Hepatitis A

Authors

Ian D. Gust, M.D.
Director, Macfarlane Burnet Centre
for Medical Research
Fairfield Hospital for Communicable
Diseases
Professor of Clinical Virology
Melbourne, Australia

Stephen M. Feinstone, M.D.
Medical Officer
Laboratory of Infectious Diseases
National Institutes of Allergy and
Infectious Diseases
National Institutes of Health
Bethesda, Maryland

CRC Press, Inc.
Boca Raton, Florida

Library of Congress Cataloging-in-Publication Data

Gust, Ian D.
Hepatitis A / Ian D. Gust, Stephen M. Feinstone
p. cm.
Includes bibliographies and index
ISBN 0-8493-5767-5
1. Hepatitis A virus I. .Feinstone, Stephen M. II. Title
[DNLM: 1. Hepatitis A. QW 170 G982h]
QR201 .H46G87 1988
616′.0194--dc19
DNLM/DLC
for Library of Congress

88-14623
CIP

Direct all inquiries to CRC Press, Inc., 2000 Corporate Blvd., N.W., Boca Raton, Florida, 33431

International Standard Book Number 0-8493-5767-5

Library of Congress Card Number 88-14623
Printed in the United States

INTRODUCTION

In this book we have tried to review current knowledge of hepatitis A and to trace the sequence of events which led to each of the major advances in this field. When we started work on hepatitis A in the early 1970s neither of us dreamed how quickly events would unfold. Since that time the etiological agent has been identified and characterized, sensitive, serological tests have been developed, the epidemiology of the disease has been defined, animal models have been developed, the genome has been molecularly cloned and sequenced, and a range of vaccines are now being evaluated. It is hard to imagine a more exciting or rewarding period.

The text was written while one of us (IDG) was a Fogarty Scholar in Residence at the National Institutes of Health. The authors would like to thank the two Bobs (Chanock and Purcell) for their support, and Peter Condliffe and his staff in the Stone House for their help and kindness, and Harry Newman and Terry Popkin for preparation of the diagrams and photographs.

THE EDITORS

Ian D. Gust, M.D., was born in Melbourne, Australia and is currently the Director of the Macfarlane Burnet Centre for Medical Research at the Fairfield Hospital for Communicable Diseases in Melbourne. Dr. Gust holds many degrees and fellowships and received his Doctor of Medicine from Melbourne University in 1974. His fellowships include the Royal College of Pathologists of Australia and the Royal Australasian College of Physicians, and he is a member of the Australian Society for Microbiology.

Dr. Gust has received many honors and other scientific recognition, he has spoken at numerous national meetings, and lectured at many international meetings. He has been a consultant for the World Health Organization and has been appointed to numerous public and university committees. He is currently a member of the editorial boards for the *Journal of Medical Virology, Asian Journal of Infectious Diseases, Journal of Infectious Diseases,* and *Journal of Clinical Microbiology.* Dr. Gust has also written many articles and technical reports.

Stephen M. Feinstone, M.D., is currently a Medical Officer at the National Institutes of Allergy and Infectious Diseases at the National Institutes of Health in Bethesda, Maryland. Dr. Feinstone received his Bachelor's degree from Johns Hopkins University in 1966 and his M.D. from the University of Tennessee Medical School in 1969. Following graduation, he served an internship at the Medical College of Wisconsin and then became a Research Associate with the National Institutes of Health. He took residency at Georgetown University Hospital before returning to the NIH in 1976.

Dr. Feinstone is a member of many societies including the American Federation of Clinical Research and the American Society for Microbiology. He has written many journal articles and technical reports. Dr. Feinstone's current research interests include virology, specifically viral hepatitis, and clinical infectious diseases. His most important scientific contributions include the first detection of hepatitis A virus which resulted in the first test for HAV, and the first description of non-A, non-B hepatitis by serologic testing. More recently, he has discovered the molecular cloning of the HAV genome.

DEDICATION

This book is dedicated to you ... and my father who doesn't understand it, my sister who doesn't believe it, and my family who have suffered through every word of it. We have tried to retell the events of the past 15 years as they happened and to produce a book which as well as being readable, would be robust enough for pressing wildflowers and thick enough to prop up tables with uneven legs.

Letters of praise, requests for sequels, and gifts should be forwarded to me; criticisms, allegations of plagarism, and writs should be forwarded to my colleague, Dr. Feinstone.

Ian D. Gust
Bethesda, January 1987

For my father, my wife and my son.

Stephen M. Feinstone
January 1988

GLOSSARY OF COMMON TERMS AND ABBREVIATIONS

A	arginine
Å	angstrom
AGM	African green monkey
AGMK	African green monkey kidney cells
AIIMS	All India Institute of Medical Sciences
ALT	alanine aminotransferase
anti-HAV	antibodies to hepatitis A virus
anti-HAV IgM	IgM antibodies to hepatitis A virus
anti-HIV	antibodies to human immunodeficiency virus
AST	aspartate aminotransferase
BGMK	Buffalo green monkey kidney
B-SC-1	continuous line of African green monkey cells
BSP	bromsulphthalein excretion test
^{14}C	radioactive carbon
C1q, C3	components of the complement pathway
C	cysteine
C.	Callithrix species of marmosets
c.c.	cubic centimeter
CCF	cephalin-cholesterol flocculation test
cc/lb	cubic centimeters per pound
CDC	Centers for Disease Control
cDNA	complimentary DNA
CF	complement fixation
cm	centimeter
CMV	cytomegalovirus
CNS	central nervous system
CO_2	carbon dioxide
CPE	cytopathic effect
CR326	Costa Rican strain of HAV
CsCl	cesium chloride
CV-1	continuous line of African green monkey kidney cells
DNA	deoxyribonucleic acid
DNAase	an enzyme which specifically destroys DNA
ds	double stranded
E.	Escherichia
EBV	Epstein-Barr virus
ECG	electrocardiogram
EIA	enzyme-linked immunoabsorbent assay
EM	electron microscopy
EMC	encephalmyocarditis virus
FDA	Food and Drug Administration
FL	continuous line of human amnion cells
FMDV	foot and mouth disease virus
FRhK-4	cloned line of fetal rhesus monkey kidney cells
FRkH-6	cloned line of fetal rhesus monkey kidney cells
FRhL-2	fetal rhesus lung cells
G	glycine
GB	serum from surgeon with hepatitis used to transmit disease to marmosets

GBG	German strain of HAV
GBM	German strain of HAV
GJA	German strain of HAV
GMT	geometric mean titre
^{3}H	radioactive hydrogen
HA	hepatitis A
HB	hepatitis B
HBIG	hepatitis B immunoglobulin
HBsAg	hepatitis B surface antigen
HBV	hepatitis B virus
HEF	human embryo fibroblasts
HEK	human embryo kidney cells
HeLa	continuous line of human epithelial cells
Hep2	continuous line of human epithelial cells
HIV	human immunodeficiency virus
HM175	Australian strain of HAV
HM790	Australian strain of HAV
NANBH	non-A, non-B hepatitis
HNIG	human normal immunoglobulin
HRV-14	human rhinovirus 14
^{125}I	radioactive isotope of iodine
IAHA	immune adherence hemagglutination
ICD	isocitric dehydrogenase
ICTV	International Committee for the Taxonomy of Viruses
ID	infectious dose
IEM	immune electron microscopy
IF	immunofluorescence
IFN	interferon
IgA	immunoglobulin A
IgG	immunoglobulin G
IgM	immunoglobulin M
IL-2	interleukin-2
IM	intramuscular
ISG	immune serum globulin
IU	International units
IV	intravenous
K30	serum containing the MS-1 strain of HAV
L	lysozome-like bodies
LAK	lymphokine activated killer cells
LFT	liver function tests
LLC-MK2	continuous line of African green monkey kidney cells
log	logarithm
MA-104	fetal rhesus monkey kidney cells
MBB	German strain of HAV
MID	marmoset infectious dose
MIT	Massachusetts Institute of Technology
mRNA	messenger RNA
MRC5	human diploid fibroblast line
MS-1	serum pool containing hepatitis A virus
MS-2	serumpool containing hepatitis B virus
MV	multivesicular bodies

MW	molecular weight
n	number
N	nucleus
NA	not applicable
ND	not done
NIAID	National Institute for Allergy and Infectious Diseases
NIH	National Institutes of Health
NK	natural killer cells
N-P	nasopharyngeal
NT	not tested
oligo-dt	oligo deoxythymidine
O	oral
OMK-210	owl monkey kidney cells
P	parenteral
P1, 2, 3	cleavage sites on picornavirus polyprotein
PA-21, PA-33	strains of HAV recovered in owl monkeys
PBL	peripheral blood leukocytes
PBS	phosphate buffered saline
PEG	polyethylene glycol
PEO	continuous line of green monkey cells
p.i.	post innoculation
PLC/PRF5	continuous line of human hepatoma cells
P/N	positive/negative ratio
poly(A)	polyadenylic acid
poly(C)	polycytidylic acid
poly(U)	polyurydylic acid
PR	polyribosomes
PV1	poliovirus type 1
RIA	radioimmunoassay
RIF	radioimmunofocus
RIFA	radioimmunofocus assy
RIFU	radioimmunofocus units
RNA	ribonucleic acid
RNAase	an enzyme which specifically destroys RNA
R.T.	room temperature
^{35}S	radioactive sulfur
S	sedimentation coefficient (Svedburg units)
S.	Saguinus species of marmosets
S/C	subcutaneous
SD11	American strain of HAV
SDS	sodium dodecyl sulfate
SDS-PAGE	sodium dodecyl sulfate polyacrylamide gel electrophoresis
SGOT	serum glutamic-oxaloacetic transaminase
SGPT	serum glutamic-pyruvic transaminase
sp.	species
SPRIA	solid phase radioimmunoassay
ss	single stranded
SV40	simian virus 40
T	thymus-dependent lymphocytes
TCID	tissue culture infectious dose
TT	thymol turbidity test

U	uridine
U-V	ultraviolet
V	vesicles
Vero	continuous line of African green monkey kidney cells
vol/vol	volume to volume
VP1, VP2, VP3, VP4	the major structural proteins of HAV
VPg	virion protein, genome
WHO	World Health Organization
WI-38	human diploid fibroblast line
WSP-5	serum pool containing HAV and HBV
wt/vol	weight to volume
WW-SS	plasma from patient with hepatitis used to transmit infection to monkeys and man
$ZnCl_2$	zinc chloride

TABLE OF CONTENTS

Chapter 1

HISTORY

I. INTRODUCTION

A. General

Today, when at least five different forms of acute viral hepatitis are recognized, it is difficult to imagine that the disease was not considered to be a distinct clinical entity until this century and that it was not until the 1940s that physicians realized that more than one form of hepatitis existed.

It is often said that hepatitis A was known at the time of Hippocrates[1] because of his description of benign epidemic jaundice which appears in the "De internis affectionibus".[2] "This type of jaundice is called epidemic because it occurs in all seasons. It is caused above all else by overindulgence, excesses of wine and after a chill. From the first moment, the body changes color and becomes yellow; the eyes become markedly jaundiced; the disease appears under the hair and under the nails. There are chills and low grade fever. The patient is weak. The head aches; the urine is yellow and thick. This form of jaundice is less dangerous and is cured if treated quickly."

While the symptoms that are described could be due to hepatitis A they are scarcely diagnostic! Caution should be exercised in ascribing this and later descriptions of epidemic jaundice to hepatitis A as a number of other diseases such as yellow fever may cause outbreaks of jaundice often associated with a high-case fatality rate. The earliest outbreaks of hepatitis which were almost certainly hepatitis A, were documented in Europe in the 17th and 18th centuries. In 1886, Hirsh compiled a list of 34 epidemics of jaundice, the earliest of which took place in Minorca in 1745.[3] A more extensive list was produced by Bachman in 1952[4] including two epidemics in the 17th century, 25 in the 18th, and 53 from 1802 to 1874. While not all of these were due to hepatitis A, some undoubtedly were. The earliest recorded outbreak in the U.S. seems to have occurred in Norfolk, Va. in 1812.[5] Brief accounts of the disease can be found over the next 50 years although it appears to have been uncommon until the outbreak of the Civil War when large numbers of cases occurred among Union troops.

B. Catarrhal Jaundice

Sporadic cases of jaundice were also recognized in the 19th century. From 1855 the disease became known as "catarrhal jaundice" as a result of a misconception about its etiology which was first suggested by Bamberger and perpetuated by the eminent pathologist Virchow.[6]

Bamberger and Virchow believed that "catarrhal jaundice" was due to inflammation of the biliary tract which resulted in the common bile duct becoming blocked by a plug of inspissated mucus. This belief was so firmly entrenched that patients were sometimes treated with magnesium sulfate which it was thought would assist in dislodging the plug.

Although many physicians, including Sir William Osler, subscribed to this theory, Bamberger and Virchow's views did not go unchallenged. Various workers suggested that epidemic and sporadic jaundice had a common etiology, that the cause of both diseases was acute inflammation of the liver and that they were caused by an infectious agent.

In 1912, Cockayne[7] published a review in which he suggested that the sporadic and epidemic forms of jaundice were probably manifestations of the same disease. Although this view was echoed by Blumer,[8] it received little attention in the medical community.

In 1922, Eppinger[9] propounded the view that the primary lesion was not obstruction but hepatocellular necrosis, a view which received support from pathologists in Britain, Scandinavia, and the U.S.[10-12] Although Eppinger felt that the most probable cause was a liver toxin, contemporaries thought that the disease was infectious, a view which gained support when Inada and his colleagues[13] showed that some outbreaks of jaundice in Japan were due to infection with leptospira. Although a number of enteric bacteria were isolated from the feces of patients with "catarrhal jaundice" no particular strain was consistently implicated which led McDonald[14] in 1908 to postulate that a virus might be involved.

C. Serum Hepatitis

The first indication of the existence of a second form of hepatitis which was spread by a different route and had a different epidemiology came from the observations of Lurmann who described an outbreak of hepatitis among shipyard workers in Bremen in 1833.[15] Lurmann observed 191 cases of jaundice in a group of 1289 shipyard workers 2 to 8 months after they had been vaccinated against smallpox with a particular batch of human glycerinated lymph. By contrast, no cases of jaundice were observed among 500 of their colleagues who were vaccinated at the same time with a different batch of vaccine. A second similar episode occurred in Germany later that year.[16] The disease, which became known as "serum hepatitis" was assumed to be due to a blood-borne infectious agent but was regarded as a medical curiosity of no general importance. Over the next 50 years however, a number of outbreaks with similar features were observed in a variety of settings. In the 1920s, outbreaks of jaundice were described in groups of American and German patients who were being treated for syphilis with injections of the arsenical preparation, salvarsan.[17] Although these outbreaks were carefully documented, the mechanism of infection was not defined and no thought appears to have been given to the possible role of syringes and needles in transmission of the disease. A few years later Flaum et al.[18] reported an outbreak of jaundice among diabetics attending a clinic to receive insulin injections. In an outstanding paper, they produced evidence that reuse of needles and syringes was responsible for transfer of the disease and suggested that in view of its prolonged incubation period "serum hepatitis" was caused by a different agent to the one responsible for "catarrhal jaundice."

In 1937, two additional outbreaks of serum hepatitis were reported by British workers. In the first, several groups of school children who had received injections of pooled convalescent serum as prophylaxis against measles developed jaundice.[19,20] Of 191 children immunized with a particular serum pool, 37 developed hepatitis and 7 died. In each instance the incubation period between receipt of the pooled serum and onset of symptoms was several months. A second outbreak of jaundice occurred among adults who had been injected with yellow fever vaccine.[21] Findlay and MacCallum,[22] who investigated the outbreak, concluded that the disease was caused by the presence of a contaminating virus, which had been introduced either in the tissue cultures in which the yellow-fever virus was propagated, or in batches of human serum used to stabilize the vaccine. In an effort to prevent contamination they changed the method of vaccine production and the strain of virus which was used. Following these changes no cases of jaundice were detected among the next 5000 vaccinees.

Although by 1939 there was excellent evidence that viral hepatitis was an infectious disease caused by two etiologically distinct agents with separate modes of transmission and distinctive epidemiologies, these concepts were outside the mainstream of medical thought. It required the impetus of World War II and massive outbreaks of disease among troops and civilians to alter this situation.

II. HEPATITIS IN THE MILITARY

A. 18th and 19th Centuries

Epidemic hepatitis is an old camp follower and has been a problem among troops for centuries. Epidemics of hepatitis were frequently described by military surgeons in accounts of battles fought in Europe during the 18th and 19th centuries. The disease occurred during the Seven Years War, both in Germany and the low countries.[23] So many epidemics occurred among German garrison and field troops between 1842 and 1856, that the disease became known as "Kriegsickterus," "Kriegsgelbsucht," or "Soldatengelbsucht" just as French soldiers in the Napoleonic Wars referred to it as "jaunisse des camps."

Hepatitis has been a problem in the American army since the Mexican War and was well documented among both white and black troops in the Union Army during the American Civil War, where it was reported to have caused more than 70,000 cases and attack rates were as high as 13 cases per 1000 troops.[25]

Although the disease was recognized to be infectious, its etiology was not known. A variety of causes were suggested, including infection with leptospira and certain strains of Salmonellae.[26,27]

B. World War I

During World War I large epidemics of hepatitis occurred in the British, French, German, and Rumanian armies.[28] Starting in July, 1915 among British troops in Alexandria, the disease spread to Gallipoli, Thessalonia, and Mesopotamia. Epidemics also occurred among French troops in the Dardenelles and German and Rumanian troops in Rumania[26,29] while the American army was largely unaffected.

Willcox,[30] who described the epidemic in British troops, observed that "the epidemic jaundice of campaigns appears to start as a gastrointestinal infection, . . . is almost certainly conveyed by the alimentary tract and contamination of food . . . probably plays an important part. Flies . . . may convey infection to food; but they are probably not the main cause of the spread of the disease". He also speculated that the disease could be transmitted by contaminated water. It was 30 years before these ideas were proven to be true.

C. World War II

Although infectious hepatitis was a major cause of morbidity among both Allied and Axis forces throughout the entire conflict, it was not until a major outbreak of serum hepatitis occurred among American troops that the magnitude of the problem was realized. As a direct result of this outbreak, army authorities in Britain and the U.S. sponsored a number of studies which were aimed at determining the etiology of the disease and developing means of controlling it.

1. The Serum Hepatitis Epidemic in American Troops

By 1940, immunization had become an accepted means of prophylaxis against yellow fever. As the problem of post-vaccination jaundice seemed to have been overcome, it seemed logical to immunize troops who might be required to fight in endemic areas. The decision to immunize American troops was put into effect in 1941 using a vaccine produced by the Rockefeller Foundation containing the 17D strain of yellow fever virus, propagated in chick embryo cells in a medium containing 1% normal human serum. To maintain vaccine production, 8 to 10 ℓ of human serum was needed per week, most of which was provided by students, interns, nurses, and laboratory technicians from Johns Hopkins School of Hygiene and Public Health.

Mass immunization of troops began in the autumn of 1941. In February of the following year, an unusual number of cases of jaundice were noted among recipients of the vaccine and it was soon apparent that an epidemic was occurring.[31] Altogether 49,233 cases of hepatitis occurred in U.S. troops in 1942, most of which followed yellow fever vaccination (see Figure 1). In some units which received the vaccine the attack rate was approximately 20%, with incubation periods ranging from 60 to 154 days (mean 96 days). A detailed investigation of the outbreak revealed that only certain lots of the vaccine were involved

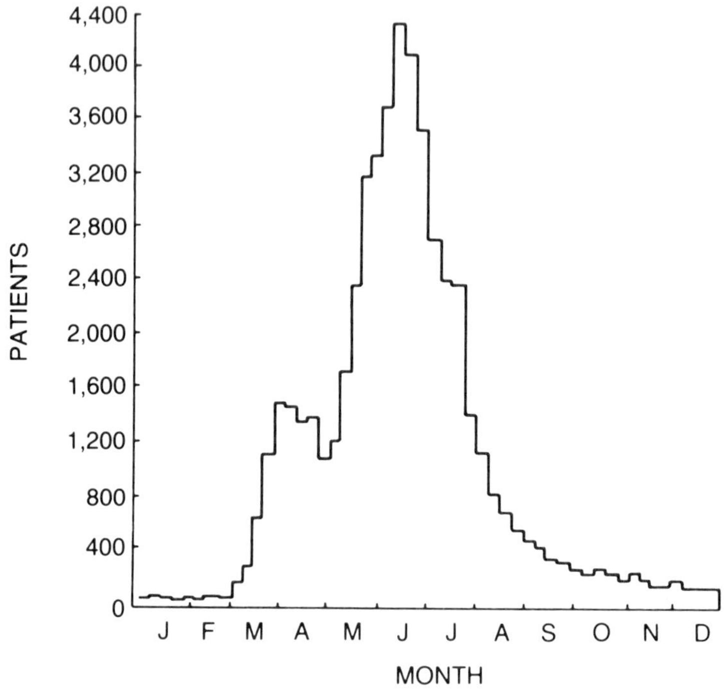

FIGURE 1. U.S. Army records of soldiers admitted to hospital with jaundice from January 1 to December 31, 1942. Most of these cases followed receipt of contaminated batches of yellow fever vaccine. (From Havens, W. P., Jr., Viral hepatitis, in *Internal Medicine in World War II,* Vol. 3, Office of the Surgeon General, Department of the Army, Washington, D.C., 1968, chap. 13. With permission.)

and cast suspicion on the role of some batches of human serum. Production procedures were rapidly altered so that the vaccine could be manufactured without human serum and within a few months the epidemic subsided.

As a result of the epidemic, physicians came to realize that some healthy adults were carriers of the serum hepatitis virus and that infection could be transmitted by inoculation of blood or blood products, or by the reuse of inadequately sterilized needles and syringes. These observations had important implications for medical practice, especially during warfare.

2. *Epidemic Hepatitis*

During the second World War, overcrowding, poor hygiene, breakdown of systems of sewage disposal, and contamination of food and water with human feces was followed by massive outbreaks of enteric disease. More than 5 million cases of hepatitis were reported in Germany among troops and civilians; and French, Commonwealth, and Axis troops suffered severely, especially during the North African campaign.[31,32]

Epidemic hepatitis, which British medical officers referred to as ''infective hepatitis,'' was a serious problem among troops in the Middle East in 1941 and 1942 (see Figure 2). In both years, epidemics coincided with severe fighting in the North African desert and at times 8 to 9% of soldiers and up to a third of officers were out of action.[33] The 1942 epidemic started in October when the British were commencing a counter offensive at El Alamein and it affected both their own troops and Italian prisoners of war. Army doctors formed the opinion that poor sanitation was responsible for the spread of infection as the disease spread most readily on battlefields which were heavily contaminated with feces and unburied corpses. In contrast, hepatitis was not a problem in hospitals, camps, and POW

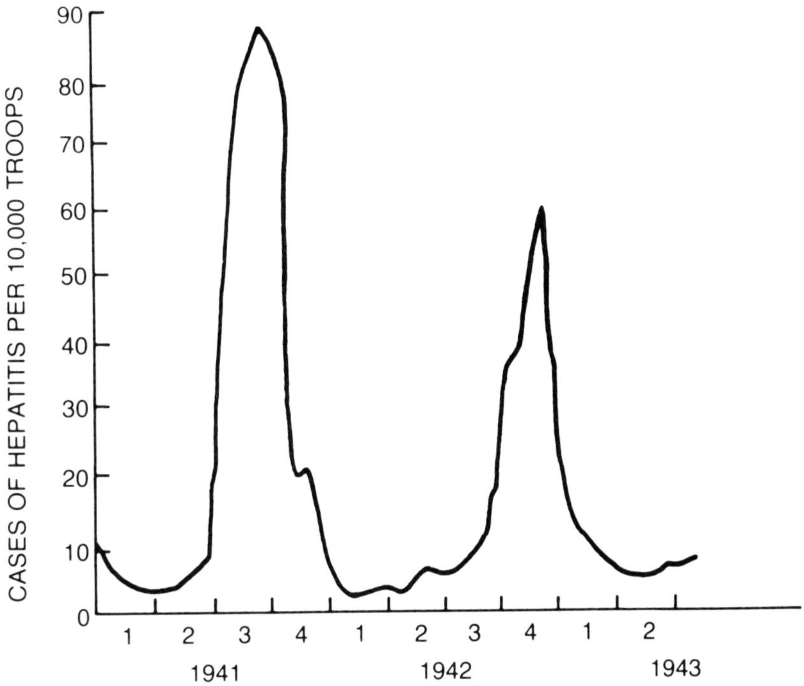

FIGURE 2. Monthly incidence of hepatitis per 10,000 British troops stationed in the Middle East between January 1941 and July 1943. (Redrawn from Paul and Gardner.[28])

compounds where standards of hygiene were higher. Some British medical officers were convinced that the disease was spread by feces and were anxious to prove this with studies in human volunteers but failed to obtain permission to perform these experiments.[34]

Hepatitis became a problem for American forces in North Africa and the Mediterranean in the second half of 1943. By October and November, a major epidemic was occurring in troops in North Africa, Egypt, and Sicily, and hospitals from Cairo to Palermo were overflowing with patients. As ways of preventing or controlling the disease were urgently needed, Brigadier General Simmons of the Army Epidemiological Board visited the region and gave the joint tasks of epidemiologic investigation recommending means of preventing the disease to the Neurological Virus Diseases Commission.

The approach of this group was colored by the recent outbreak of serum hepatitis and at first a variety of possibilities were considered. Some members thought that epidemic hepatitis was spread by droplets like influenza, while others thought that it was spread excrementally by chronic carriers like typhoid. Another school of thought held that the disease was spread by an animal or insect vector, while a fourth believed that it represented an example of biological warfare!

The Neurological Virus Diseases Commission studied the epidemiology of hepatitis in the field and made a number of important observations.[28] First, they noted that epidemic hepatitis had a much shorter incubation period than serum hepatitis with most cases occurring after an incubation period of 18 to 25 days, compared with a mean of 90 days in troops who had been infected by contaminated batches of yellow fever vaccine.

Second, a number of American soldiers who developed epidemic hepatitis in 1943 to 1944 were found to have previously suffered an attack of post-vaccination hepatitis, suggesting that the two diseases were caused by two different agents.

Third, the disease was four or five times more common among officers than among

Table 1
EVOLUTION OF THE NOMENCLATURE OF HEPATITIS A

19th Century	20th Century
Catarrhal jaundice	Campaign jaundice
Epidemic jaundice	Yellow jaundice
Benign epidemic jaundice	Infective hepatitis
Kriegsickterus	Infectious hepatitis
Soldatengelbsucht	Hepatitis type A
Jaunisse des camps	
Botkin's disease	
Bilious fever	

enlisted men, suggesting that troops who had a lower socioeconomic background had encountered the agent earlier in life and were immune.

Fourth, inquiries among local doctors revealed that although hepatitis was occasionally seen in the population, it was essentially a disease of children, suggesting that infection was endemic and that most local adults had already been exposed to the virus and were immune.

Finally, outbreaks of hepatitis among troops often followed a month or two after an outbreak of diarrhea, suggesting that both infections may have been acquired simultaneously, perhaps as a result of ingestion of fecally contaminated food or water.

With this background it was clear to the Americans, as it had been to the British, that if control measures were to be developed the mode of transmission of the disease would have to be established. To assist in this process blood, feces, urine, and other specimens were collected from patients in the epidemic area and sent to the U.S. for study. In addition, a clinical research unit was established in Naples, Italy.

While these studies were in progress, infectious hepatitis continued to be the major cause of disability among U.S. troops in Europe and North Africa and was also a serious problem in the Pacific theater where the incidence of disease in some areas exceeded 80 per 1000 per annum. In 1945, almost 40,000 cases occurred among U.S. troops and an estimated two thirds of a million days of activity were lost.[28]

III. NOMENCLATURE

The studies in human volunteers marked a turning point in understanding the etiology and mode of spread of the disease and had an immediate impact on nomenclature.

Early in the 20th century the prevalent form of acute liver disease was known as "catarrhal jaundice," or "epidemic jaundice" although in some countries local terms were used (see Table 1). In 1937, in their first report on hepatitis among recipients of yellow fever vaccine, Findlay and MacCallum[21] introduced the term "common infective jaundice" as a synonym for "catarrhal jaundice" and later suggested that the term "infective hepatitis" was preferable.[20] In 1943, the term "infectious hepatitis" was officially introduced in the U.S. to describe a form of hepatitis which had a short incubation period, spread readily from person to person, occurred in epidemic form, and seemed to be transmitted by the fecal-oral route.

The term "homologous serum jaundice" was first used in a British Ministry of Health memorandum in 1943, to describe cases of hepatitis which appeared to have a longer incubation period and to be caused by injection of blood plasma or serum.[35] From about this time the disease also became widely known as "serum hepatitis".[36]

In 1947, when hepatitis had clearly been demonstrated to be caused by two separate agents, MacCallum suggested the diseases be called hepatitis A and B and the putative etiological agents be referred to as virus A and B. This suggestion was adopted in 1952 by the First Expert Committee in Viral Hepatitis of the World Health Organization but received little support from the general scientific community.

Finally, in 1973, after the etiological agents of both forms of the disease had been identified, an Expert Committee of the World Health Organization recommended that the two diseases be officially known as hepatitis A and hepatitis B, and the etiological agents as the hepatitis A virus (HAV) and hepatitis B virus (HBV), respectively.[37] This terminology is now universally accepted.

In this chapter, the terms that were in use at the time that experiments were performed are given. Throughout the rest of the book the diseases and etiological agents will be referred to by their official names.

IV. STUDIES IN HUMAN VOLUNTEERS

A. During World War II

1. Organization of Studies in England and the U.S.

In May 1943, the British Government decided to give control of hepatitis a high priority. A Jaundice Committee was created and a research group of virologists, epidemiologists, and physicians was established in the Department of Pathology at Cambridge Univeristy.

The main objectives of the group were to determine how the disease was being spread and to devise effective means for its containment. It approached its task from two different perspectives, carrying out a series of experimental infections in human volunteers and backing this up with careful observations of the natural history of the disease in the field. To assist with these studies jaundice was made a notifiable disease in the southeast of England.[36] The studies in human volunteers concluded in November 1945 while the epidemiological observations continued until the end of 1947.

By 1943 it had become apparent that hepatitis could not be transmitted to embryonated eggs or laboratory animals and that the most suitable experimental animal available was man.[36,38] Efforts to transmit infectious hepatitis to laboratory animals continued, and in addition, groups of volunteers were recruited from conscientious objectors, prisoners, and patients with rheumatoid arthritis. This latter group was invited to participate because of reports that attacks of hepatitis were often associated with a remission in symptoms. Nasopharyngeal washings, extracts of feces, and blood specimens were collected from patients with naturally occurring infectious hepatitis and administered to selected groups of volunteers.

About the same time that the British were beginning to study the etiology and natural history of viral hepatitis, John Paul and Paul Havens established a laboratory for the study of experimental hepatitis infection in the Section of Preventive Medicine at Yale University. Two other laboratories were established in the U.S.; one at the Children's Hospital, Philadelphia, under the direction of Joseph Stokes and John Neefe, another at the school of Public Health, Ann Arbor, Michigan, under the direction of Thomas Francis. These workers obtained volunteers from groups of conscientious objectors working in state institutions near New Haven, and prisoners from two institutions, in Danbury and Wethersfield, Connecticut.

2. Detection of the Agent in Various Specimens

The first problem that workers in both countries had was obtaining specimens likely to contain the etiological agent. As previous experiments by Yoshibumi and Shigemoto in Japan[39] and Voegt in Germany[40] had suggested that epidemic hepatitis could be transmitted by ingestion of duodenal contents and blood collected during the acute phase of their illness, attention naturally turned to blood and feces as sources of the etiological agent.

Early in 1944, MacCallum and Bradley[41] succeeded in transmitting infectious hepatitis to human volunteers by enteric administration of bacteria-free suspensions of feces and by parenteral inoculation of serum which had been collected from patients early in their illness.

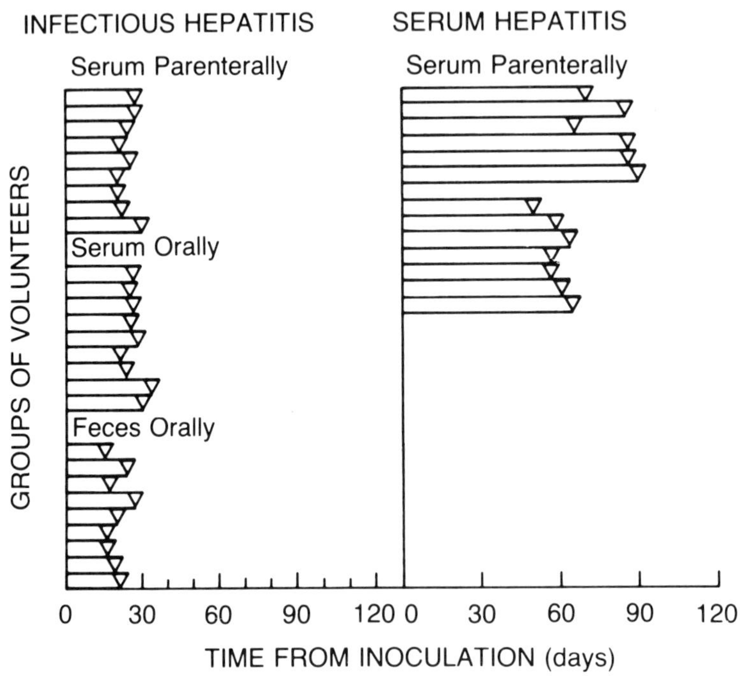

FIGURE 3. Comparison of the incubation periods observed in volunteers who developed hepatitis after administration of serum or fecal extracts collected from patients with infectious hepatitis or serum hepatitis which had been transmitted by yellow fever vaccine. (Adapted from Paul and Gardner.[28])

The success of this experiment and the lack of untoward side effects led to extensive investigations in other centers. Their findings were soon confirmed by Havens and his colleagues using specimens obtained from a soldier who had been hospitalized with a classic attack of infectious hepatitis that had been acquired in Sicily. Havens was able to induce hepatitis in 9 of 12 volunteers who received a suspension of feces by mouth and in 8 of 17 volunteers who were inoculated intravenously with acute-phase serum. These findings were extended by Neefe et al. who transmitted hepatitis to 25 of 46 volunteers by oral administration of a fecal suspension from a patient with infectious hepatitis and by oral or parenteral administration of serum collected early in the disease.[43] The incubation period in these studies ranged from 15 to 49 days.

The importance of the intestinal-oral route in infection was subsequently demonstrated by Neefe and Stokes who were able to infect volunteers with fecally contaminated water which had been collected from a well in Pennsylvania during an epidemic of the disease.[44] Volunteers who ingested the contaminated water or bacteria-free suspensions of feces from children who had become jaundiced after drinking the water, developed acute hepatitis.

Other investigators demonstrated that HAV could be transmitted parenterally as well as orally and that HAV could be detected in specimens other than feces. Havens produced evidence for the infectivity of serum obtained from patients with epidemic jaundice and demonstrated that the incubation period was independent of the route of transmission, being from 21 to 34 days when the material was given orally and 20 to 34 days when given parenterally.[45]

Irrespective of the route of inoculation, the incubation period of the disease was relatively short and considerably less than seen in volunteers inoculated with serum from patients who had developed past vaccinal hepatitis (see Figure 3).

Contradictory data was obtained on the infectivity of urine administered orally; while both

Table 2
SUMMARY OF ATTEMPTS TO TRANSMIT INFECTIOUS HEPATITIS TO ADULT VOLUNTEERS WITH SPECIMENS COLLECTED DURING THE ACUTE PHASE OF THE ILLNESS

Inoculum	Year	Route[a]	Volunteers No. inoculated (no. jaundiced)	Ref.
Duodenal fluid	1942	O	4 (1)	40
Nasopharyngeal washings	1944	NP	16 (0)[b]	41
	1945	NP	8 (0)	43
	1946	NP	3 (0)	47
Fecal extracts	1944	NP	26 (3)	41
	1944—46	O	12 (9)	42
	1944—46	O	46 (25)	43
		P	3 (0)	
	1945	O	18 (7)	46
Serum	1944	P	6 (3)	41
	1944—46	P	17 (8)	42
		O	8 (7)	
	1945	P	8 (4)	48
	1945	P	6 (1)	43
	1946	O	3 (2)	
Blood	1942	O	? (1)	40
		P	? (1)	
Urine	1942	O	? (1)	40
	1944	NP and O	19 (0)	41
	1945	O	6 (3)	46
	1945	O	7 (0)	44
	1946	O	3 (0)	47

Note: Adapted from Havens.[11]

[a] 0 = oral; P = parenteral; NP = nasopharyngeal; ? not clear from paper.
[b] Two volunteers developed "anicteric" or "inapparent" infection.

Voegt[40] and Findlay and Willcox[46] claimed to have transmitted infection, other groups were unable to repeat these observations.[41,44,47] Attempts to transmit infection by installation of nasopharyngeal washings collected early in the disease also produced contradictory results. MacCallum and Bradley, claiming to have induced anicteric or inapparent infection in 2 of 16 volunteers[41] while Havens[47] and Neefe and Stokes[44] obtained negative results in two experiments involving a total of 11 subjects. In general, this data has been interpreted as evidence against an important role for droplets in the spread of infection. The major findings of these studies are summarized in Table 2.

It became clear that epidemiological criteria were inadequate for differentiating all cases of infectious and serum hepatitis, and that specific diagnostic tests would be needed before the epidemiology and natural history of the diseases could be defined; however, another quarter of a century was to pass before these goals could be achieved.

3. Duration of Infectivity

After it was recognized that the infectious hepatitis virus was present in the feces and blood of patients during the acute phase of the disease, attempts were made to define the duration of infectivity. As relatively few experiments were undertaken and the only objective evidence of infection was the occurrence of jaundice in the recipient, positive results were more important than negative ones. While blood collected from one patient 3 days prior to

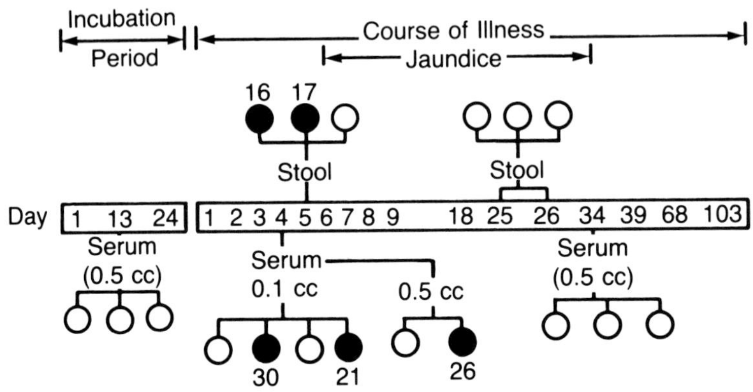

FIGURE 4. Diagnostic representation of the results obtained following administration of serum and filtrates of stool specimens, collected from a patient with experimentally induced infectious hepatitis, to volunteers. (○) Represents inoculated volunteers who failed to develop hepatitis. (●) Represents inoculated volunteers who develop hepatitis. The figure beside each circle represents the incubation period in days. (From Havens, W. P., Jr., *J. Exp. Med.*, 83, 251, 1946. With permission.)

the onset of symptoms was demonstrated to transmit infection,[48] in other experiments attempts to transmit hepatitis with blood collected halfway through the incubation period, and with blood and feces collected 1 and 3 months after the onset of symptoms were unsuccessful.[47,49] The results of one series of experiments are shown in Figure 4.

These experiments confirmed that the greatest risk of infection probably arose from contact with patients with mild forms of the disease or patients who were in the incubation period of the disease.

4. Immunity

In two sets of experiments involving a total of 19 volunteers, Havens[50] and Neefe and his colleagues[51] demonstrated that subjects who were reinoculated with the infectious hepatitis agent 6 to 9 months after their initial illness failed to develop disease.

In addition, Neefe and his colleagues[43] showed that volunteers infected with a filtered suspension of feces obtained from children involved in a water-borne outbreak of hepatitis in Pennsylvania were subsequently immune to challenge with virus recovered from the feces of a soldier who acquired his illness in Sicily. This was the first direct evidence that infectious hepatitis was caused by a single agent, a finding which had major implications for control of the disease.

In a related series of experiments, 6 volunteers who had been experimentally infected with the infectious hepatitis agent were inoculated with material known to contain the agent of serum hepatitis. Three of these people developed hepatitis.[46,51] In addition, 8 of 10 volunteers who had recovered from serum hepatitis developed illness after challenge with materials containing the infectious hepatitis agent.[51,52,53] By contrast, 29 volunteers who had developed serum hepatitis were resistant to challenge with the same inoculum.[46,51,54]

Attention then turned to the possible use of convalescent human serum for prevention of infectious hepatitis. Two groups of workers, Stokes and Neefe[55] and Havens and Paul,[56] were soon able to demonstrate that intramuscular injection of pooled normal human immunoglobulin (IG) prior to exposure could prevent or attenuate the disease.

During an outbreak of infectious hepatitis in a summer camp, Stokes and Neefe[50] immunized 53 children with 0.15 cc/lb of normal immunoglobulin and demonstrated that the attack rate over the next 32 days was much less than in uninoculated children. Only 5.7% of the children who were immunized became jaundiced compared to 45% in the control group. Havens and Paul[56] obtained similar results in children with a smaller dose of IG, 0.08 cc/lb, and subsequently it was shown that doses as low as 0.01 cc/lb could provide protection for several months.[57]

These findings had important practical implications and gamma globulin was rapidly introduced into clinical practice. In 1945 over 2700 American soldiers in the Mediterranean were immunized in the face of an epidemic. The efficacy of this approach, as determined by a reduction in the incidence of jaundice, was estimated to be 89% at 7 weeks and 73% at 12 weeks.[58]

5. Physical Characteristics of the Agent

Further studies in human volunteers by the Yale group established that infectious hepatitis was caused by an agent, presumably a virus, which was filterable through an L2 Chamberland or Seitz E.K. filter, was resistant to heating at 56°C for 30 min, and remained active after freezing for 18 months or being stored in dry ice for 32 months.[32,59]

Following their demonstration that an epidemic of infectious hepatitis in a summer camp was due to drinking fecally contaminated water,[43] Neefe et al. evaluated the effects of chlorine and precipitation of organic material on transmission of the virus.[60] In their experiments, stepwise addition of sodium carbonate, ammonium sulfate, and activated carbon successfully removed most organic material, but failed to prevent transmission of infection. Hepatitis was transmitted with suspensions of feces with a residual chlorine concentration of 1 part per million, but not by suspensions which had been exposed to a residual concentration of 15 parts per million for at least 30 min (superchlorination). As a result, from July 1945, the U.S. Army recommended that all stored water be superchlorinated before use.[61]

Subsequent studies provided the first evidence of the protective effect of extraneous protein since it was demonstrated that the virus could readily be inactivated by exposure for 30 min to 1 part per million of residual chlorine, provided that the water was first treated by coagulation settling and filtration.[62]

Thus, by the end of World War II, it had been clearly established that viral hepatitis existed in two distinct forms which were generally known as infectious and serum hepatitis. Infectious hepatitis was a disease with a short incubation period, spread predominately by the fecal-oral route. The disease occurred in both sporadic and epidemic form, was associated with secondary cases, and could be prevented or reduced in severity by administration of pooled human IG. The disease was caused by a filterable agent, presumably a virus which was relatively heat and ether stable, but could be inactivated by chlorine. Infection appeared to be associated with lifelong immunity and there was no evidence of a chronic carrier state. By contrast, serum hepatitis was a disease with a prolonged incubation period which appeared to be largely transmitted by inoculation of contaminated blood or blood products. The disease did not appear to be transmitted to close contacts, nor was there any evidence that it could be prevented by the use of IG. Infection with the etiological agent appeared to be associated with lifelong immunity and a blood-borne carrier state was assumed to occur.

A summary of the main features of the two forms of hepatitis that were known at the end of World War II is shown in Table 3.

B. After World War II

The first two decades after the war are often referred to as the golden age of virology. The development of broad-spectrum antibiotics made it possible to maintain cells in culture for prolonged periods and led to the isolation of some of the most important viruses of man — polio, Coxsackie and echoviruses, measles, mumps, and rubella. Isolation was followed by the development of specific laboratory tests, which enable the epidemiology and natural history of the infections to be defined, and ultimately by the development of killed and live vaccines. Despite rapid advances in virology, the agents responsible for infectious and serum

Table 3
THE MAJOR FEATURES OF THE TWO MAIN FORMS OF VIRAL HEPATITIS THAT WERE KNOWN IN 1947

Common name of the disease caused by	Infective hepatitis virus A	Serum hepatitis virus B
Incubation period	15—40 days	60—160 days
Source of virus in patient	Feces and blood	Blood
Route of experimental transmission	Oral and parenteral	Parenteral
Immunity	Homologous	Homologous
Value of globulin prophylaxis	Good	None
Filterable through	Seitz E.K. filter	Seitz E.K. filter
Resistance to heat	Survived 56°C for 30 min	Survived 56°C for 60 min
Resistance to diethyl ether	Survived 10% for 2 hr at 40°C	Survived triple extraction at −20°C

hepatitis could not be isolated in embryonated eggs, laboratory animals, or cell culture. Because of this, there was renewed interest in the use of human volunteers.

1. Studies at Willowbrook State School

The Willowbrook State School on Staten Island, New York, was established in 1938 to care for mentally retarded children. Completed in 1942 in a semirural setting, the school grew rapidly and at its peak provided accommodation for more than 4000 children, who were cared for by about 1000 attendants.

Infectious diseases became a major problem and large outbreaks of influenza, dysentery, and measles occurred at intervals.[63] Although the first case of infectious hepatitis was recognized in 1949, this disease did not become a serious problem until the 1950s when pressure from parents with retarded children led to a rapid increase in the number of admissions. With severe overcrowding, limited supervision, and many children who could not be toilet trained, enteric infections spread rapidly.

Between 1953 and 1957, as the population increased from 3272 to 4478, some 350 cases of hepatitis occurred among children and a further 70 among staff. During 1955, the attack rate among patients was 25 per 1000 and 40 per 1000 among staff.[64] The high incidence of second attacks of hepatitis among children suggested that not one, but two forms of the disease were circulating in the school.

About this time Drs. Robert Ward, Saul Krugman, and Joan Giles joined the staff as consultants in infectious diseases (see Figure 5). Their first task was to study the epidemiology of the disease. They were greatly assisted in these studies by access to two new assays for liver dysfunction, the detection of elevated levels of serum glutamic oxaloacetic transaminase (SGOT)*[65,66] and serum glutamic pyruvic transaminase (SGPT).**[67]

Over the next 15 years many thousands of samples were collected from children at Willowbrook State School and tested. These samples were then cataloged and stored, and proved to be a major resource for establishing the role of candidate etiological agents and defining the pattern of infection in this population.

A careful study of the distribution of cases and the sleeping and eating arrangements of the children revealed no common epidemiological factor to account for the outbreak. Hepatitis had apparently become endemic in the institution and was being transmitted from person to person presumably by close contact or by contact with feces.

It was decided to try and reduce the incidence of the disease by the wide-scale use of gamma globulin in the hope that repeated exposures to the virus, under the cover of gamma globulin, might lead to prolonged protection.

* Now called aspartate amino transferase (AST).

** Now called alanine amino transferase (ALT).

FIGURE 5. Dr. Saul Krugman, whose studies at Willowbrook State School paved the way to detection of the causative agents of both HAV hepatitis A and HBV, hepatitis B, pictured in his office during the studies. (Photo kindly supplied by Mrs. Sylvia Krugman.)

In June 1956, 1224 children, none of whom had a past history of hepatitis, were inoculated with 0.01 cc/lb gamma globulin intramuscularly; the other 2988 children acted as controls. Over the next 7 months, nine cases of hepatitis occurred in the immunized group compared with 59 in the control group. Although this was a marked reduction in incidence (from 19.7 to 7.4 cases per 1000) the results were not as impressive as those obtained earlier in adults using a higher dose of gamma globulin.[57]

The study was repeated using a larger dose. In January 1957, 1182 children were injected with 0.06 cc/lb and 1771 left as controls. Over the next 6 months only two cases of hepatitis were diagnosed in the inoculated group (a rate of 1.7 per 1000) compared with 41 in the uninoculated group (22.5 per 1000).

Having demonstrated that passive immunization would reduce the frequency of hepatitis, Ward and his colleagues wondered if it might be possible to induce "passive-active" immunity by feeding virus to children under the cover of gamma globulin.

In view of the data which had already been generated from experiments in human volunteers, the decision to deliberately infect patients at Willowbrook was not taken lightly. The group based their decision on their belief that the strain of hepatitis circulating at Willowbrook was mild, that all the children were constantly exposed to the virus, and that to be artificially infected with a controlled dose of virus in an isolated unit provided with special nursing care would not disadvantage the child in any way. Their plan was to begin with dilute inocula which were either noninfectious or produced a very mild subclinical illness, and gradually increase the dose until they found one which would regularly produce mild hepatitis in the majority of subjects.

Information about the studies, the protocols of which had been approved by the appropriate

Table 4
TITRATION OF THE FIRST WILLOWBROOK POOL OF FECES ADMINISTERED ORALLY

Group	Dose administered (g)	No. jaundiced/ no. fed	Incubation periods (days)
1	4.0	10/13	42, 42, 44, 46, 47, 54, 54, 54, 58,
2	2.0	3/5	39, 39, 46,
3	1.0	4/11	39, 39, 43, 49,
4	0.1	1/8	39,
5	0.001	0/8	—
6	0.00001	0/8	—

local and federal authorities, was conveyed to the parents by letter. Subsequently, those that expressed interest were interviewed by a social worker and the project team. Wards of the state and orphans were excluded. After a 2-week interval in which the parents had an opportunity to discuss the study with their own doctors they were contacted for a decision. Those who entered their children in the study were free to withdraw their consent at any time.

The subjects, whose ages ranged from 3 to 10 years, were admitted to a 16-bed isolation unit where they had no contact with other patients and were cared for and observed by a single physician. Nurses and attendants had minimal contact with other staff.

The Willowbrook studies built on the knowledge which had been acquired in adult volunteers during World War II and had two major advantages: the use of pedigreed inocula and access to sensitive biochemical assays which allowed the detection of subclinical infections. Ward and his colleagues began their studies by attempting to roughly quantitate the amount of virus in a clarified, bacteria-free filtrate of a pool of feces collected from several patients with acute hepatitis.[64] Before being inoculated into children this suspension had undergone extensive safety testing in monkeys, mice, and cell culture.

A dose of 4 g of feces-induced hepatitis in 10 of 13 subjects and infections were transmitted with doses as low as 0.1 g (see Table 4). The dose which would produce infection in 50% of subjects was calculated to be a little more than 1 g. During the course of this study it was noted that while most subjects become jaundiced within 35 to 40 days of inoculation, some did not develop jaundice until 60 to 70 days. It was postulated that the latter group represented secondary infections acquired by contact with children who were excreting virus in their feces during the incubation period of the disease. This was confirmed by housing six uninfected children with several others who had been inoculated. The latter group developed jaundice within 39 to 46 days of inoculation. In addition, some of those who had not been inoculated developed jaundice after incubation periods of 60 to 71 days.

Krugman et al.[68] were able to demonstrate transmission of hepatitis with extracts of feces collected during the incubation period of the disease. In three children who had developed hepatitis 39, 41, and 46 days after experimental infection by the oral route, fecal samples were collected on the 11th and 25th days. An extract of the pool of feces collected on day 11 failed to transmit hepatitis to eight volunteers whereas the pool of feces collected on day 25 (14 to 21 days prior to the onset of jaundice) caused illness in 4 of 13 recipients. They were readily able to confirm the presence of virus in the feces during the acute phase of the illness. Feces collected on the 1st and 3rd days of jaundice and from the 1st to 8th day of jaundice produced illness in volunteers whereas specimens from four patients collected between 19 and 33 days after the onset of jaundice were uniformly negative.[69] Krugman and his colleagues were able to detect viremia during the incubation period and acute phase of the illness (see Figure 6). A single specimen of serum collected 25 days before the onset

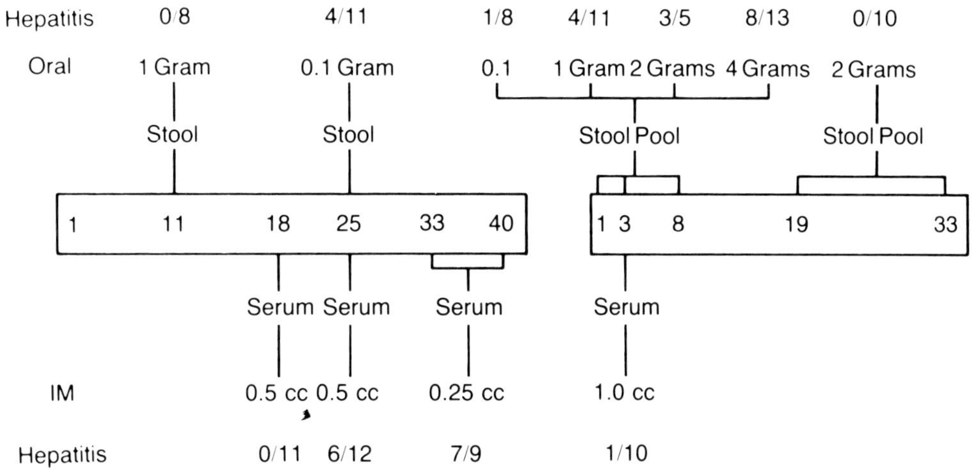

FIGURE 6. Diagnostic representation of the period of infectivity of serum and feces during the course of viral hepatitis type A. (From Krugman, S., Ward, R., and Giles, J. P., *Am. J. Med.*, 32, 717, 1962. With permission.)

of jaundice produced hepatitis in 1 of 6 volunteers.[70] By contrast, a pool of serum collected from 3 children 21 to 28 days prior to the onset of jaundice failed to transmit infection to any of 8 volunteers, whereas a second pool collected from the same children 14 to 21 days prior to jaundice was positive as were specimens collected 3 to 7 days prior to the onset of jaundice.[69] Viremia appears to end around the time that disease appears — hepatitis being transmitted to only 1 of 10 subjects with serum collected 3 days after the onset of jaundice.[69]

Subsequent studies at Willowbrook were carried out with a pool of serum known as MS-1. This and a second serum pool known as MS-2 were produced by pooling sera from children who had been infected with the serum of a patient who had developed two distinct attacks of hepatitis 6 months apart. MS-1 comprised 22 aliquots of serum collected from 10 children inoculated with serum collected from the patient during this first attack of hepatitis. These specimens were collected either just prior to the onset of jaundice (two patients) or at the time of peak transaminase activity (8 patients). The MS-1 pool was shown to be infectious after either oral or parenteral administration and regularly transmit hepatitis, with an incubation period of 30 to 40 days (see Figure 7). Disease could be induced by inoculation of as little as 0.025 $m\ell$ of serum. The MS-2 serum pool was also infectious by both the oral and parenteral route and regularly transmitted hepatitis with a long incubation period, 60 to 160 days. Krugman and his colleagues deduced that the MS-1 pool contained the agent responsible for infectious hepatitis, and the MS-2 pool the agent responsible for serum hepatitis. Conclusive evidence that the pools contained two different viruses was obtained in cross-challenge studies; children who had been infected with either MS-1 and MS-2 were shown to be resistant to challenge with the same inoculum, but fully susceptible to infection with the other inoculum.[71]

Krugman and his colleagues carried out several further studies which served to define the optimal dose of gamma globulin needed for protection against infection.[72] These experiments are discussed in detail in Chapter 10.

Perhaps the most important feature of the Willowbrook studies was the supply of pedigreed material which was obtained. These specimens were to be invaluable in proving the identity of candidate agents (see Chapter 2) and once the agent was identified, enabled the major epidemiological features to be determined rapidly.

2. *Studies at Joliet Prison*

A second series of experimental infections was carried out at the Joliet Prison, Ill. in an

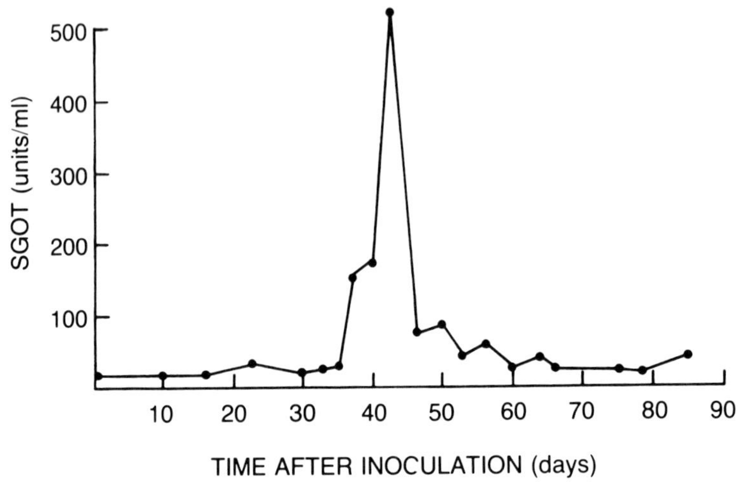

FIGURE 7. The results of serial liver function tests in a child infected with the MS-1 strain of hepatitis A. (From Krugman, S., Ward, R., and Giles, J. P., *Am. J. Med.*, 32, 717, 1962. With permission.)

attempt to generate pedigreed material of known infectivity for studies aimed at isolation of the virus in cell culture.[73,74]

The Willowbrook strain of HAV, MS-1 was used. Ten vials, each containing 0.05 mℓ of MS-1 and 0.45 mℓ of saline, and five vials of control material (0.5 mℓ of a 1:10 dilution of bovine plasma) were prepared. The contents of each vial were then increased to 40 mℓ with sterile saline and coded by an individual not involved in the study. Neither volunteers nor participants knew which preparation they had received.

The volunteers, prisoners at the Joliet Prison, were offered the opportunity to participate by an announcement on the prison radio and those who expressed interest attended question-and-answer sessions at which details of the study were discussed over several weeks. All men were warned of the risks involved and given detailed printed information about the study and any risks which might exist. No special benefits were provided. Three people unrelated to the study or the prison system witnessed the procedures for informing the volunteers and obtaining their consent.

Volunteers with a history of jaundice or drug addiction were excluded. Before admission to the isolation unit each volunteer had a complete history drawn for a variety of laboratory tests. Liver function tests were performed eight times over a 6-week period. Any abnormal findings lead to removal of the volunteer from the study.

The volunteers were isolated as a group for 2 weeks before introduction of the test material and throughout the study. Isolation included separate housing, feeding, and toilet facilities. Liver biopsies were performed 2 to 3 days prior to introduction of the test material, and prestudy fecal, urine, and plasma samples drawn and held at −70°C.

After an overnight fast and ingestion of 240 g of tap water, the coded samples were administered orally, followed by additional tap water. Subsequently, biochemical tests were performed weekly and continued for 120 days. More frequent samples were obtained where indicated. Large volumes of serum or plasma were obtained at intervals and liver biopsies repeated at the onset of illness and whenever the status of the liver was in doubt. Three of ten volunteers receiving MS-1 developed clinical, biochemical, and histological evidence of hepatitis. No illness was observed in any of the five volunteers receiving control material.

As one of the purposes of the study was to generate large volumes of material of known infectivity, a specimen of plasma obtained by plasmapheresis from a volunteer who had developed hepatitis was selected for further study. This specimen had been obtained 30 days after administration of MS-1, 8 days prior to the onset of abnormalities of liver function. A further 10 to 15 volunteers received 0.10 mℓ of this plasma — of those, five developed hepatitis after incubation periods (measured to the first SGPT > 100 units) ranging from 24 to 22 days (mean 33.4).

The Joliet studies accomplished their goals. The availability of relatively large quantities of material collected throughout the course of the infection was an impetus for renewed efforts to find the etiological agent and to develop specific diagnostic tests.[75] These efforts were successful in 1973 when Feinstone and his colleagues visualized the HAV in the feces of one of the Joliet volunteers.[76] The identification of the HAV, its adaptation to cell culture, and the development of animal models has eliminated the need for volunteer studies, other than for evaluation of candidate vaccines.

REFERENCES

1. **Zuckerman, A. J.,** *Virus Diseases and the Liver,* Butterworths, London, 1970, 16.
2. **Hippocrates,** *Ouvres Completes,* (transl), Littre, E., Ed., Balliere, Paris, 1846, 5.
3. **Hirsh, A.,** *Handbook of Geographic and Historical Pathology,* New Sydenhain Society, London, 1886, 3.
4. **Bachman, L.,** Infectious hepatitis in Europe, in *World Atlas of Epidemic Diseases,* Rodenwalt, E., Ed., Falk-Verlag, Hamburg, 1952, 1.
5. **Falkner, L.,** On epidemic jaundice, *Md. Va. Med. J.,* 15, 355, 1860.
6. **Virchow, R.,** Ueber das Vorkommen and den Nachweis des Hepatogenen, Insbesondere des Katarrhalischen Icterus, *Virchows Arch. Pathol. Anat. Physiol.,* 32, 117, 1865.
7. **Cockayne,** Catarrhal jaundice, sporadic and epidemic, and its relation acute yellow atrophy of the liver, *Q. J. Med.,* 6, 18, 1912.
8. **Blumer, G.,** Infectious jaundice in the United States. *JAMA,* 81, 353, 1923.
9. **Eppinger, H.,** Die Pathogenses des Ikterus, *Verh. Dtsch. Ges. Inn. Med.,* 34, 15, 1922.
10. **Morgan, M. T. and Brown, H. C.,** Epidemic catarrhal jaundice, some notes on the epidemiology of the disease, an account of an epidemic in the Midlands, in *Reports on Public Health and Medical Subjects, No. 42,* London, 1927.
11. **Roholm, K. and Iverson, P.,** Changes in liver in acute epidemic hepatitis (catarrhal jaundice) based on 38 aspriation biopsies, *Acta. Pathol. Microbiol. Scand.,* 16, 427, 1939.
12. **Rich, A. R.,** The pathogenesis of the forms of jaundice, *Bull. Johns Hopkins Hosp.,* 47, 338, 1930.
13. **Inada, R., Ido, Y., Hoki, R., Kaneko, R., and Ito, H.,** The etiology, mode of infection, and specific therapy of Weils disease (spirochaetosis icterohaemorrhagica), *J. Exp. Med.,* 23, 377, 1916.
14. **McDonald, S.,** Acute yellow atrophy of the liver, *Edinburgh Med. J.,* 1, 83, 1908.
15. **Lurmann, A.,** II, Eine Icterusepidemic, *Berl. Klin. Wochenschr.,* 22, 20, 1885.
16. **Jehne, A.,** Eine Icterus Epidemie in Wahrscheinlichen Zussam mit Vorausgegangenen Revaccination, *Dtsche. Med. Wochenschr.,* 11, 338, 1885.
17. **Stokes. J. H., Ruedemann, R., Jr., and Lemon, W. S.,** Epidemic infectious jaundice and its relation to the therapy of syphilis, *Arch. Intern. Med.,* 26, 521, 1920.
18. **Flaum, A., Malmros, H., and Persson, E.,** Eine Nosocomiale Ikterus-Epidemie, *Beta. Med. Scand.,* Suppl. 16, 544, 1926.
19. **McNalty, A. S.,** Acute infectious jaundice and administration of measles serum, Annual Report of the Chief Medical Officer of the Ministry of Health H.M.S.O., London, 1938.
20. **Propert, S. A.,** Hepatitis after prophylactic serum, *Br. Med. J.,* 2, 677, 1938.
21. **Findlay, G. M. and MacCallum, F. O.,** Note on acute hepatitis and yellow fever immunization, *Trans. R. Soc. Trop. Med. Hyg.,* 31, 297, 1937.
22. **Findlay, G. M., MacCallum, F. O., and Murgatroy, D. F.,** Observations bearing on the etiology of infective hepatitis (so called catarrhal jaundice), *Trans. R. Soc. Trop. Med. Hyg.,* 32, 575, 1939.
23. **Monro, D.,** *An Account of the Diseases Which Were Most Frequent in British Military Hospitals in Germany from January 1761 to the Return of the Troops to England in March 1763,* Millar, Wilson and Durham, London, 1764, 1.
24. **Gardner, H. T.,** A note on the history of epidemic viral hepatitis in Germany, *Am. J. Med.,* 8, 561, 1950.

25. **Medical and Surgical History of the War of the Rebellion. Medical History, Part 3,** U.S. Government Printing Office, Washington, D.C., 1888, 874.
26. **Sarrailhe, A., Clunet, J.,** La "Jaunisse des camps" et l' epidemie de paratyphoide des Dardenelles, *Arch. Men. Soc. Med. D. Hop. de Paris,* 40, 563, 1916.
27. **Havens, W. P., Jr. and Wenner, H. A.,** Infectious hepatitis complicated by secondary invasion with Salmonella, *J. Clin. Invest.,* 25, 45, 1946.
28. **Paul, J. R. and Gardner, H. T.,** Viral hepatitis, in *Preventive Medicine in World War II,* Vol. 4, Coates, J. B., Hoff, E. C., and Hoff, P. M., Eds., U.S. Government Printing Office, Washington, D.C., 1960, 411.
29. **Cantacuziene, J.,** Sur une Epidemie d'Icterie Observee en Roumanie Pendant la Campagne de 1917, *Presse Med.,* 26, 5411, 1918.
30. **Willcox, W. H.,** The epidemic jaundice of campaigns, *Br. Med. J.,* 1, 297, 1916.
31. **Havens, W. P., Jr.,** Viral hepatitis, in *Internal Medicine in World War II,* Vol. 3, Anderson, R. S., Ed., Office of the Surgeon General, Department of the Army, Washington, D.C., 1968, chap. 13.
32. **Gutzeit, Kr.,** Die Hepatitis Epidemica, *Muench. Med. Wochenschr.,* 92, 1161, 1950.
33. **Spooner, E. T. C.,** The 1942 epidemic of infective hepatitis in the Middle East, *Proc. R. Soc. Med.,* 37, 171, 1944.
34. **Kirk, R.,** Spread of infective hepatitis, *Lancet,* 1, 80, 1945.
35. Homologous serum jaundice, memorandum prepared by medical officers of the Ministry of Health, *Lancet,* 1, 83, 1943.
36. **MacCallum, F. O.,** Early studies on viral hepatitis, *Br. Med. Bull.,* 28, 105, 1972.
37. *WHO Technical Report Series,* 512, 1973, p. 9.
38. **Findlay, G. M., Dunlop, J. L., Brown, H. C.,** Observations on epidemic catarrhal jaundice, *Trans. R. Soc. Trop. Med. Hyg.,* 25, 7, 1931.
39. **Yoshibumi, H., Shigemoto, T.,** Human experiment with epidemic jaundice, *Acta Paediatr. Jpn.,* 47, 975, 1941.
40. **Voegt, H.,** Zur Etiologie der Hepatitis Epidemica, *Muench. Med. Wochenschr.,* 89, 76, 1942.
41. **MacCallum, F. O. and Bradley, W. H.,** Transmission of infective hepatitis to human volunteers, *Lancet,* 2, 228, 1944.
42. **Havens, J. P., Jr., Ward, R., Drill, V. A., and Paul, J. R.,** Experimental production of hepatitis by feeding icterogenic materials, *Proc. Soc. Exp. Biol. Med.,* 57, 206, 1944.
43. **Neefe, J. R., Gellis, S. S., and Stokes, J., Jr.,** Homologous serum hepatitis and infectious (epidemic) hepatitis; studies in volunteers bearing on immunological and other characteristics of the etiological agents, *Am. J. Med.,* 1, 9, 1946.
44. **Neefe, J. R. and Stokes, J., Jr.,** An epidemic of infectious hepatitis apparently due to a water borne agent, *JAMA,* 128, 1063, 1945.
45. **Havens, W. P., Jr.,** Elimination in human feces of infectious hepatitis virus parenterally introduced, *Proc. Soc. Exp. Biol. Med.,* 59, 148, 1945.
46. **Findlay, G. M. and Willcox, R. R.,** Transmission of infective hepatitis by feces and urine, *Lancet,* 1, 212, 1945.
47. **Havens, W. P., Jr.,** Period of infectivity of patients with experimentally induced infectious hepatitis, *J. Exp. Med.,* 83, 251, 1946.
48. **Francis, T., Jr., Frisch, A. W., and Quilligan, J. J., Jr.,** Demonstrations of infectious hepatitis virus in presymptomatic period after transfer by transfusion, *Proc. Soc. Exp. Biol. Med.,* 61, 276, 1946.
49. **Neefe, J. R., Stokes, J., Jr., and Reinhold, J. G.,** Oral administration to volunteers of feces from patients with homologous serum hepatitis and infectious (epidemic) hepatitis, *Am. J. Med. Sci.,* 210, 20, 1945.
50. **Havens, W. P., Jr.,** Immunity in experimentally induced infectious hepatitis, *J. Exp. Med.,* 84, 403, 1946.
51. **Neefe, J. R., Stokes, J., Jr., and Gellis, S. S.,** Homologous serum hepatitis and infectious (epidemic) hepatitis: experimental study of immunity and cross immunity in volunteers; preliminary report, *Am. J. Med. Sci.,* 210, 561, 1945.
52. **Havens, W. P., Jr.,** Experimental cross immunity between infectious hepatitis and homologous serum jaundice, *Proc. Soc. Exp. Biol. Med.,* 59, 148, 1945.
53. **MacCallum, F. O., McFarlan, A. M., and Miles, J. A. R.,** Infective hepatitis: studies in East Anglia during the period 1943-47, *Med. Res. Counc. Spec. Rep. Ser.,* No. 273, H.M.S.O. London, 1951.
54. **Oliphant, J. W., Gillam, A. G., and Larson, C. L.,** Jaundice following administration of human serum, *Public Health Rep.,* 58, 1233, 1943.
55. **Stokes, J., Jr. and Neefe, J. R.,** Prevention and attenutation of infectious hepatitis gamma globulin; preliminary note, JAMA, 127, 144, 1945.
56. **Havens, W. P., Jr. and Paul, J. R.,** Prevention of infectious hepatitis with gamma globulin, *JAMA,* 129, 270, 1945.

57. **Stokes, J., Jr., Farquuhar, J. A., Drake, M. E., Capps, R. B., Ward, C. S., Jr., and Kitts, A. W.,** Infectious hepatitis: length of protection by immune serum globulin (serum globulin) during epidemics, *JAMA,* 147, 714, 1951.
58. **Gellis, S. S., Stokes, J., Jr., Brother, G. M., Hall, W. M., Gilmore, H. R., and Beyer, E., and Morrissey, R. A.,** The use of human immune serum globulin (gamma globulin) in infectious (epidemic) hepatitis in the Mediterranean theatre of operations, *JAMA,* 128, 1062, 1945.
59. **Havens, W. P., Jr.,** Properties of the etiologic agent of infectious hepatitis, *Proc. Soc. Exp. Biol. Med.,* 58, 203, 1945.
60. **Neefe, J. R., Stokes, J., Jr., Baty, J. B., and Reinhold, J. G.,** Disinfection of water containing causative agent of infectious (epidemic) hepatitis, *JAMA,* 128, 1076, 1945.
61. **Hoff, E. C.,** *Preventive Medicine in World War II,* Vol. 2, Coates, J. B., Jr., Ed., Office of the Surgeon General, Department of the Army, Washington, D.C., 1955, 114.
62. **Neefe, J. R., Baty, J. B., Reinhold, J. G., and Stokes, J., Jr.,** Inactivation of the virus of infectious hepatitis in drinking water, *Am. J. Public Health,* 37, 365, 1947.
63. **Krugman, S.,** The Willowbrook hepatitis studies revisited: ethical aspects, *J. Infect. Dis.,* 8, 157, 1986.
64. **Ward, R., Krugman, S., Giles, J. P., Jacobs, A. M., and Bodansky, O.,** Infectious hepatitis: studies of its natural history and prevention, *New Engl. J. Med.,* 258, 407, 1958.
65. **Karmen, N.,** A note on the spectrophotometric assay of glutamic-oxaloacetic transminase in human blood serum, *J. Clin. Invest.,* 34, 131, 1955.
66. **Wroblewski, F., Jeruis, G., and LaDue, J. S.,** The diagnostic, prognostic and epidemiologic significance of serum glutamic oxaloacetic transaminase (SGOT) attenuations in acute hepatitis, *Ann. Intern. Med.,* 45, 782, 1956.
67. **Wroblewski, F. and LaDue, J. S.,** Serum glutamic pyruvic transaminase in cardiac and hepatic disease, *Proc. Soc. Exp. Biol. Med.,* 91, 569, 1956.
68. **Krugman, S., Ward, R., Giles, J. P., Bodansky, O., and Jacobs, A. M.,** Infectious hepatitis: detection of virus during the incubation period and in clincally inapparent infections, *N. Engl. J. Med.,* 261, 729, 1959.
69. **Krugman, S., Ward, R., and Giles, J. P.,** The natural history of infectious hepatitis, *Am. J. Med.,* 32, 717, 1962.
70. **Giles, J. P., Liebhaber, H., Krugman, S., and Lattimer, C.,** Early viremia and viruria in infectious hepatitis, *Virology,* 24, 107, 1964.
71. **Krugman, S., Giles, J. P., and Hammond, J.,** Infectious hepatitis: evidence for two distinctive clinical, epidemiological and immunological types of infection, *JAMA,* 200, 365, 1967.
72. **Krugman, S.,** Effect of human immune serum globulin on infectivity of hepatitis A virus, *J. Infect. Dis.,* 134, 70, 1976.
73. **Boggs, J. D., Melnick, J. L., Conrad, M. E., and Felsher, B. F.,** Viral hepatitis: clinical and tissue culture studies, *JAMA,* 214, 1041, 1970.
74. **Melnick, J. L. and Boggs, J. D.,** Human volunteer and tissue culture studies of viral hepatitis, *Can. Med. Assoc. J.,* 106, 461, 1972.
75. **Decker, R. H., Overby, L. R., Ling, C.-M., Frosner, G., Deinhardt, F., and Boggs, J.,** Serologic studies of transmission of hepatitis A in humans, *J. Infect. Dis.,* 139, 74, 1979.
76. **Feinstone, S. M., Kapikian, A. Z., and Purcell, R. H.,** Hepatitis A: detection by immune electron microscopy of a virus-like antigen associated with acute illness, *Science,* 182, 1026, 1973.

Chapter 2

IDENTIFICATION OF HEPATITIS A VIRUS

I. INTRODUCTION

Although infectious hepatitis (hepatitis A) was distinguished from homologous serum hepatitis (hepatitis B) in the 1940s,[1] the specific etiologic agent resisted identification for another three decades.[2] Human volunteer studies performed during this interval generated proven infectious materials and helped to define some of the characteristics of the causative agent (see Chapter 1).

Classic virological methods were first applied to the isolation and identification of the etiological agent; these included small animal inoculation, inoculation of embryonated chicken eggs, and inoculation of in vitro cell cultures.[3] Such studies were either negative, confusing, or resulted in the isolation of contaminating viruses. Later, serological methods were applied to the search for the agent; this approach was eventually successful in 1973, but isolation of the virus in tissue culture did not occur for another 6 years.

Epidemiologic studies and human volunteer experiments performed during and after World War II showed that hepatitis A was caused by a filterable agent that survived ether extraction. The disease was shown to be spread by the fecal-oral route, and contaminated water was often implicated as the cause of point source outbreaks. Furthermore, in distinction to hepatitis B, a carrier state did not seem to exist. Later, studies performed at the Willowbrook School provided information on the duration of viremia and shedding of virus into the feces as well as providing a well-characterized collection of infectious materials such as the MS-1 inoculum.[4] This inoculum was used to generate larger quantities of reagents in adult volunteers, including preinfection, acute and convalescent sera, and serial fecal specimens[5] (see Chapter 1). The existence of this knowledge and the availability of pedigreed reagents ultimately led to the identification of the etiologic agent.

II. VIROLOGIC METHODS

A. Cell Culture Studies

During the expansion of cell culture techniques in the later 1940s and 1950s, many of the most important human viruses were isolated and identified. As there was no reason to believe that the hepatitis agents would resist identification by these techniques, a large number of cell culture studies were initiated.

1. The Detroit-6 System

The most important of these was undertaken by Rightsel, McLean, and coworkers at the Parke-Davis Company.[6] Though several cell lines were investigated, the most extensive studies were undertaken with Detroit-6 cells, a continuous line of epithelial cells that had been derived from the bone marrow of a patient with lung cancer. These investigators described a cytopathic effect (CPE) that developed in Detroit-6 cells after inoculation with a variety of clinical specimens from patients with putative infectious or serum hepatitis (see Figure 1). The CPE could be serially passaged and appeared to be neutralized by late convalescent sera from patients with infectious hepatitis. Neutralization tests with sera from patients convalescing from what was thought to be serum hepatitis, produced inconsistent results, while sera from patients in the early convalescent stage of infectious hepatitis occasionally produced a CPE rather than neutralization.

Numerous specimens were tested by the group including coded panels provided by other

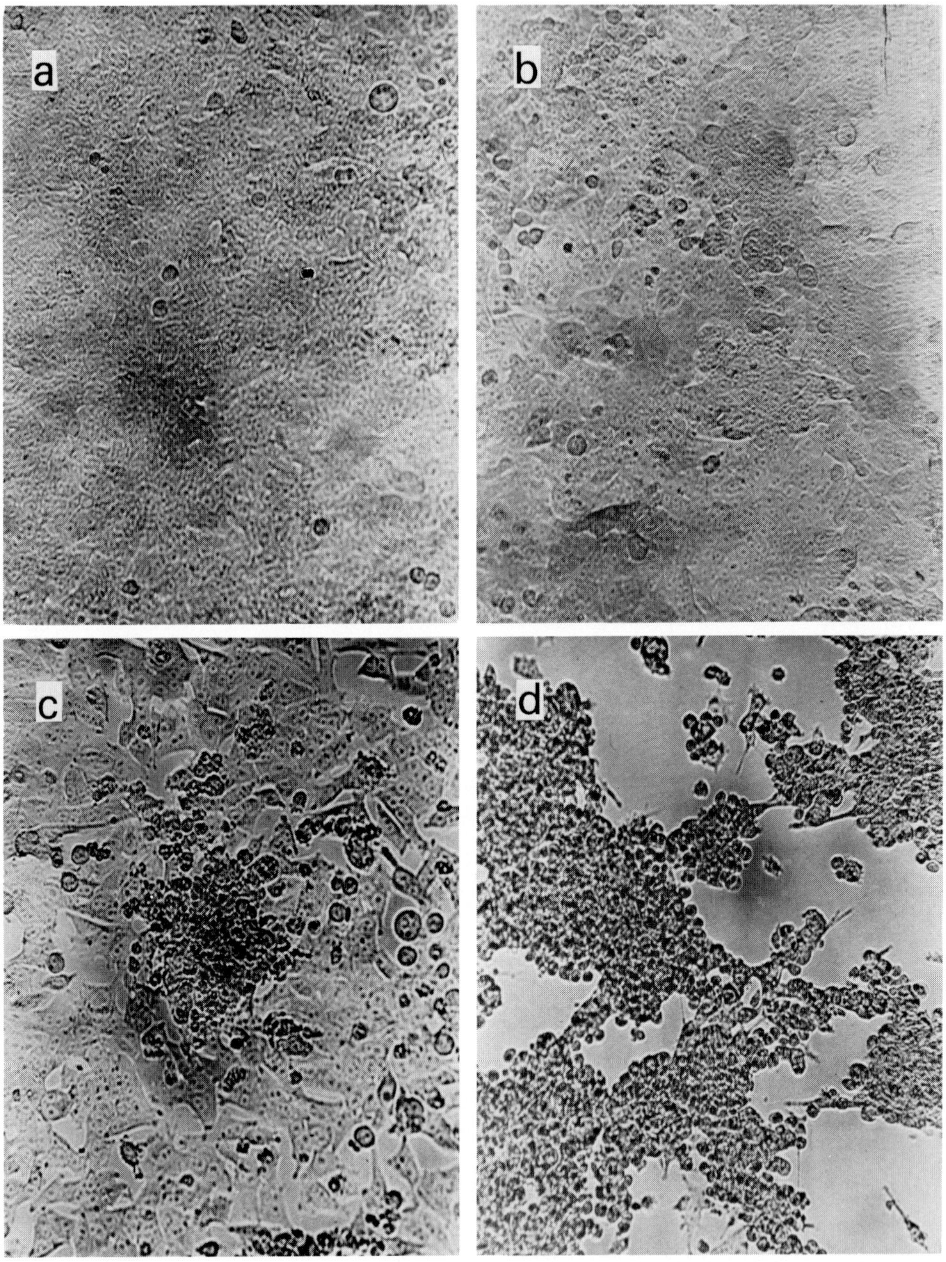

FIGURE 1. Detroit-6 cells inoculated with serum from a patient with infectious hepatitis (A) uninfected cell control, (B) cells inoculated with serum from a healthy individual, (C) cells inoculated with serum from a patient with infectious hepatitis showing a 1+ CPE, and (D) the same specimen several days later showing a 4+ CPE. (From a series of experiments in the author's laboratory.)

laboratories; most of the specimens tested were sera but some fecal specimens were also shown to produce a CPE. Many other cell lines were analyzed by the Parke-Davis group but only the Detroit-6 cells appeared to produce promising results.

Other investigators had a difficult time reproducing the results of the Parke-Davis group. The Detroit-6 cells could only be maintained if they were cultured in media containing human serum and seemed to spontaneously lose their sensitivity to the CPE-producing agents.

Some human sera either failed to support the growth of the cells or altered their sensitivity to the CPE-producing agents, while others appeared to neutralize the CPE. Although it was clear that a higher proportion of specimens that should have contained the infectious hepatitis virus produced a CPE, some specimens that on epidemiological grounds should not have contained the virus, also produced a CPE.

The Detroit-6 system was also established in Australia by Cole, a member of the original Parke-Davis team. In a separate set of experiments the Australian group claimed to isolate a CPE-producing agent from liver biopsy specimens obtained from a baby with biliary atresia and from three babies with neonatal hepatitis.[7] In a second report Cole inoculated Detroit-6 cells with the sera from patients with infectious hepatitis and other forms of liver disease.[8] While all 24 sera from patients with infectious hepatitis caused a CPE, a similar effect was also detected in cells inoculated with serum from one patient with serologically confirmed infectious mononucleosis, one with cholelithiasis and another with a pancreatic cyst, as well as from sera from 3 of 20 normal blood donors. Although CPE-producing agents were sometimes recovered from the serum of patients with acute hepatitis but not from controls tested under code, these results were not reproducible. Because of concern about the specificity of the results being obtained, subsequent studies were performed under code and independently evaluated by two observers. Under these conditions when six sera from patients with infectious hepatitis and a similar number of sera from controls were inoculated into Detroit-6 cells, a CPE was detected in 11 of 12 specimens.[9] Cross and Marmion, who were asked to independently evaluate the specificity of the observations, found no correlation between the source of inoculum and the development of CPE.[10] In addition, they observed poor reproducibility between the results obtained with duplicate aliquots of the same specimen, control inocula often being negative in one test but strongly positive in another. It was concluded from these studies that the CPE observed in Detroit-6 cells was not caused by replication of the HAV and its origin has never been satisfactorily explained.

Due to internal inconsistencies of the system as well as the failure of other workers to reproduce the original observations, interest in the Detroit-6 system gradually disappeared until 1970 when these cells were again used in an attempt to isolate HAV from specimens collected from volunteers infected with the MS-1 strain of HAV.[5]

Plasma collected from a volunteer K, 30 days after inoculation (K30) was shown to be infectious for other volunteers. This plasma was inoculated onto a cloned line of Detroit-6 cells and a CPE noted after one subpassage.[11] The CPE-producing agent was passaged more than 20 times in Detroit-6 cells and developed a titer of about 10^4 $TCID_{50}/m\ell$ of cell culture harvest. The agent was demonstrated to pass through a 50 nm Millipore filter but not 25 nm or 10 nm Wright-Flemming gradacol membranes, and to be resistant to ether. Electron microscopy of inocula containing the CPE-producing agent revealed small spherical virus-like particles measuring 18 to 20 nm in diameter. Both full and empty particles were observed but some preparations had a predominance of empty particles. Convalescent serum from volunteer K and one of two other volunteers who developed hepatitis after inoculation with MS-1, neutralized the CPE to titers of 1:64 and 1:16, respectively, but the third volunteer's convalescent serum had no neutralizing activity.

Further studies on the agent isolated from the K-30 plasma suggested that it was a parvovirus.[12] In addition to its size and resistance to ether, the virus was shown to resist both acid pH and heating to 50°C, and to grow more rapidly in the presence of adenovirus. The origin of this agent is not clear, it seems most likely to have been a contaminating virus present in minute quantities in the original cell cultures.

B. Animal Experiments

1. The GB Agent

In 1967, Deinhardt and colleagues at the University of Chicago reported that marmoset monkeys could serve as models of human hepatitis.[13] One of the inocula used in their studies was a specimen of plasma obtained from a surgeon, GB, who had developed acute hepatitis without a known exposure to infection. Plasma from this person transmitted hepatitis to four of four marmosets and was passed five times from animal to animal. The GB agent was initially thought to be the causative agent of hepatitis A and was extensively characterized.[14]

Data gradually accumulated to show that while GB was capable of inducing hepatitis in some experimental animals it was not the HAV. Parks and Melnick at Baylor College of Medicine[15] confirmed that the GB inoculum, after five passages in marmosets, regularly produced hepatitis in cotton-topped (*Saguinus oedipus*) and white-lipped marmosets (*S. nigricollis*) but not in the common marmoset (*Callithrix jacchus*). Using an inoculum with a titer of $>10^4$ $<10^5$ marmoset ID/mℓ, they demonstrated that the infectivity of GB could be greatly reduced by treatment with ether (for 18 hr at 4°C) and was completely eliminated by heating to 56°C for 30 min, treatments that should have had no effect on HAV. In filtration studies, the virus was not retained by a 50-nm filter, but the titer was reduced at least 2.5 $\log_{10}$ by passage through a 25-nm filter. Interestingly, three of nine animals inoculated with GB serum which had been passed through a 10-nm filter developed mild hepatitis, with a prolonged incubation period. On the basis of these filtration studies the size of the GB agent was estimated to be about 32 nm.

The infectivity of the GB inoculum was not affected by preincubation with pooled human gamma globulin or with convalescent marmoset sera. In fact, a marmoset serum pool that was obtained from GB-infected animals 60 to 80 days after their liver function tests had returned to normal, proved to be infectious unless heated to 56°C. Despite the absence of neutralizing antibody in convalescent sera, 18 out of 20 marmosets that developed hepatitis after primary inoculation were resistant to challenge with the original inoculum.

In contrast to their observations with the GB agent, Parks and Melnick were unable to prove that marmosets were susceptible to infection with pedigreed inocula such as the MS-1 serum pool known to contain HAV.[15] Of 97 cotton-topped marmosets inoculated with 16 different inocula, only 4 animals developed definite hepatitis and another 20 questionable hepatitis. As controls the authors inoculated 15 marmosets with normal human sera, 7 with pooled sera from normal marmosets and left a further 48 uninoculated. Of the 70 control animals, 6 developed definite and 20 questionable hepatitis! Thus, in these studies there appeared to be little difference in the frequency with which hepatitis developed in controls or animals which had been inoculated with material from patients with HAV.

Sera from two marmosets that were inoculated with sera collected from patients with hepatitis A transmitted hepatitis to other marmosets. In addition, sera from several uninoculated controls also contained transmissible agents. In both situations challenge studies suggested that transmissible agents were related to GB.

In summary, Parks and Melnick found that while the GB agent regularly induced hepatitis in marmosets, bona fide inocula from patients with hepatitis A did not. In addition a GB-like agent could sometimes be recovered in marmosets which were inoculated with human or marmoset sera. GB hepatitis could not be neutralized by pooled human gamma globulin preparations or by convalescent sera from infected animals. Finally, some of the characteristics of the GB agent, namely sensitivity to ether and heat, differed from those reported for HAV from experiments in human volunteers.[14] It was therefore concluded that GB was probably a marmoset virus which was not related to human disease.

In retrospect, the most puzzling aspect of the Baylor studies was their failure to demonstrate transmission to marmosets with inocula known to contain HAV. Perhaps some of the animals that were used were already immune to infection or alternatively, the presence of another form of hepatitis in the animal colony, presumably a GB-like agent, may have made the results impossible to interpret. In addition it is now felt that *S. mystax* and *S. labiatus* species of marmosets are a better model for HAV infections than the *S. oedipus* and *S. nigricollis* species used at Baylor.

Table 1
CHARACTERISTICS OF THE GB AGENT REVEALED BY STUDIES IN MARMOSETS

Morphology	?	Neutralization	
EM	Two particles seen 22 nm and 35 nm	Convalescent serum	No
Filtration	<20—30 nm	Pooled immunoglobulin	Partial
Density in CsCl	1.19—1.23 g/mℓ	Results of challenge with:	
Nucleic acid	?	GB	Resistant
Stability		MS-1	Susceptible
Ether		Host range	
20% 4°C, 18 hr	Stable	Chimpanzees	No
50% 4°C, 18 hr	Partially inactivated	Marmosets	
Temperature (°C)		*Saguinus* species	Yes
−20°	Stable > 1 month	*Callithrix jacchus*	No
4°	Stable > 1 month	Galagos	No
50°, 30 min	Stable	Squirrel monkeys	No
56°, 30 min	Partially inactivated	Susceptibility	
60°, 30 min	Inactivated	IV or IM	Yes
60°, 60 min	Inactivated	Oral	Yes
100°, 5 min	Inactivated		

The major characteristics of the GB agent which were revealed by animal studies are shown in Table 1. Despite the wealth of information which has been gathered about GB, its relationship to human hepatitis remains unknown. Since it was originally derived from a surgeon with no known hepatitis exposure, it has been speculated that it may be one of the agents responsible for non-A, non-B hepatitis. The surgeon was shown retrospectively to have no serologic evidence of infection by either HAV or HBV and therefore by definition had non-A, non-B hepatitis. Other strains of non-A, non-B hepatitis can be readily transmitted to chimpanzees but marmosets appear to be more difficult to infect. By contrast GB does not appear to infect chimpanzees[16] and there is evidence that marmosets that have recovered from infection with GB are still susceptible to at least one strain of non-A, non-B hepatitis (unpublished observations).

Two morphologically distinct types of particles have been detected by immune electron microscopy in infectious inocula and specimens obtained from animals injected with GB. Almeida and colleagues reported a 22 nm particle in the acute-phase serum of a marmoset with GB hepatitis[17] (see Figure 2A). While the particles appeared to be aggregated by antibody present in the convalescent serum of some infected animals, immune complexes were also detected after addition of normal marmoset serum and in the inoculum alone. Appleton detected particles with a similar size and morphology in the serum and stool of marmosets infected with the serologically related Berlin agent.[18]

By contrast, Dienstag observed particles measuring 35 nm in diameter in homogenates of liver from marmosets (*S. nigricollis* and *S. mystax*) with acute GB hepatitis[19] (see Figure 2B). These particles were aggregated by antibody found in convalescent serum but also appeared to be aggregated by preinoculation serum. It was not stated if these particles could be seen as preformed immune complexes if no serum was added. Neither of these particles meets all the criteria required for acceptance as the etiologic agent.

2. *The Berlin Agent*

In 1968 Kohler reported the occurrence of hepatitis in some marmosets which had been inoculated with pig cell cultures that had, in turn, been inoculated with sera from patients with infectious hepatitis.[20] The Berlin and GB agents seemed to have many properties in common and were later shown to be antigenically related[21] (see Chapter 4).

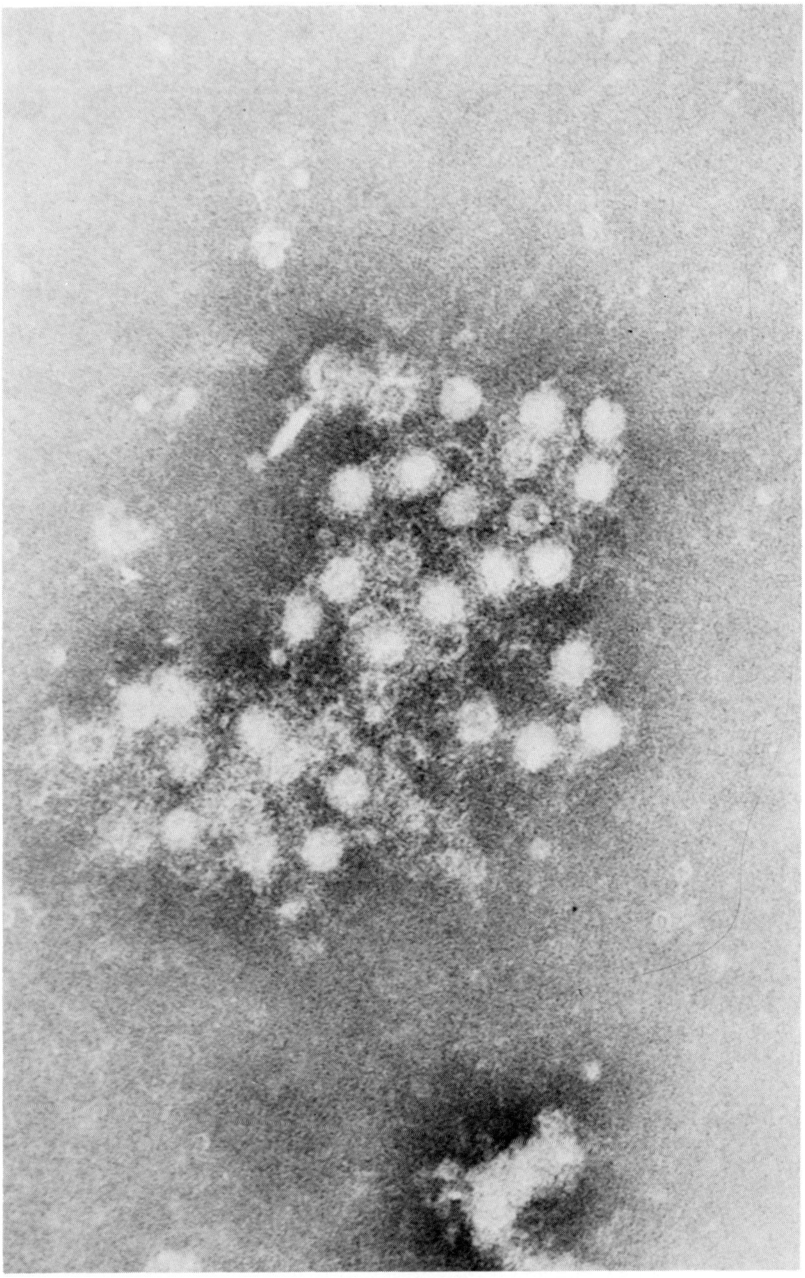

FIGURE 2. (A) An immune aggregate of 22 nm particles obtained by reacting serum collected from a marmoset during the acute phase of infection with the GB agent with serum collected during the convalescent phase. (Magnification × 250,000.) (From the author's collection.) (B) An immune aggregate of 35 nm particles obtained by reacting a gradient-purified liver homogenate from a marmoset infected within the GB agent with convalescent serum from another infected animal. (Magnification × 300,000.) (From Dienstag, J. L., Wagner, J. A., Purcell, R. H., London, W. T., and Lorenz, D. E., *Nature (London)*, 264, 260, 1976. With permission.)

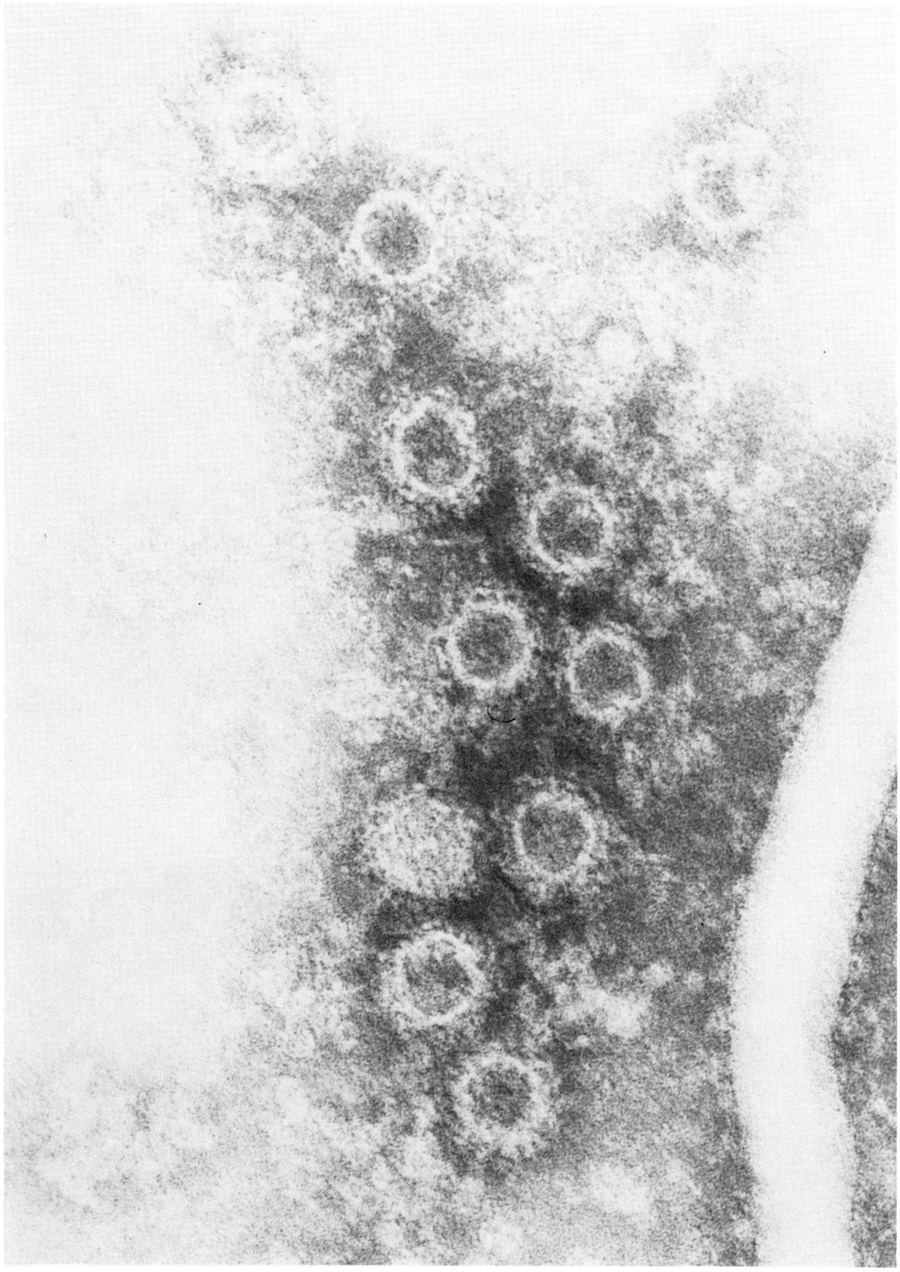

FIGURE 2B.

3. Other Studies

In a coded study with inocula derived from volunteers who had received the MS-1 strain of HAV, Deinhardt and his colleagues[22] demonstrated that sera collected during the acute phase of illness regularly transmitted hepatitis to both white-lipped and cotton-topped marmosets, but that serum from the same volunteers collected prior to infection, failed to produce hepatitis. In a later study, it was shown that while MS-1 and GB hepatitis both induced homologous immunity neither strain induced immunity against the other.[22-24] These studies were of critical importance as they clearly showed that GB was not related to classical hepatitis A (see Chapter 4) and at the same time demonstrated that marmosets were susceptible to HAV. The value of marmosets for the study of hepatitis A was rapidly confirmed in other laboratories[25,26] (see Chapter 3).

III. SEROLOGIC METHODS

While tissue culture, animal, and human volunteer studies continued, the identification of a serum antigen specifically associated with infection with the HBV (the Australia antigen) provided impetus to the use of serologic techniques to search for a HAV-associated antigen in patients with hepatitis A.

A. Milan Antigen

In 1970, Del Prete and his colleagues in Milan described an antibody in the serum of a multiply transfused patient with aplastic anemia, which appeared to react with both the Australia antigen and with a second antigen present in the serum of some patients with epidemic hepatitis.[27] The Milan antigen was detected in 65% of patients involved in three different epidemics of short incubation, (Australia antigen negative) hepatitis, in Italy and in England. Ninety percent of patients in these outbreaks who were tested within 2 weeks of the onset of symptoms had antigen in their serum, with the detection rate falling to 10% by the fourth week. The antigen proved to be labile on storage, with less than 10% of initially positive sera remaining reactive after 65 days at −20°C.

The antigen was detected in approximately 1% of 6000 normal sera, but in distinction to the Australia antigen, it was detected only transiently. The presence of the antigen in some apparently healthy individuals was suggested to be due to a low level of subclinical disease in the general population. Antibody to the Milan antigen was not found in any of the 6000 normal sera or in any of 3000 regular volunteer blood donors.[28]

Preliminary data on the biophysical properties of the antigen[28] suggested that it had a sedimentation coefficient of 11 to 14 S, migrated electrophoretically with β_2-globulin, and would pass through a Sephadex® G-200 column in the void volume. Taylor and colleagues sought to further characterize the Milan antigen and to determine its morphology by electron microscopy.[29] They found that antigen-antibody complexes in agar gels stained with fat red 7B and naphthalene black but not with Sudan black, that the density of the antigen was 1.05 g/cm^3 in sucrose and 1.08 g/cm^3 in cesium chloride (CsCl) and that its activity could be destroyed by ether treatment. Electron microscopy of negatively stained pellets of sera containing the Milan antigen, revealed disk-like structures about 40 to 50 nm in diameter that could be aggregated by specific antibody (see Figure 3). It was deduced from these data that the Milan antigen was probably an abnormal serum lipoprotein that had been previously described in patients with cholestatic jaundice.[30,31]

While the antigen was detected more frequently in patients with hepatitis A it was also found in the sera of some patients with hepatitis B and in sera from patients with other forms of liver disease such as halothane hepatitis, primary biliary cirrhosis, and Crohn's disease. The published reports do not mention whether the Milan antigen was sought or detected in feces or if patients with hepatitis A developed antibody to the antigen during the convalescent phase of their illness.[29]

B. Fecal Antigen

Recognizing that hepatitis A is generally transmitted by the fecal-oral route, Ferris and colleagues in Melbourne decided to study feces from patients with this disease for the presence of specific viral antigens.[32] In their initial studies sera from 37 multiply-transfused hemophiliacs were used to test fecal extracts from six patients with HBsAg-negative hepatitis and

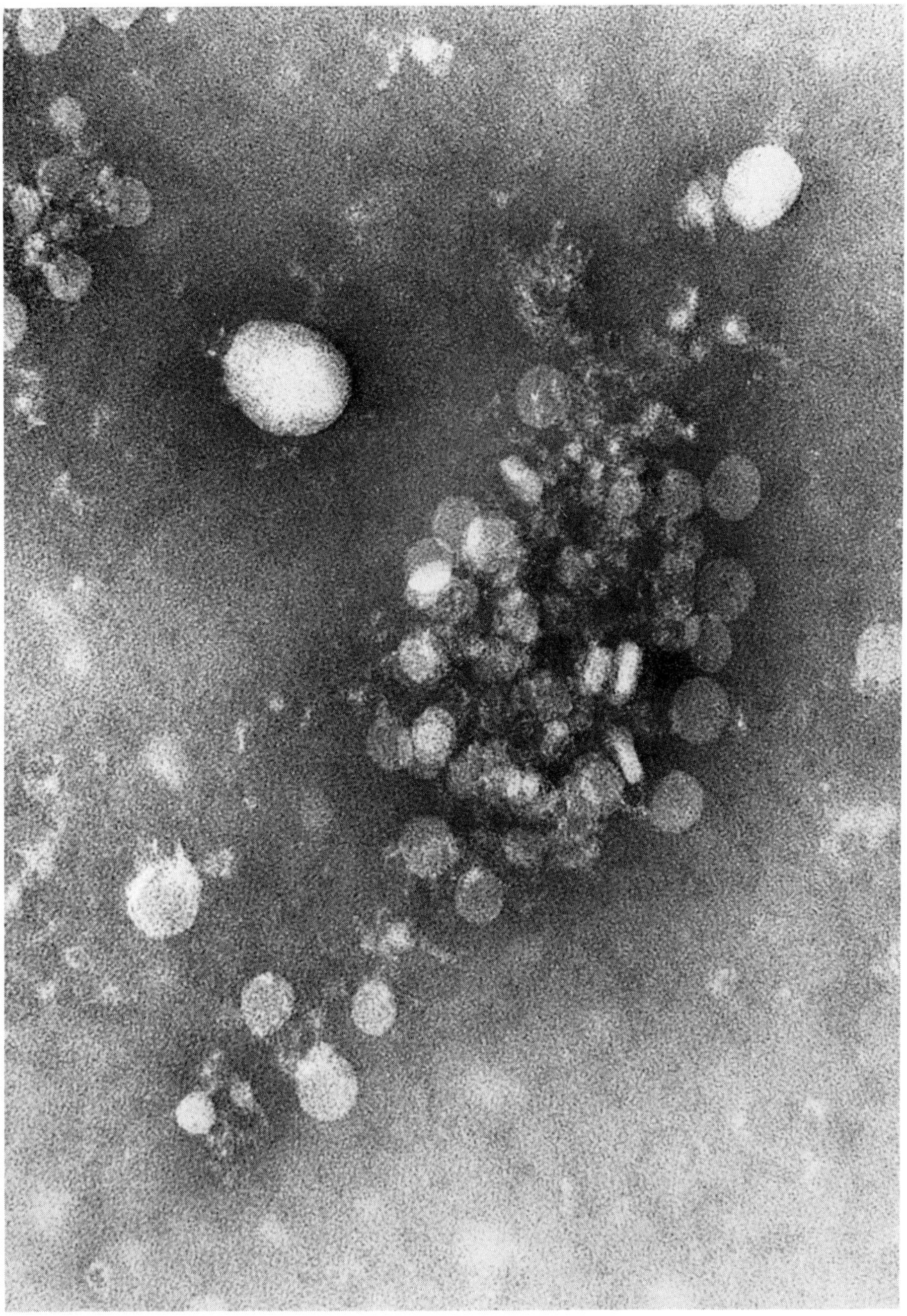

FIGURE 3. An aggregate of Milan antigen showing two distinctive forms: a roughly spherical, pleomorphic particle with a diameter of approximately 40 to 50 nm and brightly staining rod-like structures with a long axis of 40 to 50 nm and a short axis of 10 to 13 nm. (Magnification ×181,000.) (From a series by June Almeida, Patricia Taylor, and A. J. Zuckerman. With permission.)

seven patients with diseases other than hepatitis. In simple agar diffusion tests two of these sera produced single-precipitin bands with specimens from four of six patients with hepatitis and with only one of the other specimens. Neither of the two hemophiliac sera reacted with HBsAg by gel diffusion. Because only limited volumes of these two detector sera existed, the positive fecal extracts and antigen containing sucrose-gradient fractions of these extracts, were used to immunize rabbits. After appropriate absorption, some of the rabbit sera also produced precipitin bands with the fecal extracts from patients with hepatitis.

A larger study was performed with one of the rabbit sera. Single precipitin bands were detected in fecal extracts from 90 of 220 patients with acute viral hepatitis compared with 5 of 158 patients with other diseases. In patients with hepatitis, fecal antigen was usually present in the first specimen collected after the patient was admitted to hospital, becoming undetectable an average of 18 days after the onset of dark urine. Of the 220 patients with hepatitis, 13 had HBsAg detected in sera collected during the acute phase of their illness and were thus either acutely or chronically infected with HBV; three of these patients were also positive for fecal antigen.

Immune electron microscopy (IEM) of clarified fecal extracts containing the fecal antigen revealed aggregates of two types of particles:[33] large numbers of 15 to 25 nm spherical particles that resembled HBsAg and small numbers of 40 to 45 nm particles that were said to resemble the hepatitis B virion (Dane particle) although they were more pleomorphic and usually had no internal structure (see Figure 4). The resemblance of these large particles to HBV appeared to be even greater, if HBV was first treated with sodium deoxycholate. Particles resembling the elongated or filamentous form of HBsAg were not observed in fecal antigen-positive specimens. Although antibody to HBsAg did not react with fecal antigen, antibody produced against preparations containing large number of mature hepatitis B virions (Dane particles) appeared to aggregate the larger fecal antigen particles and antibody to fecal antigen appeared to react with Dane particles but not HBsAg particles.

Madden et al. evaluated the fecal antigen test in an outbreak of hepatitis that occurred at a school for mentally retarded children in Virginia.[34] They were only able to detect the antigen in 2 of 34 fecal specimens from 22 patients with hepatitis. They concluded that fecal antigen was probably not related to this outbreak of hepatitis.

In summary, fecal antigen was detected originally in the feces of patients with HBsAg-negative hepatitis, using sera from multiply transfused hemophiliacs. Rabbit antiserum to this antigen was used to detect the antigen in about 40% of hospitalized patients with HBsAg-negative hepatitis in Australia. Particles that resembled HBsAg and HBV were occasionally observed in positive fecal extracts by electron microscopy (EM) and there appeared to be some cross-reactivity between these particles and the hepatitis B virion. The results were not reproducible in other laboratories and the precise nature of the fecal antigen has not been established.

In retrospective evaluation of electron micrographs of immune aggregates of fecal antigen, it is clear that most of the particles originally described were not viruses, however some of the aggregates contain particles that may well be HAV (see Figure 4). When three of the animal antisera to fecal antigen were subsequently evaluated with specific tests, it was found that two had low levels of anti-HAV, indicating that the fecal extracts used to immunize rabbits had indeed contained HAV.[35] This study illustrated some of the difficulties of performing EM and IEM on fecal material and underlined the importance of determining the specificity of antigen-antibody reactions by the use of carefully chosen controls.

IV. DISCOVERY OF THE VIRUS

In 1972, workers in the Laboratory of Infectious Diseases at the National Institute of Allergy and Infectious Diseases (NIAID) of the National Institutes of Health (NIH) began

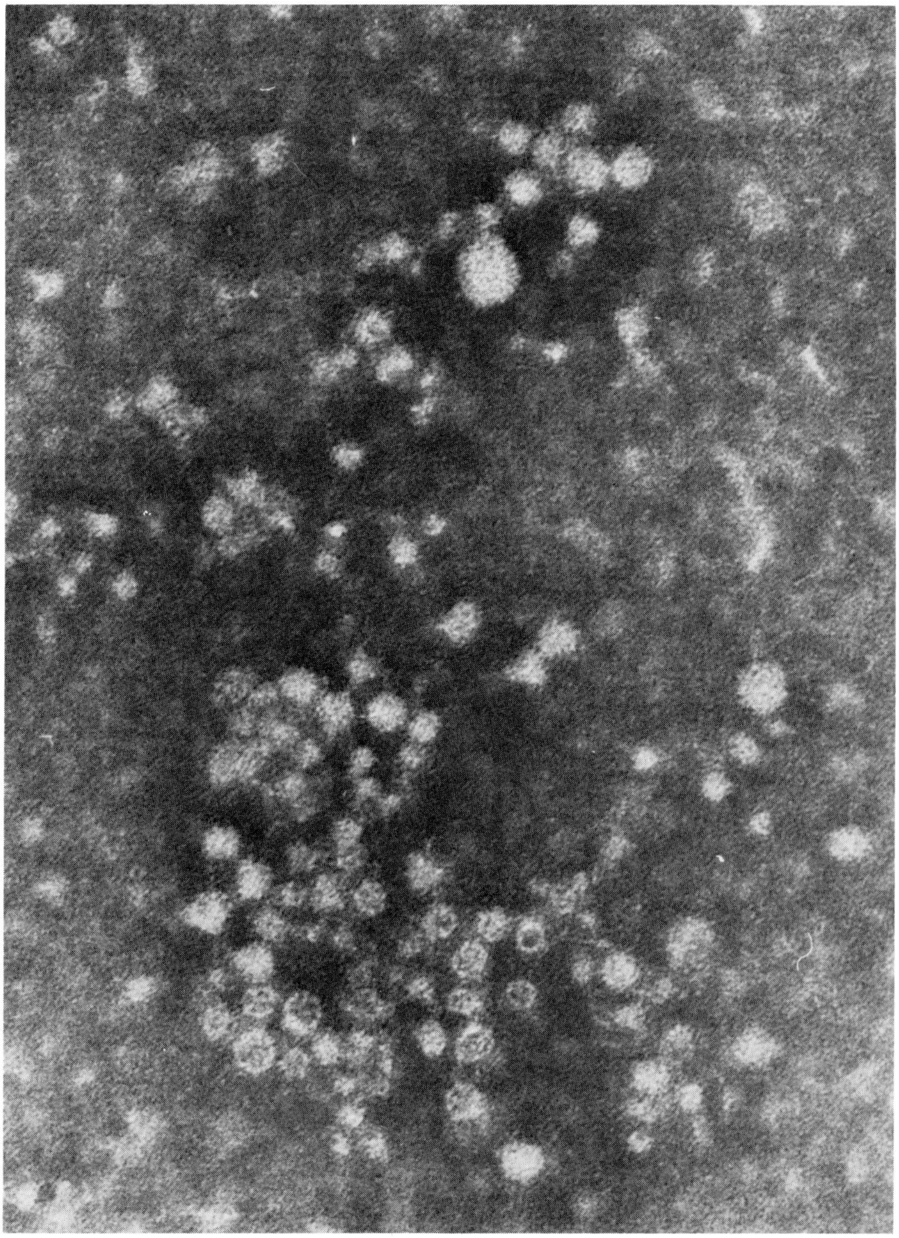

FIGURE 4. Immune aggregates obtained by reacting a partially purified extract of feces from a patient with hepatitis A containing "fecal antigen," with serum from a rabbit which had been immunized with purified antigen. Particles of different sizes can be seen. (Magnification ×200,000.) (From the authors' collection.)

to apply the technique of immune electron microscopy to the problem of hepatitis A. Albert Kapikian had learned the technique from June Almeida[36] while on sabbatical leave at the Wellcome Laboratories in England, and had applied it to the search for fastidious viruses which might be involved in the etiology of upper respiratory-tract infections. By reacting secretions from patients with upper respiratory-tract infections (from whom no virus could be isolated by conventional tissue culture techniques) with convalescent serum from one particular patient, he was able to identify coronavirus-like particles and to show that the patient from whom the specimen had been obtained developed a specific antibody response.[37] Later, Kapikian and his colleagues used IEM to detect the first human gastroenteritis virus, the Norwalk agent, in the feces of patients with diarrhea.[38] Since IEM is a serologic test, they were also able to use it to detect antibody to the virus and demonstrate that infected subjects developed increasing quantities of antibody to this particle.

It was reasoned that since hepatitis A was an enteric infection, it might yield to a similar approach. The disease was known to be spread by the fecal-oral route and that although a brief viremia occurred during the incubation period, feces were more highly infectious and virus was shed into them for a prolonged period of time. It seemed reasonable therefore to follow the Australian approach and search for virus particles in fecal samples from patients with naturally acquired and experimentally induced infections. While IEM can be an extremely powerful technique, there are many difficulties in employing it to identify an unknown virus in a specimen containing as much particulate matter as feces. Since the morphology of the virus was not known and human feces contains many particulate objects, some of which resemble, or are, viruses (human or bacterial) the importance of any particles seen would depend on demonstrating a reaction with specific antibody. When a virus particle reacts with its own antibody, antibody molecules bind to the particles and may cause them to aggregate. These antibody-coated and antibody-aggregated particles are recognizable under the electron microscope from particles without antibody. Even with this marker the search for previously unrecognized viruses is very labor intensive, time consuming, and requires a considerable level of experience and expertise. As differences in the quantity of virus and antibody in sequential specimens from the same patients may be small, it is essential that studies involving putative etiological agents be carried out under code and that appropriate controls be included at each stage.

Acute-phase fecal specimens from several outbreaks of hepatitis A were examined, using both convalescent serum and commercial gamma globulin prepared from large pools of human serum as a source of antibody.

In specimens collected during an outbreak among young adults on Palau Island, Micronesia,[39] Feinstone and his colleagues observed an unusual particle in the feces of 2 of 17 patients (see Figure 5). This particle resembled a complex virus with a core and outer envelope but was highly pleomorphic and variable in size. While 8 out of 15 patients involved in the outbreak had antibody to the particle, a serologic rise could only be demonstrated in two. No serologic response to the particle could be demonstrated in volunteers who received the MS-1 strain of HAV although three appeared to have antibody in sera collected prior to infection. In addition, two of two lots of commercial gamma globulin tested contained large quantities of antibody to the particle. The morphology of these particles was so unusual and the serologic data so unconvincing that it was felt that specimens from other sources should be studied.

The Armed Forces Institute of Pathology then provided the NIH group with several specimens of feces and serum from adult volunteers that had been infected with the MS-1 strain of HAV.[5] Feinstone and his colleagues (see Figure 6) had examined several of these fecal specimens, including one from volunteer K obtained 30 days after inoculation, a time at which his serum was known (from studies in marmosets) to be infectious. They were initially unable to identify particles in any of these specimens by IEM, using either convalescent serum or commercial gamma globulin as a source of antibody. However, a systematic search of fecal specimens from four volunteers which had been provided under code, eventually yielded positive results. Single particles, doublets, and triplets heavily coated with antibody were observed when the acute-phase feces of one of the volunteers was reacted with gamma globulin. The particles had a diameter of approximately 27 nm and resembled

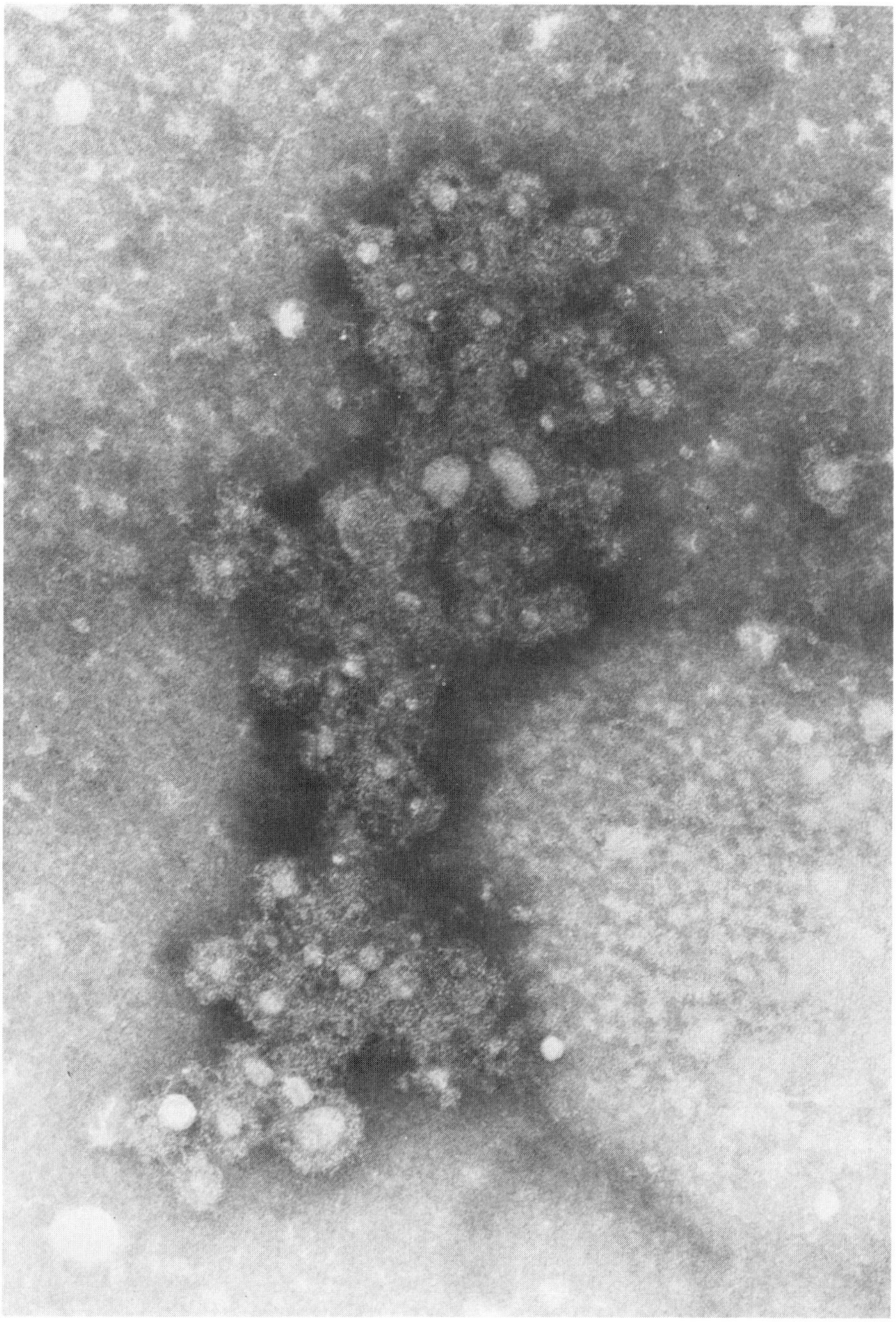

FIGURE 5. (A) Immune aggregates of unusual pleomorphic particles produced by reacting an extract of feces obtained from a patient in Palau during the acute phase of an attack of hepatitis, with convalescent serum obtained from another patient involved in the outbreak. (B) Pleomorphic particles, some of which appear to contain core-like structures, purified from the same specimen of feces by gradient centrifugation. (Magnification ×120,000.) (From the authors' collection.)

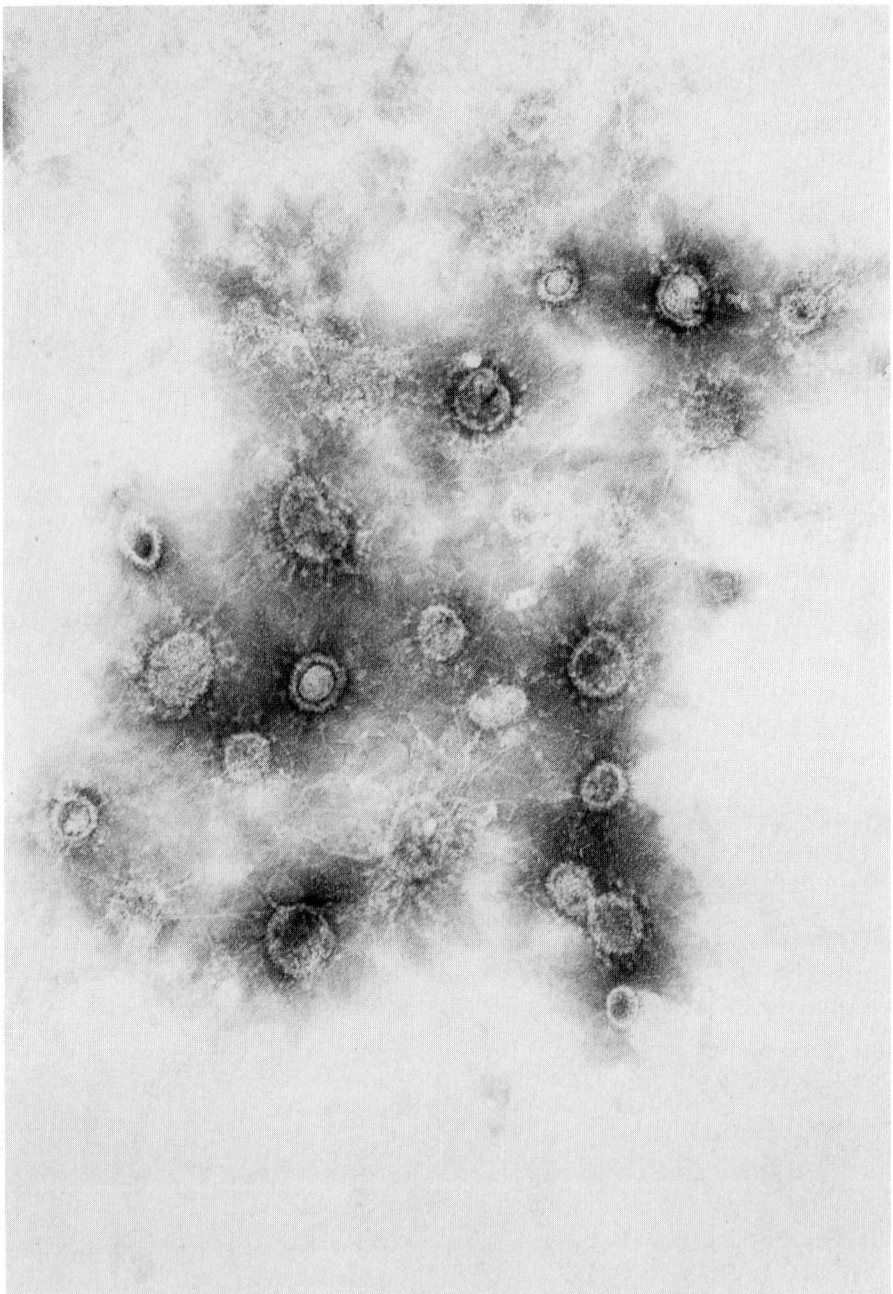

FIGURE 5B.

many other small nonenveloped viruses.[2] The volunteer study provided preinfection serum that could be used as negative controls. The NIH group then tested paried (preinfection and convalescent) sera under code against the specimen that had been shown to contain particles. No antibody-coated or aggregated particles could be identified when fecal extracts containing the particles were reacted with preinoculation sera, however convalescent sera contained antibody which invariably reacted with the particles (see Figure 7).

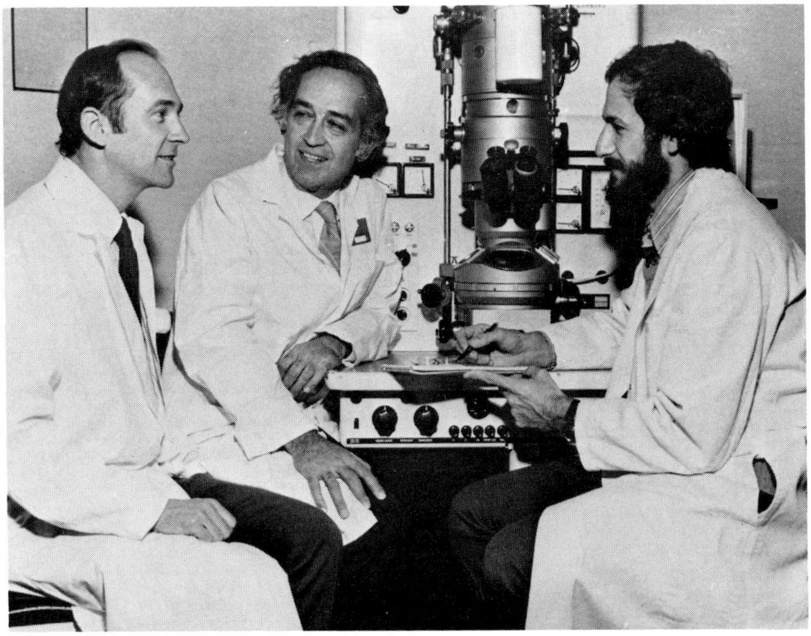

FIGURE 6. L—R: Drs. Robert Purcell, Albert Kapikian, and Stephen Feinstone of the National Institute of Allergy and Infectious Diseases, NIH, seated around the electron microscope used to identify HAV. (Photograph courtesy of NIH.)

The specificity of the reaction was proven by demonstrating seroconversions to the particle in six out of six experimentally infected volunteers and seroconversion or antibody rises in six out of six patients with naturally acquired hepatitis A infections occurring in Massachusetts and American Samoa. No serologic relationship was demonstrated between the 27 nm particle and the HBV or with the Norwalk agent which is morphologically similar to HAV and also found in the feces (see Table 2).

By the time this study was undertaken the virus which had been isolated from patients with hepatitis A in Detroit-6 cells had been shown to be a parvovirus, antigenically related to the Kilham rat virus complex.[12] In coded experiments, convalescent sera from volunteers infected with hepatitis A were found to have no reactivity against the Kilham virus nor did hyperimmune serum against the rat virus react with the 27 nm hepatitis A particle. Similarly, no antigenic relationship was demonstrated between the 27 nm particle and the GB agent. Finally, using hyperimmune rabbit sera prepared by Geoff Cross, the NIH group demonstrated that there was no relationship between fecal antigen and the particle they had detected.

In retrospect the evidence that the 27 nm virus-like particle was the hepatitis HAV was overwhelming, however, the long history of failure in this field caused some workers to initially doubt NIH data and a vigorous debate ensued. Almeida et al.[40] expressed concern about the specificity of the IEM findings pointing out that an increase in antibody levels to various antigens in stool can often be observed following acute hepatitis. These doubts were soon overcome when it was demonstrated that shedding of particles in the stool was temporally related to disease, that patients with hepatitis A developed a specific antibody response to the 27 nm particle and the results were confirmed in other centers using materials obtained from a wide range of epidemiological situations including water-borne, food-borne, intrafamily, and institutional outbreaks as well as experimentally transmitted infections.[35,41,42]

Soon after the initial report, 27 nm virus-like particles were detected in the feces of eight of nine Australian patients with naturally acquired hepatitis A, but not in the feces of three

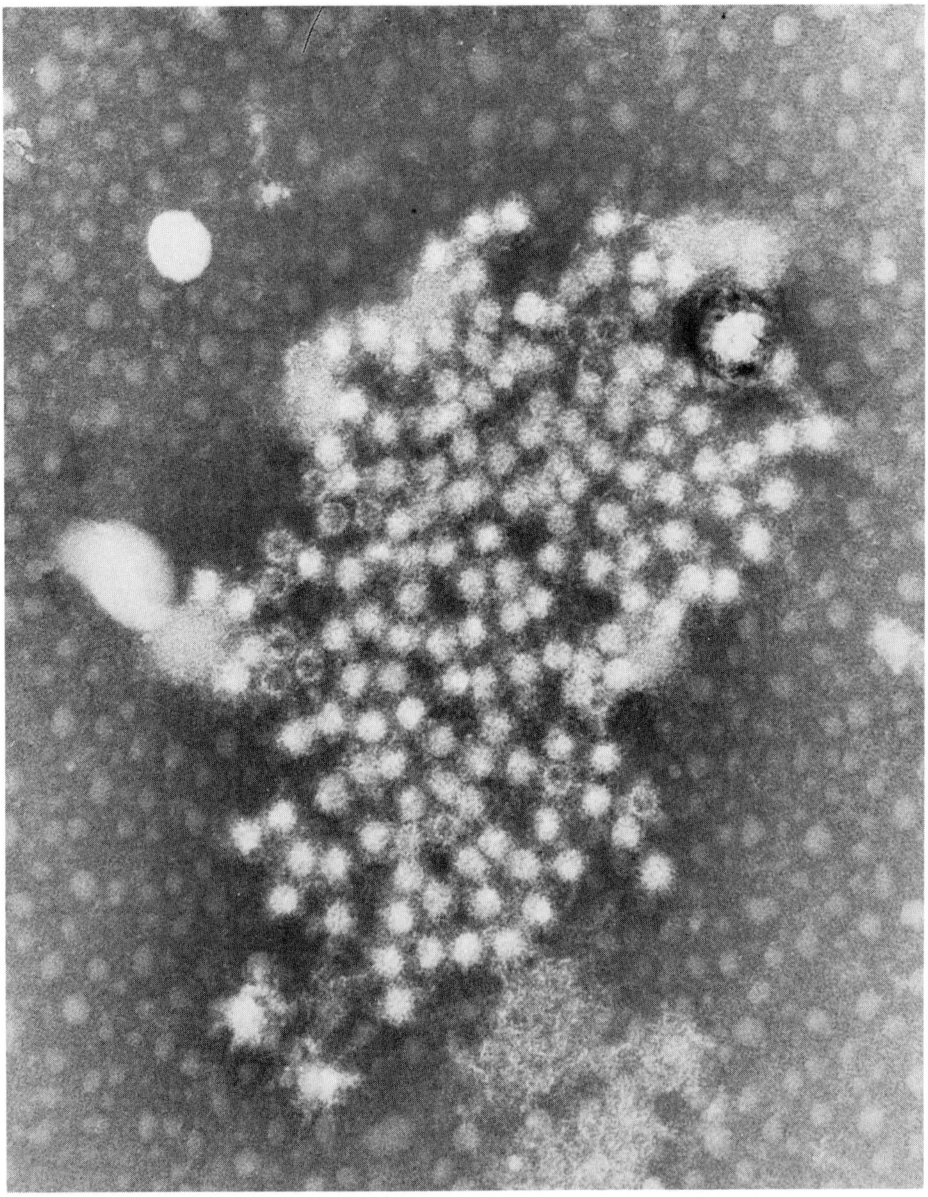

FIGURE 7. An immune aggregate of 27 nm particles obtained by reacting a gradient purified extract of feces collected from a patient with MS-1 type hepatitis, with convalescent serum from a patient with naturally acquired hepatitis A. The particles appear to have cubic symmetry and both full and empty particles can be seen. (Magnification ×136,000.) (From the authors' collection.)

patients with hepatitis B or nine patients with other diseases.[43] When acute and convalescent sera from patients with a variety of different types of liver disease were tested for the presence of antibody to these particles under code, seroconversions were demonstrated only in patients with hepatitis A. The serologic identity of the Australian particles and the particles observed at NIH was confirmed by an exchange of reagents between the laboratories. Shortly after serologically and morphologically identical particles were also observed in the feces of patients involved in a common-source epidemic of hepatitis in Arizona.[44]

Table 2
DETECTION OF ANTIBODY TO HAV IN ACUTE AND CONVALESCENT SERA FROM PATIENTS WITH HEPATITIS A, HEPATITIS B AND NORWALK GASTROENTERITIS

Patient	Acute	Convalescent	Acute	Convalescent
	Experimentally induced hepatitis A		Naturally acquired hepatitis A	
1	0	1—2[a]	0—1	3—4
2	0	1—2	0	3
3	0	3—4	0	3
4	0	3—4	0	2
5	0	1—2	0	3
6	0	3—4	1—2	3—4
	Naturally accquired hepatitis B		Experimentally induced gastroenteritis	
1	0	0	0	0
2	0	0	2—3	2—3

[a] Antibody detected by IEM and quantitated on a 0—4+ scale.

In 1973, Mascoli and her colleagues at the Merck Institute for Therapeutic Research transmitted hepatitis to marmosets by inoculating them with fecal extracts from Costa Rican patients with hepatitis A.[45,46] Virus-like particles measuring 27 nm in diameter were observed in liver homogenates of infected animals and extracts of liver tissue could be used as a source of antigen for complement fixation and immune-adherence hemagglutination tests for hepatitis A antigen and antibody.[47,48]

After Feinstone and his colleagues used the technique of IEM to identify seronegative chimpanzees, workers at the NIH, the Food and Drug Administration (FDA), and the Centers of Disease Control (CDC), demonstrated that these animals could be infected with the MS-1 and Phoenix strains of HAV.[49,50] Following intravenous or oral inoculation, the animals developed enzyme elevations after an incubation period of 3 to 4 weeks. Hepatitis A particles appeared in the feces prior to the onset of abnormal liver function tests, rose to peak levels, and disappeared within a few days of the first enzyme elevations. Circulating antibody to the virus became detectable about the time of onset of biochemical and histological hepatitis and rose gradually, reaching peak levels within 4 to 6 weeks.[51] Although humans were not reinoculated with the chimpanzee material, Koch's postulates were essentially fulfilled and the 27 nm hepatitis A particle was finally accepted as the etiological agent of hepatitis A.

The development of sensitive diagnostic tests for HAAg and anti-HAV greatly facilitated studies on the virus and within a short period of time its buoyant density was shown to be about 1.34 g/cm^3 in CsCl, its sedimentation coefficient 160 S, its genome to be single-strand plus sense RNA, and its capsid to consist of four structural proteins (see Chapter 3). All these characteristics indicated that HAV was a member of a relatively common virus family (the picornaviridae) which had been known for many years and had been extensively studied. Why then had HAV proved so difficult to identify and why had it resisted such strenuous efforts by the best laboratories in the world? While in retrospect the problem seems quite simple, identification of HAV was delayed because the virus replication could not be recognized in the cell culture systems available at the time. Identification of the virus was eventually possible because of the availability of a new technique (IEM) and of well characterized materials (both feces and sera) which had been generated by studies in human volunteers.

In reviewing the work of the 1950s and 1960s, it is clear that several mistakes were made. While it was known that HAV was excreted in the feces for prolonged periods and that viremia was transient and of low titer, many workers confined their studies to serum specimens, presumably because these were easier to obtain and more pleasant to work with! In addition, as the serum hepatitis epidemic in American troops in the 1940s colored the views of clinicians and epidemiologists, the discovery of the Australia antigen in the 1960s colored the view of virologists. Many people assumed that HAV and HBV would have a similar structure even though it was known that the two diseases had distinctive epidemiological, clinical, and virologic features.

As with many new developments in science, the identification of HAV was based on a large amount of preceding work, and occurred at the time when new techniques capable of solving the problem had been developed. The final element was the availability of young scientists who were prepared to undertake the unpleasant task of processing and studying human and animal feces.

REFERENCES

1. **Neefe, J. R., Stokes, J., Jr., and Gellis, S. S.,** Homologous serum hepatitis and infectious (epidemic) hepatitis: experimental study of immunity and loss immunity in volunteers; preliminary report, *Am. J. Med. Sci.,* 210, 561, 1945.
2. **Feinstone, S. M., Kapikian, A. Z., and Purcell, R. H.,** Hepatitis A: detection by immune electron microscopy of a virus-like antigen associated with acute illness, *Science,* 182, 1026, 1973.
3. **MacCollum, F. D.,** Early studies on viral hepatitis, *Br. Med. Bull.,* 28, 105, 1972.
4. **Krugman, S., Giles, J. P., and Hammond, J.,** Infectious hepatitis: evidence for two distinctive clinical, epidemiological and immunological types of infection, *JAMA,* 200, 365, 1967.
5. **Boggs, J. D., Melnick, J. G., Conrad, M. E., and Felsher, B. F.,** Viral hepatitis: clinical and tissue culture studies, *JAMA,* 214, 1041, 1970.
6. **Rightsel, W. A., Keltsch, R. A., Taylor, A. R., Boggs, J. D., and McLean, J. W., Jr.,** Status report on tissue-culture cultivated hepatitis virus. I. Virology laboratories studies, *JAMA,* 177, 678, 1961.
7. **Cole, R. A., Danks, D. M., and Campbell, P. E.,** Hepatitis virus in neonatal liver disease, *Lancet,* 1, 1368, 1965.
8. **Cole, R. A.,** Viral agents from patients with infective hepatitis, *Aust. J. Exp. Biol. Med. Sci.,* 43, 497, 1965.
9. **Ferris, A. A. and Cole, R. A.,** Detroit-6 cells and infective hepatitis, *Med. J. Aust.,* 2, 1213, 1966.
10. **Cross, G. F. and Marmion, B. P.,** Cell culture and infectious hepatitis, *Med. J. Aust.,* 2, 1214, 1966.
11. **Melnick, J. L., Boucher, D. W., Craske, J., and Boggs, J.,** Properties of a virus isolated from patients with MS-1 infectious hepatitis, *J. Infect. Dis.,* 124, 76, 1971.
12. **Mirkovic, R. R., Adamova, V., Boucher, D. W., and Melnick, J. L.,** Identification of the kirk "Hepatitis" virus as a member of the parvovirus (picodnavirus) group, *Proc. Soc. Exp. Biol. Med.,* 138, 626, 1971.
13. **Deinhardt, F., Holmes, A. W., Capps, R. B., and Popper, H.,** Studies on the transmission of human viral hepatitis to marmoset monkeys, *J. Exp. Med.,* 125, 673, 1967.
14. **Parks, W. P., Melnick, J. L., Voss, W. R., Singer, D. B., Rosenberg, H. S., Alcott, J., and Casazza, A. M.,** Characterization of marmoset hepatitis virus, *J. Infect. Dis.,* 120, 548, 1969.
15. **Parks, W. P. and Melnick, J. L.,** Attempted isolation of hepatitis viruses in marmosets, *J. Infect. Dis.,* 120, 539, 1969.
16. **Tabor, E., Peterson, D. A., April, M., Seeff, L. B., and Gerety, R. J.,** Transmission of human non-A, non-B hepatitis to chimpanzees following failure to transmit GB agent hepatitis, *J. Med. Virol.,* 5, 103, 1980.
17. **Almeida, J. D., Deinhardt, F., Holmes, A. W., Peterson, D. A., Wolfe, L., and Zuckerman, A. J.,** Morphology of the GB hepatitis agent, *Nature (London),* 261, 608, 1976.

18. **Appleton, H.,** Virus particles in marmoset hepatitis, *Nature (London),* 267, 729, 1977.
19. **Dienstag, J. L., Wagner, J. A., Purcell, R. H., London, W. T., Lorenz, D. E.,** Virus-like particles and GB agent hepatitis, *Nature (London),* 264, 260, 1976.
20. **Kohler, H. and Apodaca, J.,** Hepatitis bei Affen Nach Verimpfung von Adenovirun aus Menshen mit Virushepatitis. I. Bestimmung der Enzymaktivitaten im Verlauf der Erkrankung, *Zentralbl. Bakteriol.,* 206, 1, 1968.
21. **Appleton, H.,** Hepatitis transmission in marmosets, *J. Infect. Dis.,* 132, 500, 1975.
22. **Deinhardt, F., Holmes, A. W., Wolfe, L., and Junge, V.,** Transmission of viral hepatitis to nonhuman primates, *Vox Sang.,* 19, 261, 1970.
23. **Holmes, A. W., Wolfe, L., Deinhardt, F., and Conrad, M.,** Transmission of human hepatitis to marmosets: further coded studies, *J. Infect. Dis.,* 124, 520, 1971.
24. **Deinhardt, F., Wolfe, L., Junge, V., and Holmes, A. W.,** Viral hepatitis in nonhuman primates, *Can. Med. Assoc. J.,* 106, 468, 1972.
25. **Lorenz, D., Barker, L., Stevens, D., Peterson, M., and Kirchstein, R.,** Hepatitis in the marmoset *Saguinus mystax, Proc. Soc. Exp. Biol. Med.,* 135, 348, 1970.
26. **Mascoli, C. C., Ittensohn, O. L., Villarejos, V. M., Arguedas, J. A., Provost, P. J., and Hilleman, M. R.,** Recovery of hepatitis agents in the marmoset from human cases occurring in Costa Rica, *Proc. Soc. Exp. Biol. Med.,* 142, 276, 1973.
27. **Del Prete, S., Costantino, D., and Doglia, M.,** Detection of new serum-antigen in three epidemics of short-incubation hepatitis, *Lancet,* 2, 579, 1970.
28. **Del Prete, S., Costantino, D., Doglia, M., Fabiani, M. P., Borin, R., and Jacini, A.,** Epidemic hepatitis-associated antigen (Milan Antigen), *Am. J. Dis. Child.,* 123, 326, 1972.
29. **Taylor, P. E., Almeida, J. D., Zuckerman, A. J., and Leach, J. M.,** Relationship of Milan Antigen to abnormal serum lipoprotein, *Am. J. Dis. Child.,* 123, 329, 1972.
30. **Eder, H. A., Russ, E. M., Pritchett, R. A. R., Wilber, M. M., and Barr, D. P.,** Proteinlipid relationships in human plasma: in biliary cirrhosis, obstructive jaundice, and acute hepatitis, *J. Clin. Invest.,* 34, 1147, 1971.
31. **Hamilton, R. L., Havel, R. J., and Kane, J. P.,** Cholestasis: lamellar structure of the abnormal human serum lipoprotein, *Science,* 172, 475, 1971.
32. **Ferris, A. A., Kaldor, J., Gust, I. D., and Cross, G.,** Fecal antigen in viral hepatitis, *Lancet,* 2, 243, 1970.
33. **Cross, G. F., Waugh, M., Ferris, A. A., Gust, I. D., and Kaldor, J.,** Virus-like particles associated with a fecal antigen from hepatitis patients and with Australia antigen, *Aust. J. Exp. Biol. Med. Sci.,* 49, 1, 1971.
34. **Madden, D. L., Mathew, E. B., Dietzman, D. E., Sever, J. L., and Nagler, B.,** Search for FERRIS's fecal antigen in stool samples obtained during a large outbreak of infectious hepatitis, *Proc. Soc. Exp. Biol. Med.,* 139, 1028, 1972.
35. **Purcell, R. H., Dienstag, J. L., Feinstone, S. M., and Kapikian, A. Z.,** Relationship of hepatitis A antigen to viral hepatitis, *Am. J. Med. Sci.,* 270(1), 61, 1975.
36. **Almeida, J. D. and Waterson, A. P.,** The morphology of virus-antibody interaction, *Ad. Virus Res.,* 15, 307, 1969.
37. **Kapikian, A. Z., James, H. D., Jr., Kelly, S. J., and Vaughn, A. L.,** Detection of coronavirus strain 692 by immune electron microscopy, *Infect. Immun.,* 7, 111, 1973.
38. **Kapikian, A. Z., Wyatt, R. G., Dolin, R., Thornhill, T. S., Kalica, A. R., and Chanock, R. M.,** Visualization by immune electron microscopy of a 27 nm particle associated with acute infectious nonbacterial gastroenteritis, *J. Virol.,* 10, 1075, 1972.
39. Trust Territory of the Pacific Islands Monthly Reports, Morbidity and Mortality 3: June to August, 1967.
40. **Almeida, J. D., Gay, F. W., and Wreghitt, T. G.,** Pitfalls in the study of hepatitis A, *Lancet,* 2, 748, 1974.
41. **Locarnini, S. A., Ferris, A. A., Stott, A. C., and Gust, I. D.,** Pitfalls in hepatitis A?, *Lancet,* 2, 1007, 1974.
42. **Dienstag, J. L., Feinstone, S. M., Kapikian, A. Z., and Purcell, R. H.,** Immune electron microscopy and hepatitis A, *Lancet,* 1, 102, 1975.
43. **Locarnini, S. A., Ferris, A. A., Stott, A. C., and Gust, I. D.,** The relationship between a 27 nm virus-like particle and hepatitis A as demonstrated by immune electron microscopy, *Intervirology,* 4, 110, 1974.
44. **Maynard, J. E., Bradley, D. W., Gravelle, C. R., Hornbeck, C. L., and Cook, E. H.,** Virus-like particles in hepatitis A, *Lancet,* 2, 1207, 1974.
45. **Mascoli, C. C., Ittensohn, O. L., Villarejos, V. M., Arguedas, J. A., Provost, P. J., and Hillman, M. R.,** Recovery of hepatitis agents in the marmoset from human cases occurring in Costa Rica, *Proc. Soc. Exp. Biol. Med.,* 142, 276, 1973.

46. **Provost, P. J., Ittensohn, O. L., Villarejos, V. M., Arguedas, J. A., and Hilleman, M. R.,** Etiologic relationship of marmoset-propagated CR326 hepatitis A virus to hepatitis in man, *Proc. Soc. Exp. Biol. Med.,* 142, 1257, 1973.
47. **Provost, P. J., Ittensohn, O. L., Villarejos, V. M., and Hilleman, M. R.,** A specific complement-fixation tests for human hepatitis A employing CR326 virus antigen, *Proc. Soc. Exp. Biol. Med.,* 148, 962, 1975.
48. **Miller, W. J., Provost, P. J., McAleer, W. J., Ittensohn, O. L., Villarejos, V. M., and Hilleman, M. R.,** Specific immune adherence assay for human hepatitis A antibody. Application to diagnostic and epidemiologic investigations, *Proc. Soc. Exp. Biol. Med.,* 149, 254, 1975.
49. **Dienstag, J. L., Feinstone, S. M., Purcell, R. H., Hoofnagle, J. H., Barker, L. F., London, W. T., Popper, H., Peterson, J. M., and Kapikian, A. Z.,** Experimental infection of chimpanzees with hepatitis A virus, *J. Infect. Dis.,* 132, 532, 1975.
50. **Maynard, J. E., Bradley, D. W., Gravelle, C. R., Ebert, J. W., and Krushak, P. H.,** Preliminary studies of hepatitis A in chimpanzees, *J. Infect. Dis.,* 131, 194, 1975.
51. **Maynard, J. E., Lorenz, D., Bradley, D. W., Feinstone, S. M., Krushak, D. H., Barker, L. F., and Purcell, R. H.,** Review of infectivity studies in nonhuman primates with virus-like particles associated with MS-1 hepatitis, *Am. J. Med. Sci.,* 270(1), 81, 1975.

Chapter 3

THE PHYSIOCHEMICAL PROPERTIES OF HEPATITIS A VIRUS

I. INTRODUCTION

In contrast to most other viruses of man, a great deal was known about the physical and chemical properties of HAV prior to its isolation in cell culture. Studies in human volunteers had demonstrated that the agent would pass through both a Seitz E.K. and a Chamberland No. 2 filter, that it was stable to ether and acid, capable of surviving in the frozen state for more than 1 year, but could be inactivated by boiling, autoclaving, or contact with 1 ppm residual chlorine (see Chapter 1).

Little additional information was obtained until the virus had been identified[1] and tests developed which were capable of detecting viral antigens in clinical specimens and purified material.[2,3] Although HAV was isolated in cell culture in 1979[4] it was some time before high yields could be obtained in cell lysates or supernatant fluid. As a result much of our knowledge of the physical and chemical properties of HAV has been derived from studies of particles purified from fecal specimens collected from patients with naturally acquired hepatitis A and from experimentally infected animals. In some instances the quantities of virus available were so small that it was necessary to pool several specimens to obtain sufficient material for analysis.

Despite these difficulties a great deal of information was obtained on the structure and organization of the virus, which has subsequently been confirmed by the use of tissue culture-derived virus and extended by the application of modern molecular biological techniques (see Chapter 7).

II. THE PARTICLE

A. Size and Morphology

In 1973 Feinstone et al.[1] used the technique of immune electron microscopy (IEM) to visualize small spherical virus-like particles in the feces of patients with hepatitis A. Measurements of these particles, the edges of which were partly obscured by antibody, revealed an average particle size of 27 nm (see Figure 1). Morphologically identical particles were soon recognized in the feces of patients with sporadic or epidemic hepatitis A in Australia[5] and in other parts of the U.S.[6] (see Figure 2). Hepatitis A particles have also been detected in the feces, liver, and bile of experimentally infected chimpanzees and marmosets,[7-9] and in the supernatant fluid and lysate of infected cell cultures.[4] In each case EM revealed naked, approximately spherical particles, with a diameter of 27 to 28 nm, sometimes as individual particles surrounded by a halo of antibody, sometimes as large immune aggregates, and occasionally as crystalline arrays or clusters of particles enclosed by a membrane-lined vesicle (see Figure 3). In these early studies the most detailed morphology and the most accurate measurements of particle size were made by Locarnini and his colleagues[5] on clusters of particles which were shed into the feces in lattice-like arrays (see Figure 3C). In some specimens both empty and full particles can be visualized (see Figure 1) and electron-dense cores have been noted by some workers (see Figure 4). Siegl and Eggers,[10] using the Markham rotation technique, have produced evidence that the capsid of HAV has 32 capsomeres arranged as either a rhombic tricontrahedron or a pentacis dodecahedron, although this is in conflict with data on the structure of other picornaviruses, all of which appear to have 60 structural units.

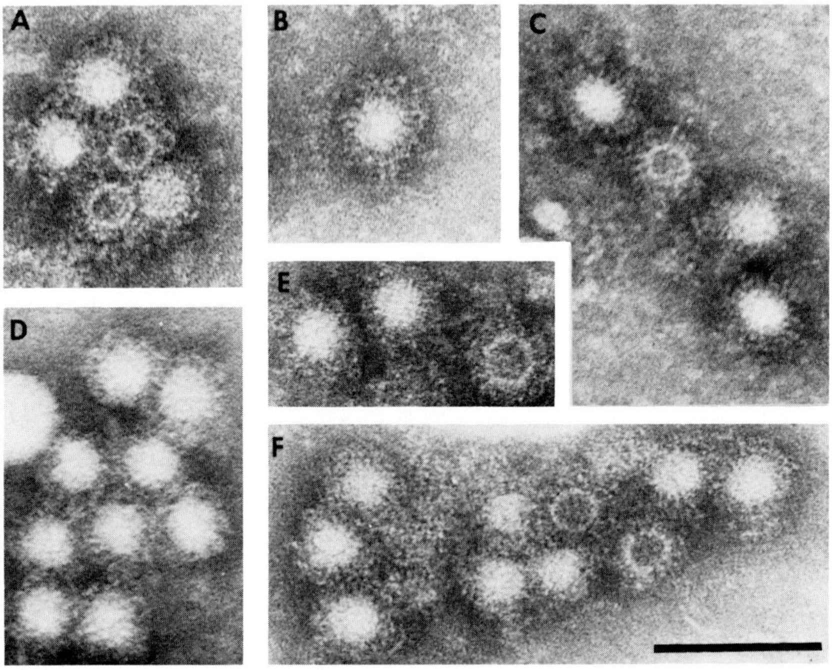

FIGURE 1. A representative sample of HAV-like particles observed after 0.9 mℓ of the 2% stool suspension was incubated with 0.1 mℓ of diluted serum, and then centrifuged and prepared for EM. The particles appeared to have cubic symmetry and resemble picornaviruses morphologically. (A and B) An aggregate and a single particle observed after incubation of the stool filtrate with a serum from a convalescent patient from the Joliet study. The aggregate is composed of three "full" and two "empty" particles heavily coated with antibody. The single particle is also heavily coated with antibody. The quantity of antibody in this serum to the hepatitis A antigen was scored as 3 to 4+; (C and D) aggregates after incubation of the stool filtrate with the serum of a patient convalescing from hepatitis contracted during a naturally occurring outbreak. This serum was also rated 3 to 4+; (E and F) aggregates and particles after incubation of the stool filtrate with immune serum globulin; they are heavily coated with antibody. The bar represents 100 nm. (From Feinstone, S. M., Kapikian, A. Z., and Purcell, R. H., *Science,* 182, 1026, 1973.)

B. Buoyant Density and Sedimentation Coefficients

Initial studies on the buoyant density of HAV particles purified from the feces of patients with naturally acquired infections and the feces of experimentally infected chimpanzees produced confusing results with values ranging from 1.24 to 1.50 g/mℓ in CsCl.[9,11-17] While some of the variations appear to be due to different purification procedures and different methods of determining buoyant density[18] there is convincing evidence that purified preparations of HAV contain three distinct populations of particles[19] (see Figures 5 to 7).

Mature HAV virions band at approximately 1.32 to 1.34 g/mℓ in CsCl and sediment at approximately 160 S which is similar to the properties exhibited by members of the Enterovirus and Cardiovirus genera of the family picornaviridae. Siegl et al.[20] have shown that, when poliovirus type 2 and HAV are compared in the same centrifugation experiments their buoyant densities and sedimentation coefficients are almost identical, with HAV having slightly lower values (1.332 g/mℓ and 156 S; 1.340 g/mℓ and 160 S). The buoyant density of mature particles derived from infected cell cultures using native virus or following internal labeling of the viral protein,[18,21] is in agreement with data obtained using particles derived from human or chimpanzee feces.[14,15,17]

Another population of HAV particles which band at 1.27 to 1.32 g/mℓ in CsCl and

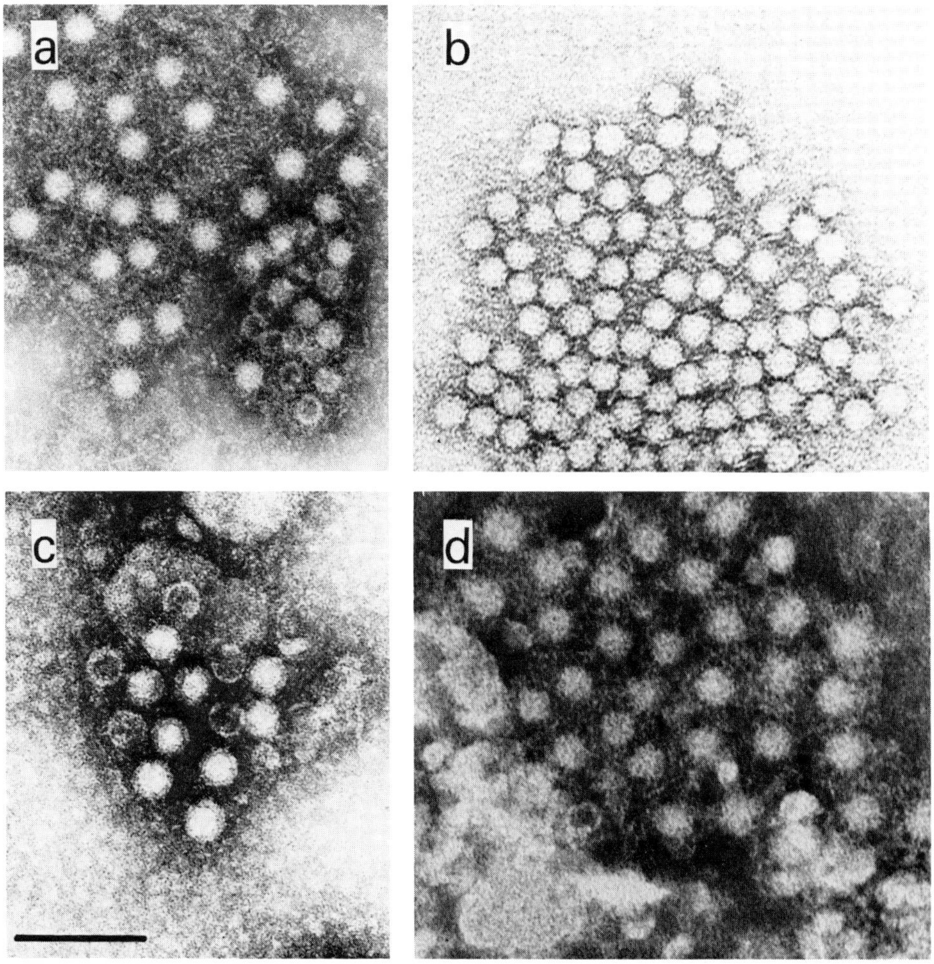

FIGURE 2. A composite electron micrograph demonstrating that strains of HAV recovered from different locations following natural and experimental infections, are morphologically indistinguishable. (a) The MS-1 strain recovered from the feces of an adult volunteer who had been infected by intramuscular injection of serum; (b) the HM175 strain recovered from the feces of an adult male in Melbourne, Australia. This man was part of an outbreak of hepatitis in which a family member acquired the disease after contact with an infected child; (c) the Phoenix strain which was recovered from the feces of an individual involved in a common source epidemic in Phoenix, Arizona; (d) the CR326 strain which was obtained from the feces of an individual with acute sporadic hepatitis A in Costa Rica. (A and C — From the author's collection; B — from Locarnini, S. A., Ferris, A. A., Stott, A. C., and Gust, I. D., *Intervirology,* 4, 110, 1974. With permission; D — from Provost, P. J., Wolanski, B. S., Miller, W. J., Ittensohn, O. L., McAleer, W. J., and Hilleman, M. R., *Proc. Soc. Exp. Biol. Med.,* 148, 532, 1975. With permission.)

sediment at 70 to 80 S can be detected in many preparations. The concentration of these particles, initially thought to be empty capsids, (which contain no nucleic acid and are hence noninfectious), varies markedly from specimen to specimen.[22] Recently, Lemon and his colleagues[21] have demonstrated that gradient fractions containing this population of particles are infectious and have recovered normal-length RNA from them. It is possible, however, that these results reflect contamination of the low-density particles with mature virions which cannot be completely separated under the conditions used.

The third population of particles has a density of 1.39 to 1.50 g/mℓ in CsCl, due it has been suggested, to a more open virion structure which either binds CsCl or allows it to

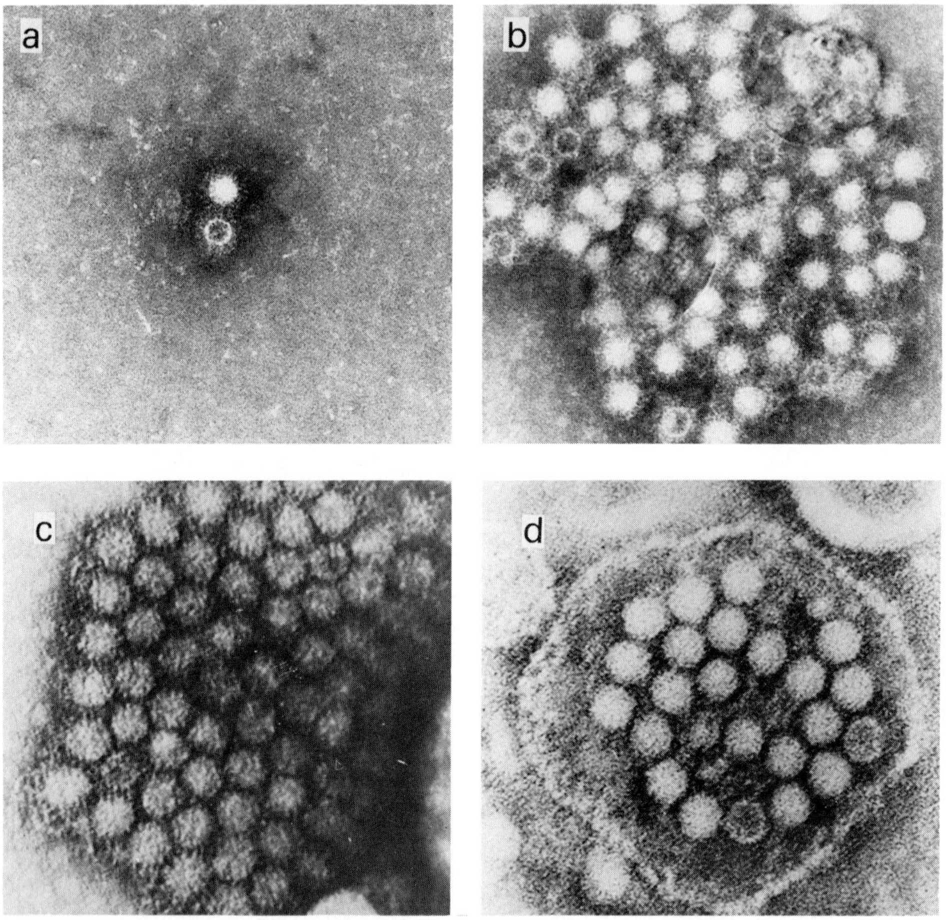

FIGURE 3. A composite of electron micrographs showing a variety of different ways in which HAV particles may be visualized in fecal extracts. (a) Scattered particles coated with antibodies; (b) large immune complexes; (c) crystalline arrays; (d) clusters of particles contained in a membrane-lined vesicle. (Varying magnifications used.) (a, b, d — From the authors; collection; c — from Locarnini, S. A., Ferris, A. A., Stott, A. C., and Gust, I. D., *Intervirology*, 4, 110, 1974. With permission.)

penetrate into the virion, a situation which had been noted with poliovirus[23,24] and some other members of the picornaviridae.[25] These particles contain RNA[21] but tend to be less stable than the mature virion and upon recentrifugation dissociate into components which sediment at 50, 90, 160, and 230 S.[17,26]

C. Resistance to Physical and Chemical Agents

The successful transmission of HAV infection to marmosets and chimpanzees and adaptation of some strains to growth in cell culture, has provided new and more sensitive means of studying the susceptibility of HAV to various physical and chemical agents.

1. Heat

While it has been known for many years that HAV will remain viable if stored at −20°C or below, studies in human volunteers and experimental animals shed little light on its sensitivity to heat, other than an apparent resistance to treatment at 56°C. In 1970 Krugman

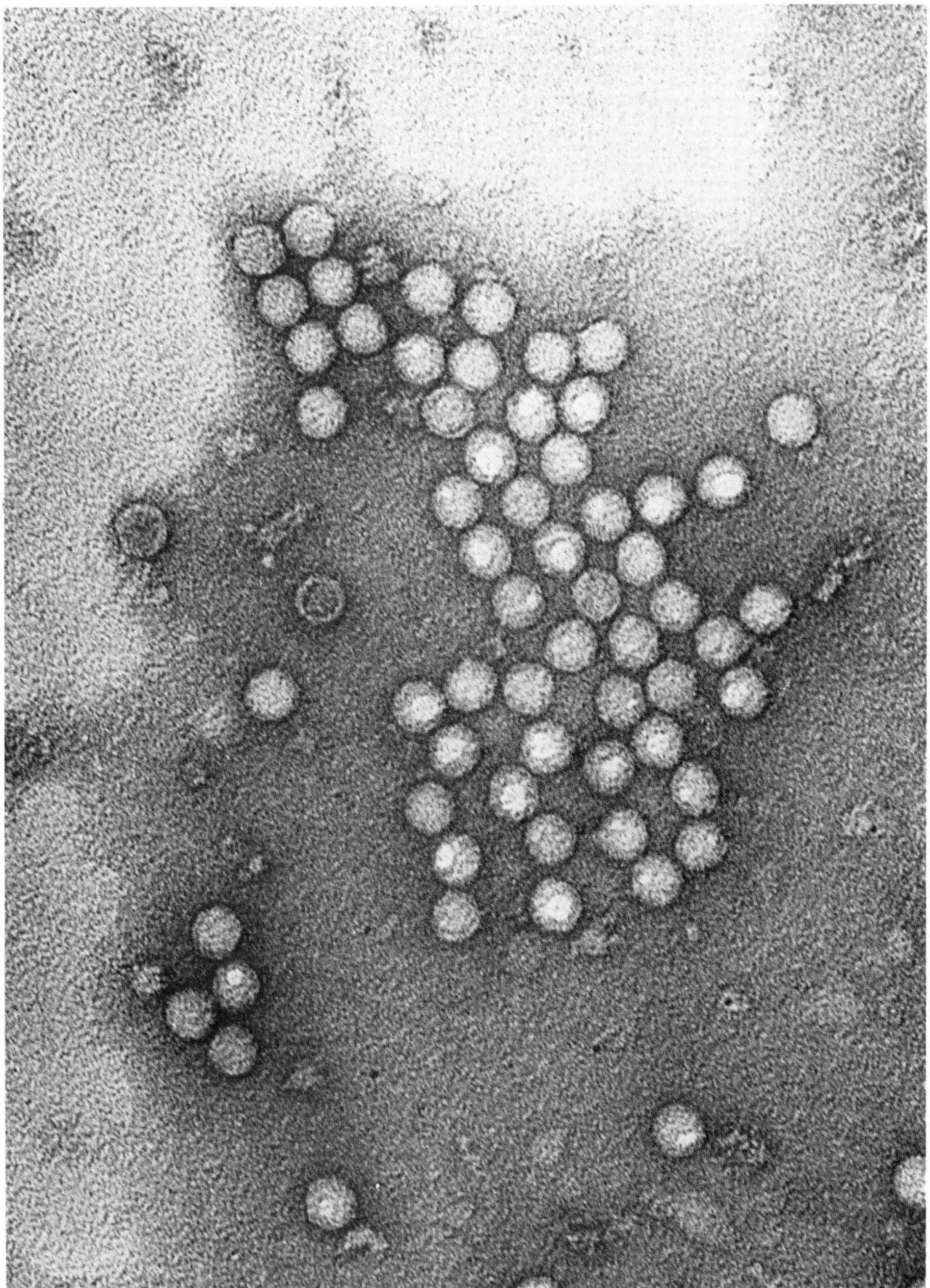

FIGURE 4. Electron micrographs revealing full and empty capsids and particles with electron dense cores. Bar equals 100 nm. (From Coulepis, A. G., Locarnini, S. A., Ferris, A. A., Lehmann, N. I., and Gust, I. D., *Intervirology,* 10, 24, 1978. With permission.)

et al.[27] demonstrated that inoculation of volunteers with aliquots of the MS-1 serum pool which had been heated to 98°C for 1 min failed to transmit disease, suggesting that HAV could be destroyed by boiling. Some years later Provost and his colleagues were able to confirm these observations in marmosets using the CR326 strain which had been heated to 100°C for 5 min.[9]

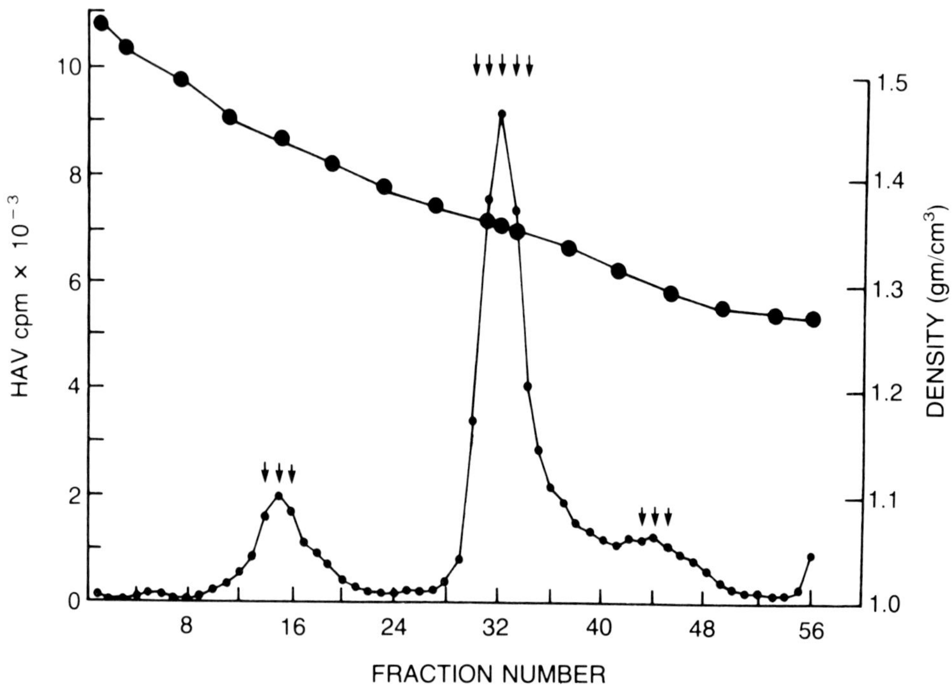

FIGURE 5. Detection of HAV particles with heavy, intermediate, and light-buoyant densities in the supernatant fluid of infected cell cultures. HAV was precipitated from clarified supernatant fluid by addition of 10% PEG and 2.3% NaCl, and subjected to isopyknic centrifugation in 5 mℓ gradients at 200,000 for 22 hr at 4°C in a Beckman SW 40 motor. Aliquots of each fraction were tested for HAAg by radioimmunoassay and density using an Abbe-type refractometer. The arrows indicate the fractions which were rebanded in CsCl (see Figure 6). (From Lemon, S. M., Jansen, R. W., and Newbold, J. E., *J. Virol.*, 54, 78, 1985. With permission.)

The ability to propagate the virus in cell culture has allowed scientists in several countries to evaluate its sensitivity to heat and a variety of other agents.

HAV appears to be more heat stable than any other picornavirus: treatment at 60°C for 1 hr does not destroy the virus and 60°C for 10 to 12 hr results in only partial inactivation. On the other hand, heating to 85°C for 1 min completely inactivated preparations containing more than 10^4 infectious doses of virus. Probably the most extensive study of the heat sensitivity of HAV is that of Parry and Mortimer,[28] who tested the effect of various conditions on a phosphate buffered saline (PBS) suspension containing 10^5 $TCID_{50}$ of cell culture-derived virus. Treated suspensions were inoculated into susceptible cells (FRhK4) which were tested 5 weeks later for the presence of HAAg by RIA. A 2 × $\log_{10}$ reduction in titer was observed after treatment at 65°C for 2 min and a 4 × $\log_{10}$ reduction when the time was doubled. The virus was completely inactivated after 4 min at 70°C, 30 sec at 75°C, 5 sec at 80°C, and virtually instantaneously at 85°C.

Fecal suspensions of HAV are more resistant to heat than virus isolated from cell culture, presumably because virus may be sequestrated within small particles of fecal material. For example, Frosner[29] reported no loss of activity when fecal suspensions containing HAV were treated at 60°C for 2 hr and Scheid and Deinhardt[30] reported only a 100-fold loss in titer when suspensions were held at 37°C for 7 days, or at room temperature for 4 weeks. Similarly Parry and Mortimer[28] demonstrated that HAV would remain viable in milk after treatment for 30 min at 62.8°C or 15 sec at 71.6°C. Since these are the conditions set by statute for pasteurization of milk in Great Britain, this finding raised concern about the efficacy of the pasteurization process. However, as milk-borne epidemics of hepatitis A are extremely rare, the current procedures are probably effective for all but gross contamination.

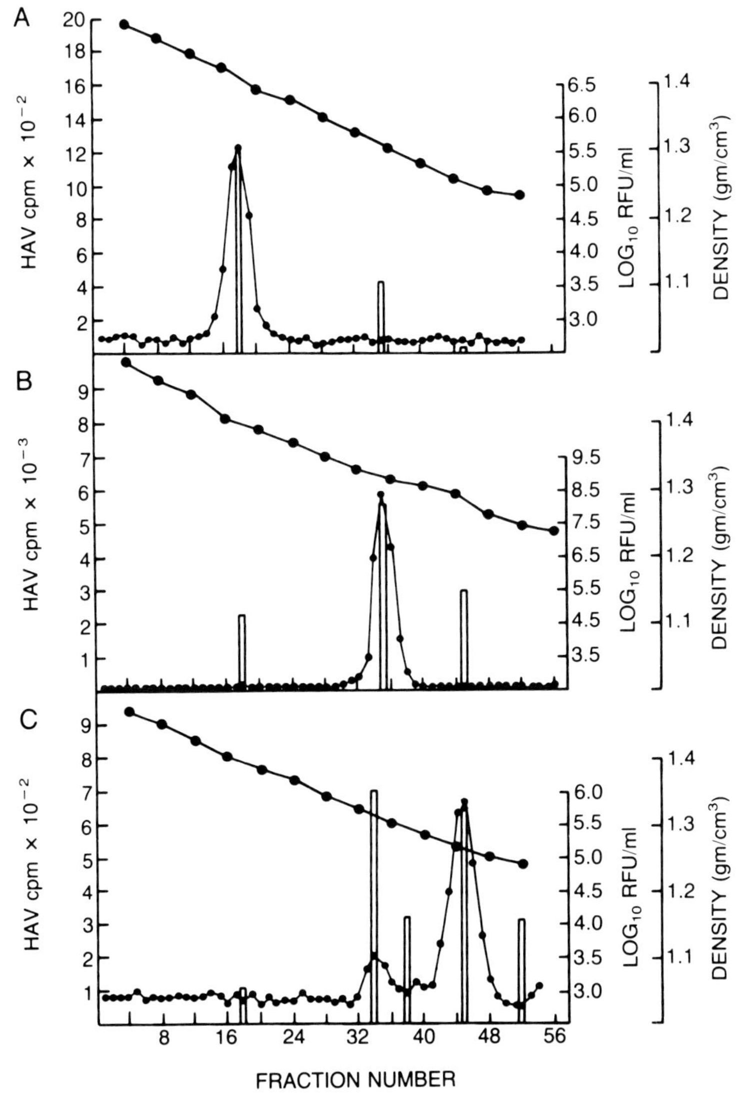

FIGURE 6. Rebanding of HAV particles from the gradient shown in Figure 5. (A) Dense particles from fractions 14 to 16; (B) the major component obtained by rebanding fractions 30 to 34; (C) light particles from fractions 43 to 45. The results of detection of HAAg by radioimmunoassay and infectious particles by radioimmunofocus assay are shown. (From Lemon, S. M., Jansen, R. W., and Newbold, J. G., *J. Virol.*, 54, 78, 1985. With permission.)

Outbreaks of hepatitis A have been reported following ingestion of raw or partially cooked shellfish harvested from waters contaminated with human feces,[31] suggesting that even steaming shellfish or heating them until the shells open may be insufficient to destroy the virus.[32]

Infectivity of HAV suspensions appears to be lost after incubation at 37°C for 2 weeks or room temperature for 5 weeks, although dried specimens containing extraneous protein can be stored at 25°C for a least 30 days and remain viable.[33] The ability of the virus to survive in the dried state highlights the need for meticulous care when handling feces in the laboratory and for disinfecting potentially infected surfaces.

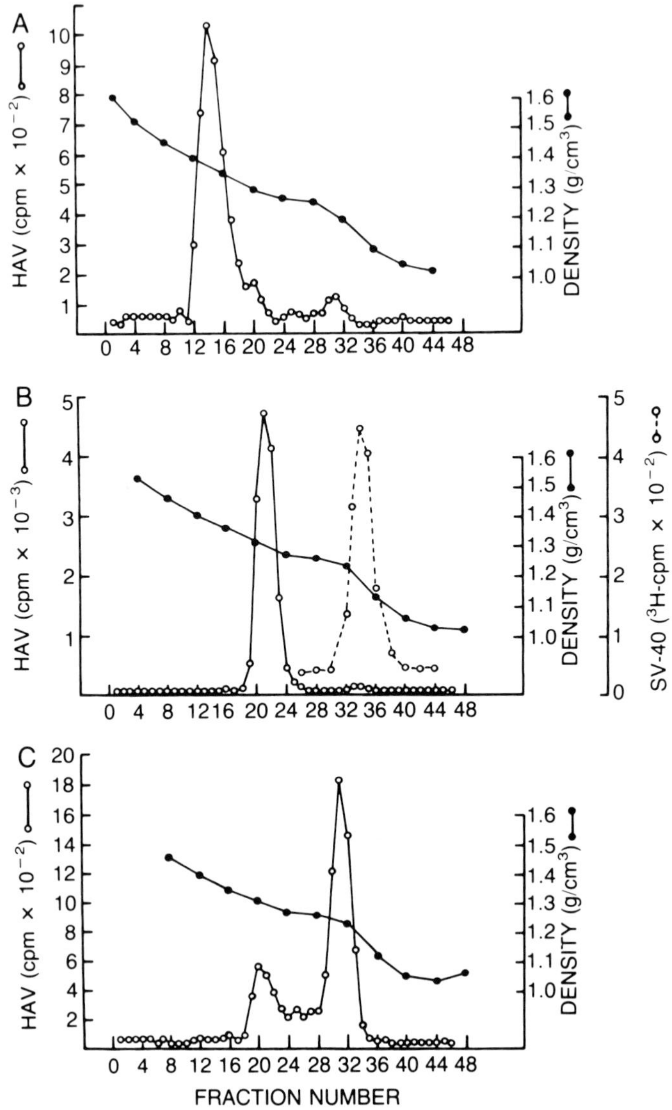

FIGURE 7. Isopyknic banding in metrizamide of (A) dense, (B) intermediate, (C) light-density HAV particles recovered from CsCl gradients. The buoyant density of ^{3}H-labeled SV40 is drawn in (B). (From Lemon, S. M., Jansen, R. W., and Newbold, J. E., *J. Virol.*, 54, 78, 1985. With permission.)

Siegl et al.[34] have demonstrated that HAV is significantly more resistant to heat than poliovirus and suggested that it should not be classified with any of the current picornavirus genera but should be given its own genus within the picornaviridae (see Chapter 7).

2. pH

As with other enteroviruses HAV is stable at pH 3,[30,33,34] although there is some evidence that the particles are less resistant than poliovirus and begin to lose their characteristic morphological appearance if left at low pH at room temperature for several hours[26] or at 60°C for 1 hr.[34]

3. Lipid Solvents

HAV lacks a lipid-soluble outer envelope and is thus resistant to treatment with 20% diethyl ether,[9] chloroform, and up to 50% trichlorotrifluoroethane.[17] Treatment of fecal suspensions with chloroform or trichlorotrifluoroethane is a useful preliminary step in purification of the virus[35] and the preparation of bacteria-free inocula for cell culture.

4. Other

The data on inactivation of HAV by chlorine, which was obtained in studies on human volunteers, has been extended by Peterson et al.[36] who treated fecal extracts with various concentrations of chlorine and assayed for the presence of live virus in susceptible marmosets. They found infectivity was reduced by doses of 0.5 to 1.5 mg free-residual chlorine per liter at pH7 for 60 min., and abolished within 15 min by concentrations of 2 to 2.5 mg/ℓ. Although chlorine is the most common means of treating contaminated water (and is used in many laboratories for decontaminating spills containing HAV), iodine (3 mg/ℓ) and potassium permanganate (30 mg/ℓ) have been shown to be equally effective.

Ethanol, at a concentration of 70% for 1 hr at room temperature, has been demonstrated to reduce the titer of HAV preparations by 5.5 $\log_{10}$.[37]

Treatment with high dilutions of formalin for prolonged periods has been demonstrated to inactivate HAV (1:4000 for 72 hr at 37°C or 96 hr at 35.5°C),[9] although higher concentrations for shorter periods of time at lower temperatures (1:350 at 20°C for 1 hr) are only partially effective. Workers at the Merck Institute[38] have prepared an experimental hepatitis A vaccine by purifying HAV from the livers of infected marmosets, and treating it with formalin at a final concentration of 1:4000 for 4 days at 35.5°C. The vaccine, which contained approximately 10^{10} particles per milliliter, retained antigenicity but was not infectious when administered intravenously to susceptible marmosets.

Finally, Provost et al.[9] have demonstrated that suspensions of HAV can be readily inactivated by ultraviolet (UV) irradiation (at 1-1 W for 60 sec and Frosner[29] has shown that the virus is very sensitive to UV irradiation at 60°C.

The susceptibility of HAV to chemical and physical agents is very similar to that of the prototype member of the enterovirus group, poliovirus, with the exception that HAV is considerably more stable to heat.

III. STRUCTURAL PROTEINS

Until recently most of our knowledge about the structure and composition of the capsid of HAV came from studies on virus which had been purified from the feces of patients with naturally acquired disease. As peak shedding of virus particles tends to occur prior to the patient being admitted to the hospital, adequate specimens were difficult to obtain and the purification procedures which were developed to maximize virus recovery were very laborious.[35]

Coulepis and his colleagues[39,40] were the first to determine that HAV was made up of four major polypeptides and that these were of similar size to the structural polypeptides of poliovirus. Virus was purified from the feces of patients with hepatitis A by a process of chloroform extraction, differential centrifugation, column chromatography, and rate-zonal ultracentrifugation, disrupted by boiling in 0.1% sodium dodecylsulfate (SDS) and 1% dithiothreitol and the proteins separated by SDS-polyacrylamide gel electrophoresis (SDS-PAGE) and identified by staining with Coomassie blue. Originally three structural proteins were detected, which by reference to standard external molecular weight (MW) markers were estimated to have MWs of 34,000, 25,500, and 23,000, respectively[38] (see Figure 8). By analogy with other picornavirus these proteins were referred to as VP1, VP2, and VP3. A fourth, low-MW polypeptide, characteristic of other members of the picornavirus, was not detected.

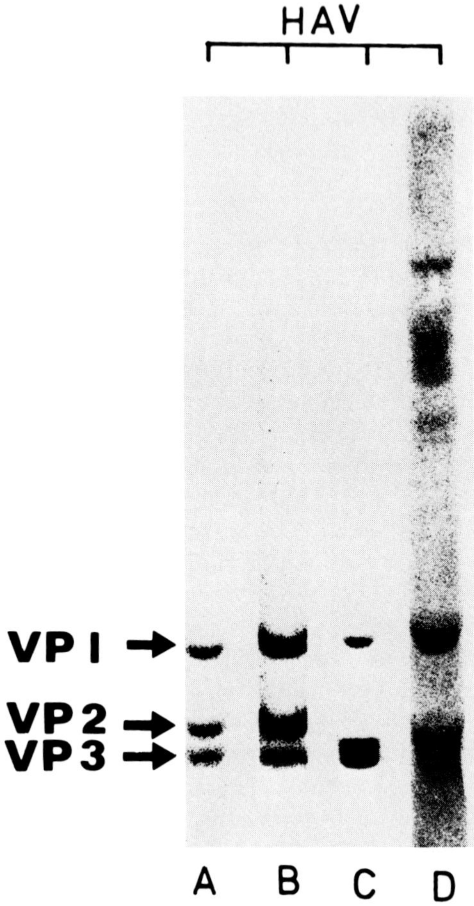

FIGURE 8. Discontinuous SDS-PAGE analysis of four different preparations of HAV in column gels. Migration was from top to bottom and the polypeptide bands were identified by staining with Coomassie brilliant blue. The four preparations were purified from fecal specimens obtained from four separate patients with hepatitis A. The four gels were run separately and are placed side by side to show the reproducibility of the results. Gels A, B, and D contained 10% (w/v) acrylamide, while gel C contained 7% (w/v) acrylamide; the sample applied to gel D was prepared by immune precipitation. The different conditions applied to gels C and D resulted in poorer resolution of VP2 and VP3; however, the MWs of these polypeptides when compared to the external molecular weight markers, electrophoresed under the same conditions, were the same as those in gels A and B. The high molecular MW-protein contaminants shown in gel D probably represent the breakdown products of the IgG used in the immune precipitation reaction. (From Coulepis, A. G., Locarnini, S. A., Westaway, E. G., Tannock, G. A., and Gust, I. D., *Intervirology*, 18, 107, 1982. With permission.)

Subsequently these workers repeated the procedure using purified virus which had been labeled with ^{125}I, then disrupted and analyzed by SDS-PAGE and autoradiography[40] (see Figure 9). This study provided more accurate estimates of the MWs of the three major structural polypeptides (VP1, 33,000; VP2, 26,500; VP3, 22,500) and revealed the existence of a fourth polypeptide, VP4, with a MW of about 14,000.

These findings were rapidly confirmed,[22,41,43] although each group obtained slightly different values for the MWs so that the reported MWs for VP1 vary from 30 to 34,000, for VP2 from 24 to 26,500, for VP3 from 21 to 23,000 and for VP4 from 7 to 14,000. These figures are very similar to those reported for members of the picornaviridae.[25]

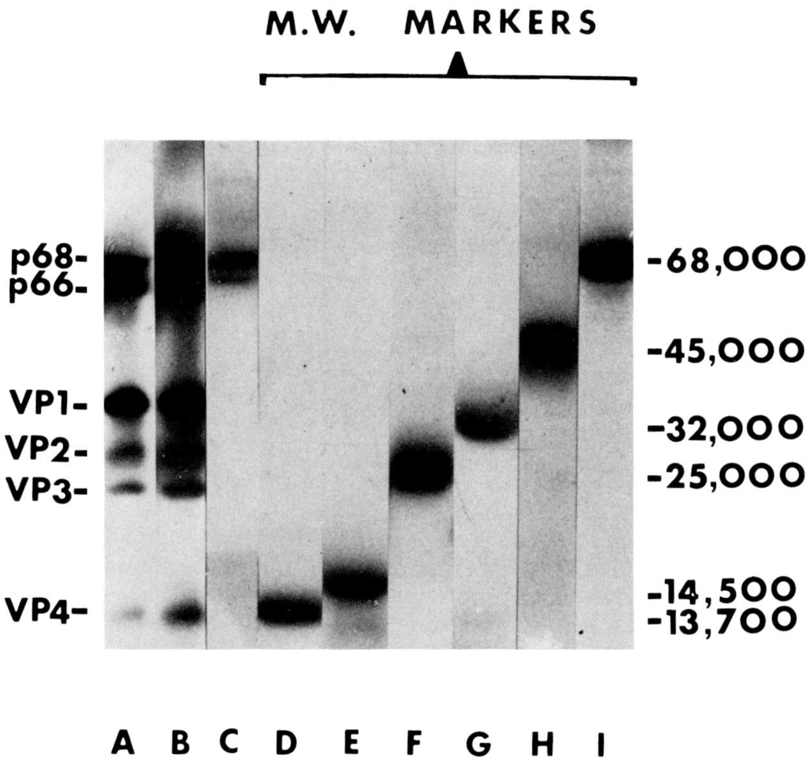

FIGURE 9. Autoradiogram of two HAV preparations which were disrupted with 10% SDS then iodinated for 45 sec and subjected to discontinuous SDS-PAGE in a 13% (w/v) slab gel together with a series of iodinated-MW markers. Tracks A and B represent two preparations of HAV; tract C the laboratory-grade bovine serum albumin (BSA) used throughout the purification protocol; track D, ribonuclease; track E, lysozyme; track F, chymotrypsinogen; track G, carbonic anhydrase; track H, ovalbumin; track I, BSA (analytical grade). The MWs of the marker proteins are shown on the right of the figure. The period of film exposure was insufficient to detect VP4 in tract A. VP4 was detected in track B by overexposure. P66 and P68 in tracks A and B appear to be BSA contaminants which were not separated from HAV during purification. Analysis of the laboratory grade BSA (track C) revealed only the presence of P66 and P68. (From Coulepis, A. G., Locarnini, S. A., Westaway, E. G., Tannock, G. A., and Gust, I. D., *Intervirology*, 18, 107, 1982. With permission.)

In addition to the four structural proteins, some workers[17,42] have reported detection of an additional polypeptide with a MW of approximately 40,000, which was thought to represent a precursor polypeptide equivalent to VP0 of poliovirus. Molecular cloning and nucleic acid sequencing has enabled the precise size of the capsid proteins to be predicted. These predictions indicate the MWs of VP1 to be 32.8; VP2, 24.8; VP3, 27.3, and VP4 to be only 2.5 KD (see Chapter 7).

The precise location of each of the polypeptides is uncertain although there is some evidence that, as with other picornaviruses, VP1, VP2, and VP3 are located on the surface of the virion.

IV. NUCLEIC ACID

Early work with HAV established its size and morphology but failed to resolve whether the agent was a picornavirus or a parvovirus.[11] Indirect evidence that the genome consisted of RNA was provided by Provost et al.[9] who noted that HAV particles and HAAg are found almost exclusively in the cytoplasm of infected hepatocytes. These particles could be stained with acridine orange, indicating the genetic material was either single-stranded DNA or RNA. Furthermore suspensions of HAV purified from infected liver tissue were partially inactivated by treatment with RNAase at 60°C for 1 hr but not by DNAase.

While confirmation of the intracytoplasmic location of the virus was rapidly obtained,[44,45] Bradley and his colleagues[26] provided stronger evidence that the genome was RNA by demonstrating that the sedimentation rate of HAV in sucrose was altered by treatment at pH 10 to 11 at 40°C for 1 hr and incubation with RNAase, whereas treatment with DNAase had no effect.

Siegl and Frosner[17] disrupted highly purified HAV by treatment with 4 *M* urea and 90% formamide and studied the nucleic acid which was released by EM and ultracentrifugation. They observed a poulation of linear molecules with two predominant species whose molecular weights were calculated to 1.9×10^6 and 1.3×10^6, considerably smaller than the genomes of other picornaviruses (2.3 to 2.8×10^6).[25] These findings were soon confirmed by Moritsugu and his colleagues (reported by Feinstone[41]) who demonstrated that nucleic acid released from disrupted particles was sensitive to RNAase but not to DNAase.

Direct evidence of the nature of the HAV genome was provided by Coulepis et al.[46] who extracted nucleic acid from purified virus by treatment with SDS and labeled it with ^{125}I (see Figure 10). Under nondenaturing conditions, sucrose gradient centrifugation showed that the genome sedimented at 33 S and had a buoyant density of 1.64 g/cm^3 suggesting that it was a linear single-stranded molecule with a MW of 2.25×10^6 (see Figure 11).

Siegl et al.[18] confirmed those findings using virus which had been grown in cell culture and labeled internally. They found the sedimentation coefficient of the genome, under nondenaturing conditions, to be 32.5 S corresponding to a MW of 2.3×10^6. Rather different values were obtained when the experiment was repeated under fully denaturing conditions suggesting that the genome contains extended regions with a secondary structure.

Coulepis et al.[46] also provided the first evidence that the HAV genome contain sequences of polyadenylic acid (poly A) by demonstrating that radiolabeled HAV RNA was bound to a column of sepharose 4B charged with polyuridylic acid (see Figure 12). Comparable results were obtained with internally labeled virus: after digestion of the genome with RNase and chromatography of the products on oligo (dt) cellulose, a poly A tract estimated at 40 to 50 nucleotides was identified. In common with all picornaviruses this sequence has been shown to be located at the 3′ end of the genome[47] (see Chapter 7).

The presence of a poly A tract suggested that HAV possessed a genome of the same polarity as its intracellular messenger RNA (that is, positive stranded) and that like other picornaviruses, under appropriate conditions the nucleic acid alone would be infectious. Both Locarnini et al.[48] and Siegl[18] have provided indirect evidence for this proposition by demonstrating the production of novel, apparently virus-coded polypeptides or the development of intracytoplasmic HAAg in cell cultures inoculated with purified viral RNA, although neither was able to detect the production of mature virus particles. Sequencing of the cloned DNA of HAV has confirmed that the virion genome is positive sense.[47,49]

V. REPLICATIVE STRATEGY

Although HAV has been adapted to growth in cell culture, by comparison with other picornaviruses, such as poliovirus, it replicates poorly. As relatively little is known about the replicative strategy of HAV, the reasons for this difference are unknown.

Locarnini et al.[48] inoculated continuous cell lines of monkey kidney origin with HAV and looked for evidence of production of viral-coded proteins and nucleic acid. Although no effect on host-cell macromolecular synthesis was observed, viral-coded proteins could be detected (by the double-label, co-electrophoresis, and subtraction technique) 3 to 6 hr

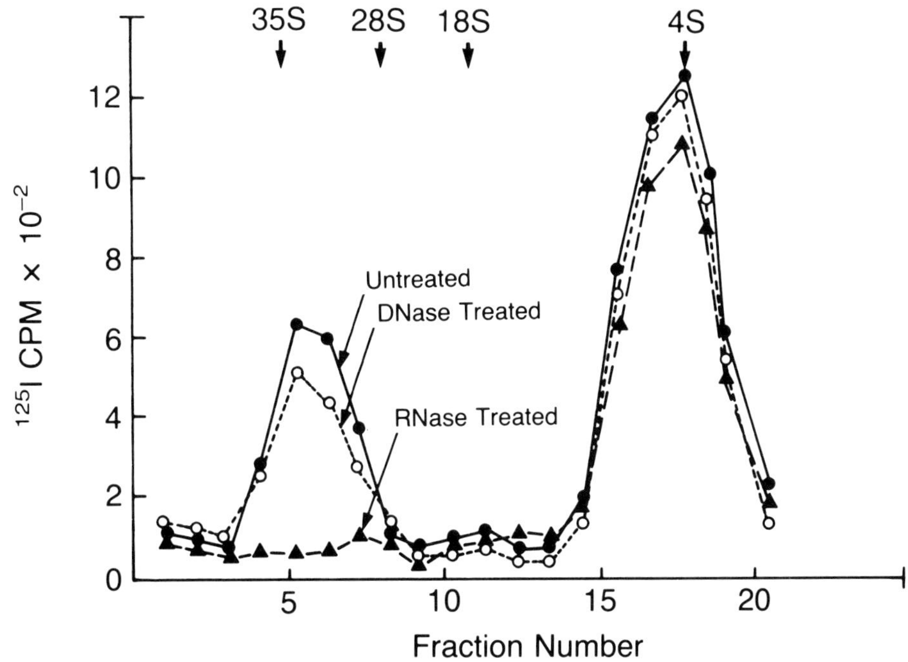

FIGURE 10. Rate-zonal ultracentrifugation analysis and nuclease digestion of nucleic acid released from HAV and iodinated using the chloramine-T method. A 500-μℓ sample of iodinated HAV nucleic acid was centrifuged at 310,000 *g* through 4.0 mℓ 15 to 30% (wt/v) sucrose gradient prepared in NET buffer (100 m*M* NaCl, 10 m*M* EDTA, 10 m*M* Tris) for 2.33 hr at 7°C. Such gradients conserve double-stranded RNA structures while rendering single-stranded RNA susceptible to RNase digestion. The profiles of acid-insoluble radioactivity in fractions for untreated (●—●) aliquots and aliquots treated with RNase (▲—▲) and DNase (○---○) are shown. The arrows represent the positions of 35 S poliovirus RNA and 28 S, 18 S, and 4 S rRNAs, which were determined in parallel gradients. The 33 S peak appears to be DNase-resistant and RNase-sensitive, while the 4 S peak is both DNase- and RNase resistant. The 33 S peak was found to be pronase resistant, while the 4 S peak was completely degraded by this treatment. These results indicate that the 33 S peak is the HAV genome and the 4 S peak probably consists largely of small amounts of iodinated protein which were released from the RNA but remained in the aqueous phase after phenol-chloroform extraction. (From Coulepis, A. G., Locarnini, S. A., Westaway, E. G., Tannock, G. A., and Gust, I. D., *Intervirology*, 18, 107, 1982. With permission.)

after infection and continued to be produced for at least 48 hr. In addition, evidence of viral RNA replication was obtained within 2 to 6 hr of infection with both a 33 S peak of single-stranded RNA and a 22 S peak of what appeared to be double-stranded RNA being detected. The 33 S RNA was found to have the same MW as virion RNA and to contain a poly A tract while the 22 S RNA was thought to represent the replicative intermediate.

Similar patterns of protein and nucleic acid production were obtained by these workers when cells were infected with RNA which had been extracted from purified virions, suggesting that the genome of the virus was positive stranded. They were unable to define the defect in the replicative cycle but suggested that a restriction might be operating at the level of RNA synthesis.

Subsequently Anderson et al.[50] have infected B-SC-1 cells with HAV at a low multiplicity of infection and demonstrated the development of specific viral antigens, infectious RNA, and mature virions. Under single-cycle growth conditions, a 12-hr lag phase was demonstrated followed by 12 hr of logarithmic growth. Using dot-blot hybridization and radioim-

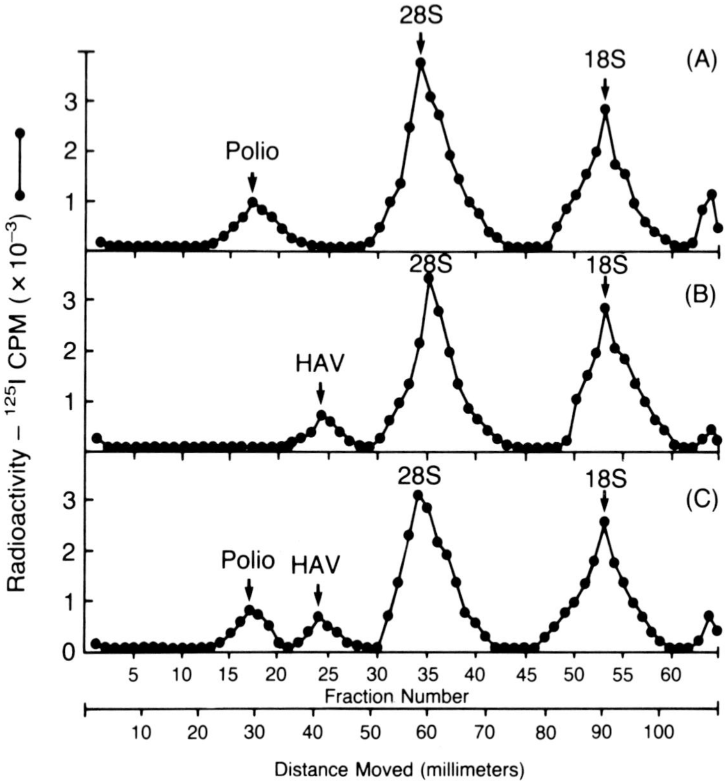

FIGURE 11. Continuous SDS-PAGE profiles of iodinated (A) poliovirus and ribosomal RNAs, (B) HAV and ribosomal RNAs, and (C) poliovirus, HAV, and ribosomal RNAs run in parallel cylindrical gels containing 2.2% (w/v) acrylamide and 0.5% (v/v) agarose under nondenaturing conditions. Following electrophoresis the gels were frozen at $-70°C$ for 1 hr and then cut into 65 equal slices using stacked razor blades. The slices were put directly into tubes and the radioactivity present in each was determined on a gamma counter. The genome of HAV appears to have a MW greater than 28 S ribosomal RNA but less than that of the 35 S poliovirus RNA. Measurements of MWs of the HAV genome were determined by comparing its electrophoretic mobility with the mobilities of standard RNAs. The mobility of the HAV genome thus yielded a MW of $(2.25 \text{ to } 0.15) \times 10^6$. (From Coulepis, A. G., Locarnini, S. A., Westaway, E. G., Tannock, G. A., and Gust, I. D., *Intervirology,* 18, 107, 1982. With permission.)

munofocus assays, they found that HAV RNA begins to accumulate from 6 to 12 hr postinfection, about the time that infectious virus became detectable. While other groups[51] have suggested that the failure of HAV to grow rapidly may be related to problems with penetration or uncoating of the virus, Anderson et al. were unable to demonstrate a significant difference in the penetration and uncoating of HAV and poliovirus. The site of restriction of HAV replication is unknown but may be at the level of RNA replication.

VI. CLASSIFICATION

A. Classification of Picornaviruses

The International Committee on Taxonomy of Viruses (ICTV) which is responsible for the classification of viruses has published two major reports on the picornaviridae,[25,52] a family of very small RNA-containing viruses, united by a common physical structure and

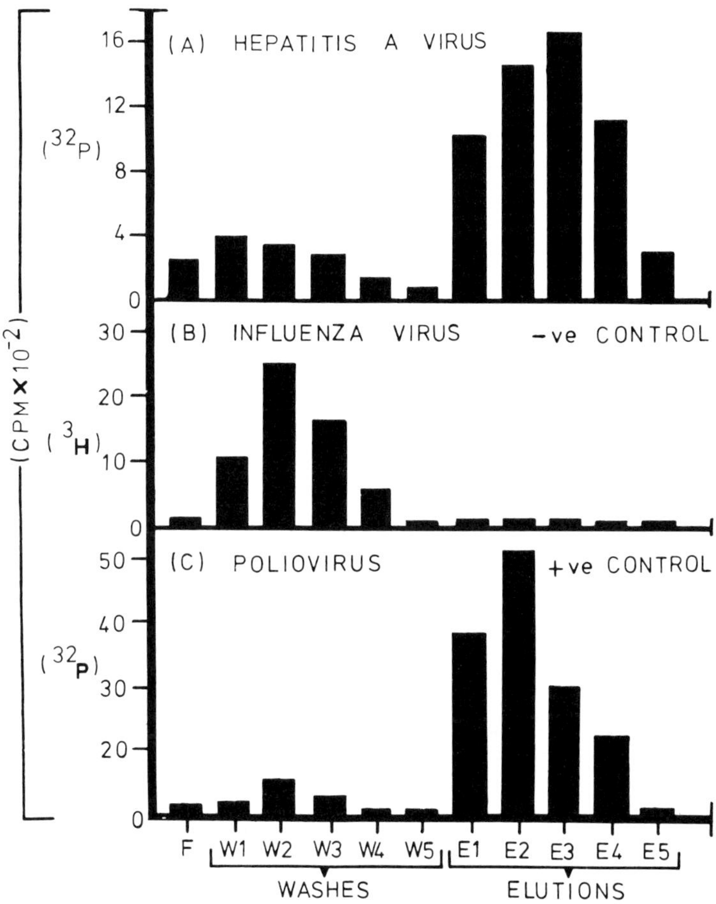

FIGURE 12. Elution profile of (A) HAV RNA, (B) influenza A strain PR8 RNA, and (C) poliovirus type 1 RNA in poly(U)-Sepharose 4B columns. The 33 S peak of ^{32}P-labeled HAV RNA from a sucrose gradient was collected and the RNA present precipitated in the presence of 100 g of carrier RNA. The RNA was resuspended in 0.5 mℓ NET buffer and applied to a 0.5 × 2 cm column. The initial filtrate (F) was collected and nonadherent RNAs were removed in five steps (W1 to W5), each with 0.5 mℓ of washing buffer (1 part formamide:3 parts NET). Poly A-containing RNAs were then eluted in five steps (E1 to E5) with 0.5 mℓ of elution buffer (9 parts formamide:1 part NET). The radioactivity in the filtrate, washes, and eluates is shown in (A). A similar procedure was followed using RNAs prepared by SDS-phenol:chloroform extraction of purified influenza A strain PR8 virus (B) and for poliovirus type 1 (C). (From Coulepis, A. G., Locarnini, S. A., Westaway, E. G., Tannock, G. A., and Gust, I. D., *Intervirology,* 18, 107, 1982. With permission.)

(in those members that have been studied) a common genome strategy. Members of this family are characterized by a naked ether-resistant, icosohedral capsid, measuring 22 to 30 nm in diameter, with no obvious surface features. The capsid is composed of approximately equal molar amounts of four major polypeptides whose aggregate MW is between 80 to 120 kdalton — typically three of the polypeptides have MWs of between 20 and 40 kdalton, the other being much smaller, around 5 to 10 kdalton. The genome of these viruses is linear single-stranded RNA with a MW of about 2.5×10^6, which is infectious, positive-stranded, and carries a polyadenylate tract at the 3′ end. Virus replication occurs in the cytoplasm.

The family can be separated into four genera — Enterovirus, Cardiovirus, Rhinovirus,

Table 1
DISTINGUISHING FEATURES OF PICORNAVIRUS GENERA

Property	Enterovirus	Cardiovirus	Rhinovirus	Apthovirus
Major members	Polio, echo, coxsackie	Encephalo-myelitisvirus	Human and animal rhino viruses	Foot and mouth disease viruses
Serotypes	>70	1	>110	7
Main hosts	Man, vertebrates	Man, rodents, vertebrates	Man, vertebrates	Most cloven-footed animals
Host range and tissue tropism	Narrow to wide	Wide	Narrow	Wide
Major habitat	Gut	CNS, heart	Upper respiratory tract	Generalized
Optimal growth temperature	36—37°C	36—37°C	33—34°C	36—37°C
Virion size	22—28 mm	24—30 mm	24—30 mm	23—25 mm
Acid sensitivity infectivity				
pH3	Stable	Stable	Labile	Labile
pH6	Stable	Labile	Becoming labile	Labile at low ionic strength; stable at high ionic strength
Density in CsCl g/cm³	1.33—1.35	1.34 also 1.44	1.38—1.41	1.43—1.45
% RNA	29	31	30	31.5

and Apthovirus — on the basis of certain properties such as acid stability, buoyant density, and clinical manifestations.[25]

The main features used for distinguishing the gerera are shown in Table 1.

B. Classification of HAV

In 1980, in response to increasing knowledge about the structure of HAV, the ICTV established an ad hoc study group to consider the taxonomic position of HAV. This group presented its first report in August 1981 and recommended that HAV be classified as a picornavirus within the genus Enterovirus.[53] This suggestion was accepted by the Executive Council and membership of ICTV and the suggestion by Dr. Melnick that it receive the next enterovirus number (number 72) was noted.

While it is clear that on the basis of its size, buoyant density, the composition of its capsid, and the size and nature of the genome, HAV can readily be classified as a picornavirus, it has a number of features which set it apart from each of the four established genera. These include its heat stability (which is greater than any known animal RNA virus), a relatively high degree of secondary structure in the genome, and perhaps its ability to induce persistent nonlytic infections in some cell lines.

Recent data from studies of the molecular biology of picornaviridae[53] (see Chapter 7) indicate the HAV has little nucleotide sequence homology with other picornaviruses, distinctive peptide cleavage sites, and a smaller VP4, suggesting that it should probably be classified in a separate genus. A comparison of these major characteristics of picornaviruses is presented in Table 2.

Table 2
COMPARISON OF HAV WITH OTHER PICORNAVIRIDAE

General Features

	Example	Density (CsCl)	Acid pH	Heat 60°C	Capsid proteins (kdalton) VP1	VP2	VP3	VP4
HAV	HM-175	1.32—1.34	Stable	Stable	31.9	24.8	27.8	2.5
Enterovirus	PV-1	1.34	Stable	Labile	33.5	30.0	26.4	7.4
Rhinovirus	HRV-14	1.39—1.42	Labile	Labile	31.7	29.0	25.1	7.2
Cardiovirus	EMC	1.34	Stable	Labile	23.3	24.7	24.3	8.5
Apthovirus	143	1.45	Labile	Labile	32.4	28.5	26.2	7.2

Composition of Genome RNA

	HAV	Enterovirus	Rhinovirus	Cardiovirus	Apthovirus
C	16	20—24	20	24—26	28
A	29	27—30	34	25—27	26
G	22	23—28	19	23—24	24
U	33	23—25	27	24—27	22
G and C	38	~46	39	~49	52
Poly C	No	No	No	Yes	Yes
Homology	None	Rhinovirus	Enterovirus	None	None

REFERENCES

1. **Feinstone, S. M., Kapikian, A. Z., and Purcell, R. H.,** Hepatitis A: detection by immune-electronmicroscopy of a virus-like antigen associated with acute illness, *Science,* 182, 1026, 1973.
2. **Hollinger, F. B., Bradley, D. W., Maynard, J. R., Dreesman, G. R., and Melnick, J. R.,** Detection of hepatitis A viral antigen by radioimmunoassay, *J. Immunol.,* 115, 1464, 1975.
3. **Purcell, R. H., Wong, D. C., Moritsugu, Y., Dienstag, J. L., Routenberg, J. A., and Boggs, J. D.,** A microtitre solid-phase radioimmunoassay for hepatitis A antigen and antibody, *J. Immunol.,* 116, 349, 1976.
4. **Provost, P. J. and Hilleman, M. R.,** Propagation of human hepatitis A virus in cell culture, *Proc. Soc. Exp. Biol. Med.,* 149, 254, 1979.
5. **Locarnini, S. A., Ferris, A. A., Stott, A. C., and Gust, I. D.,** The relationship between a 27 nm virus-like particle and hepatitis A as demonstrated by immune electron microscopy, *Intervirology,* 4, 110, 1974.
6. **Gravelle, C. R., Hornbeck, C. L., Maynard, J. E., Schable, C. A., Cook, E. H., and Bradley, D. W.,** Hepatitis A: report of a common source outbreak with recovery of a possible etiologic agent. II. Laboratory studies, *J. Infect. Dis.,* 131, 167, 1975.
7. **Maynard, J. E., Bradley, D. W., Gravelle, C. R., Ebert, J. W., and Krushak, D. H.,** Preliminary studies of hepatitis A in chimpanzees, *J. Infect. Dis.,* 131, 194, 1975.
8. **Schulman, A. N., Dienstag, J. L., Jackson, D. R., Hoofnagle, J. H., Gerety, R. J., Purcell, R. H., and Barker, L. F.,** Hepatitis A antigen particles in liver, bile and stool of chimpanzees, *J. Infect. Dis.,* 134, 80, 1976.
9. **Provost, P. J., Wolanski, B. S., Miller, W. J., Ittensohn, O. L., McAleer, W. J., and Hilleman, M. R.,** Physical, chemical and morphological dimensions of human hepatitis A virus strain CR326, *Proc. Soc. Exp. Biol. Med.,* 148, 532, 1975.
10. **Siegl, G. and Eggers, H. J.,** Failure of guanidine and 2-(α hydroxybenzyl) benzimidazole to inhibit replication of hepatitis A virus in vitro, *J. Gen. Virol.,* 61, 111, 1982.
11. **Feinstone, S. M., Kapikian, A. Z., Gerin, J. L., and Purcell, R. H.,** Buoyant density of the hepatitis A virus-like particle in cesium chloride, *J. Virol.,* 13, 1412, 1974.
12. **Bradley, D. W., Hornbeck, C. L., Gravelle, C. R., Cook, E. H., and Maynard, J. E.,** CsCl-banding of hepatitis A-associated virus-like particles, *J. Infect. Dis.,* 131, 304, 1975.

13. **Bradley, D. W., Gravelle, C. R., Cook, E. H., Fields, H. A., and Maynard, J. E.,** Cyclic excretion of hepatitis A virus in experimentally infected chimpanzees: biophysical characterization of the associated HAV particles, *J. Med. Virol.*, 1, 133, 1977.
14. **Bradley, D. W., McCaustland, K. A., Schreeder, M. T., Cook, E. H., Gravelle, C. R., and Maynard, J. E.,** Multiple buoyant densities of hepatitis A virus in cesium chloride gradients, *J. Med. Virol.*, 1, 219, 1977.
15. **Moritsugu, Y., Dienstag, J. L., Valdesuso, J., Wong, D. C., Wagner, J., Routenberg, J. A., and Purcell, R. H.,** Purification of hepatitis A antigen from feces and detection of antigen and antibody by immune adherence hemagglutination, *Infect. Immun.*, 13, 898, 1976.
16. **Rakela, J., Fay, O. H., Stevenson, D., Gordon, I., and Mosley, J. W.,** Similarities of two hepatitis A virus strains, *Bull. WHO*, 54, 561, 1976.
17. **Siegl, G. and Frosner, G. G.,** Characterization and classification of virus particles associated with hepatitis A. II Type and configuration of nucleic acid, *J. Virol.*, 26, 48, 1978.
18. **Siegl, G.,** The biochemistry of hepatitis A virus, in *Hepatitis A*, Gerety, R. J., Ed., Academic Press, Orlando, Fla., 1984, 12.
19. **Coulepis, A. G., Locarnini, S. A., Westaway, E. G., Tannock, G. A., and Gust, I. D.,** Biophysical and biochemical characterization of hepatitis A virus, *Intervirology*, 18, 107, 1982.
20. **Siegl, G., Frosner, G. G., Gauss-Muller, V., Tratschin, J. D., and Deinhardt, F.,** The physicochemical properties of infectious hepatitis A virion, *J. Gen. Virol.*, 57, 331, 1981.
21. **Lemon, S. M., Jansen, R. W., and Newbold, J. E.,** Infectious hepatitis A particles produced in cell culture consist of three distinct types with different buoyant densities in CsCl, *J. Virol.*, 54, 78, 1985.
22. **Tratschin, J. D., Siegl, G., Frosner, G. G., and Dienhardt, F.,** Characterization and classification of virus particles associated with hepatitis A. III. Structural proteins, *J. Virol.*, 38, 151, 1981.
23. **Yamaguchi-Koll, V., Weigers, K. J., and Drzeniek, R.,** Isolation and characterization of dense particles from poliovirus infected HeLa cells, *J. Gen. Virol.*, 26, 307, 1975.
24. **Weigers, K. J., Yamaguchi-Koll, J., and Drzeniek, R.,** Differences in the physical properties of dense and stardard poliovirus particles, *J. Gen. Virol.*, 34, 465, 1977.
25. **Cooper, P. D., Agol, V. I., Bachrach, H. L., Brown, F., Ghendon, Y., Gibbs, A. J., Gillespie, J. H., Longberg-Holm, K., Mandel, B., Melnick, J. L., Mohanty, S. B., Pouey, R. C., Rueckart, R. R., Schaffer, F. L., and Tyrell, D. A. J.,** Picornaviridae: a second report, *Intervirology*, 10, 165, 1978.
26. **Bradley, D. W., Fields, H. A., McCaustland, K. A., Cook, E. H., Gravelle, C. R., and Maynard, J. E.,** Biochemical and biophysical characterization of light and heavy density hepatitis A virus particles: evidence HAV is an RNA virus, *J. Med. Virol.*, 2, 175, 1978.
27. **Krugman, S., Giles, J. P., and Hammond, J.,** Hepatitis virus: effect of heat on the infectivity and antigenicity of the MS-1 and MS-2 strains, *J. Infect. Dis.*, 122, 432, 1970.
28. **Parry, J. V. and Mortimer, P. P.,** The heat sensitivity of hepatitis A virus determined by a single tissue culture method, *J. Med. Virol.*, 14, 277, 1984.
29. **Frosner, G. G.,** Zuchtang des Hepatitis A Virus in Gewebekultur: Moglich Keit zur Virus Produktion fur Impstoffe und Testzwecke zur Untersuchung von Patienten auf Infektiositat und zur Prufung von Disinfektionsmitteln, *Offentl. Gesund. Wes.*, 44, 370, 1982.
30. **Scheid, R. and Deinhardt, F., quoted by Siegl, G.,** The biochemistry of hepatitis A virus, in *Hepatitis A*, Gerety, R. J., Ed., Academic Press, Orlando, Fla., 1984, 14.
31. **Dienstag, J. L., Gust, I. D., Lucas, C. R., Wong, D. C., and Purcell, R. H.,** Mussel-associated viral hepatitis, type A: serological confirmation, *Lancet*, 1, 561, 1976.
32. **Koff, R. S. and Sear, H. S.,** Internal temperature of steamed clams, *N. Engl. J. Med.*, 276, 737, 1967.
33. **McCaustland, K. A., Bond, W. W., Bradley, D. W., Ebert, J. W., and Maynard, J. E.,** Survival of hepatitis A virus in feces after drying and storage for 1 month, *J. Clin. Microbiol.*, 16, 957, 1981.
34. **Siegl, G., Weitz, M., and Kronauer, G.,** Stability of hepatitis A virus, *Intervirology*, 22, 218, 1984.
35. **Locarnini, S. A., Coulepis, A. G., Ferris, A. A., Lehmann, N. I., and Gust, I. D.,** Purification of hepatitis A virus from human feces, *Intervirology*, 10, 300, 1978.
36. **Peterson, D. A., Hurley, T. R., Hoff, J. C., and Wolfe, L. G.,** Effect of chlorine treatment on infectivity of hepatitis A virus, *Appl. Environ. Microbiol.*, 45, 223, 1983.
37. **Scheid, R., Deinhardt, F., Frosner, G. G., Abb, J., Zachoval, R., and Siegl, G.,** *Viral Hepatitis, 1981 International Symposium*, Szmuness, W., Alter, H. J., and Maynard, J. E., Eds., Franklin Institute Press, Philadelphia, 1981, 627.
38. **Provost, P. J. and Hilleman, M. R.,** An inactivated hepatitis A vaccine prepared from infected marmoset liver, *Proc. Soc. Exp. Biol. Med.*, 159, 201, 1978.
39. **Coulepis, A. G., Locarnini, S. A., Ferris, A. A., Lehmann, N. I., and Gust, I. D.,** The polypeptides of hepatitis A virus, *Intervirology*, 10, 24, 1978.
40. **Coulepis, A. G., Locarnini, S. A., and Gust, I. D.,** Iodination of hepatitis A virus reveals a fourth structural polypeptide, *J. Virol.*, 35, 572, 1980.

41. **Feinstone, S. M., Moritsugu, Y., Shih, R. W.-K., Gerin, J. L., and Purcell, R. H.,** Characterization of HAV, in *Viral Hepatitis*, Vyas, G. N., Cohen, S. N. and Schmid, R., Eds., Franklin Institute Press, Philadelphia, 1978, 41.
42. **Siegl, G., Frosner, G. G., Gauss-Muller, V., Tratschin, J. D., and Deinhardt, F.,** The physicochemical properties of infectious hepatitis A virions, *J. Gen. Virol.*, 57, 331, 1981.
43. **Hughes, J. V., Stanton, L. W., Tomassini, J. E., Long, W. J., and Scolnick, E. M.,** Neutralizing monoclonal antibodies to hepatitis A virus: partial localization of a neutralizing antigen site, *J. Virol.*, 52, 465, 1984.
44. **Mathieson, L., Feinstone, S. M., Purcell, R. H., and Wagner, J.,** Detection of hepatitis A antigen by immunofluorescence, *Infect. Immun.*, 18, 524, 1977.
45. **Murphy, B. L., Maynard, J. E., Bradley, D. W., Ebert, J. W., Mathieson, L. R., and Purcell, R. H.,** Immunofluorescence of hepatitis A virus antigen in chimpanzees, *Infect. Immun.*, 21, 663, 1978.
46. **Coulepis, A. G., Tannock, G. A., Locarnini, S. A., and Gust, I. D.,** Evidence that the genome of hepatitis A virus consists of single-stranded RNA, *J. Virol.*, 37, 473, 1981.
47. **Ticehurst, J. R., Racaniello, V. R., Baroudy, B. M., Baltimore, D., Purcell, R. H., and Feinstone, S. M.,** Molecular cloning and characterization of hepatitis A virus cDNA, *Proc. Nat. Acad. Sci. U.S.A.*, 80, 5885, 1983.
48. **Locarnini, S. A., Coulepis, A. G., Westaway, E. G., and Gust, I. D.,** Restricted replication of human hepatitis A virus in cell culture: intracellular biochemical studies, *J. Virol.*, 37, 216, 1981.
49. **Najarian, R., Caput, D., Gee, W., Potter, S. J., Renard, A., Merryweather, J., Van Nest, G., and Dina, D.,** Primary structure and gene organization of human hepatitis A virus, *Proc. Nat. Acad. Sci. U.S.A.*, 82, 2627, 1985.
50. **Anderson, D. A., Locarnini, S. A., Ross, B. C., Coulepis, A. G., Anderson, B. N., and Gust, I. D.,** Single-cycle growth kinetics of hepatitis A virus on B-SC-1 cells, in *Positive Stranded RNA Viruses*, Reuckart, R., Ed., Alan R. Liss, New York, 1986, in press.
51. **Wheeler, C. M., Fields, H. A., Schable, C. A., Meinke, W. J., and Maynard, J. E.,** Adsorption, purification, and growth characteristics of hepatitis A virus strain HAS-15 propagated in fetal rhesus monkey kidney cells, *J. Clin. Microbiol.*, 23, 434, 1986.
52. **Melnick, J. L., Agol, V. I., Bachrach, H. L., Brown, F., Cooper, P. D., Fiers, W., Gard, S., Gear, J. A. S., Ghendon, Y., Kasza, L., Laplaca, M., Mandel, B., McGregor, S., Mohanty, S. B., Plummer, G., Rueckart, R. R., Schaffer, F. L., Tagaya, I., Tyrell, D. A. J., Voroshilova, M., and Wenner, H. A.,** Picornaviridae, *Intervirology*, 4, 303, 1974.
53. **Gust, I. D., Coulepis, A. G., Feinstone, S. M., Locarnini, S. A., Moritsugu, Y., Najera, R., and Siegl, G.,** Taxonomic classification of hepatitis A virus, *Intervirology*, 20, 1, 1983.

Table 1
OUTBREAKS OF HEPATITIS IN HANDLERS OF NONHUMAN PRIMATES FROM 1958 TO 1974

	No. of clusters	No. of cases
Chimpanzees	37	165
Gorillas	1	4
Celebes apes	1	4
Wooley monkeys	2	9
Mixed groups	5	38
Total	46	220

most of the animals which were used were jungle caught and had been held in crowded and unhygienic conditions prior to transport to the laboratory, ample opportunity existed for them to acquire infection from other animals, their attendants, or from contaminated food or water. Acquisition of human infections seemed to occur quite commonly — for example, rhesus monkeys often become infected with measles virus between capture and arrival in the laboratory.[10]

In 1957, Deinhardt and his colleagues[11] began an extensive study of hepatitis infection in nonhuman primates in the Belgian Congo. Chimpanzees were tested frequently to define their usual pattern of liver function and attempts were made to transmit hepatitis to recently captured animals. Normal values of bilirubin, CCF, thymol turbidity (TT), SGOT, and BSP excretion were established, the range of fecal parasites normally encountered determined, and serial liver biopsies (obtained with a Vim-Silverman needle) examined to establish the normal pattern of liver histology.

Animals were inoculated with a filtrate of human feces and a pool of serum which had transmitted infection to human volunteers at the Willowbrook State School (see Chapter 1) or with a pool of feces collected from patients involved in an epidemic of hepatitis in Toronto.[12] Two thirds of 37 chimpanzees inoculated with these materials developed abnormalities of liver function and histological evidence of acute liver damage compared with 1 of 15 uninoculated controls. Some of the animals developed a biphasic illness suggesting that the serum pool may have contained more than one agent. Unfortunately these studies were terminated by political upheavals in the country and all the specimens which had been collected were lost.

B. Outbreaks in Primate Handlers

Attempts to transmit hepatitis to nonhuman primates had been virtually abandoned when Hillis in 1961 reported an outbreak of hepatitis among chimpanzee handlers at an American Air Force base.[13] A search of the literature revealed no less than 18 similar episodes involving between one and eight cases.[14] Over the next 10 years or so, a further 46 episodes resulting in 200 cases of hepatitis in individuals in close contact with nonhuman primates were recorded[12] (see Table 1).

The outbreak described by Hillis occurred in 1953 among Air Force personnel handling chimpanzees and monkeys in a research laboratory in New Mexico.[13] Of 21 people who had repeated close contact with a newly acquired batch of juvenile chimpanzees, 11 developed acute hepatitis. Hillis subsequently reported on nine icteric and two anicteric cases which occurred at the Holloman Air Force Base over a 29-month period. In that outbreak, veterinarians and other handlers of newly imported young chimpanzees became ill within 28 to 43 days of first contact with the animals. By contrast, no cases of hepatitis were observed among handlers of adult chimpanzees housed in another part of the base.[15]

Table 2
RESULTS OF TESTS FOR EVIDENCE OF HAV INFECTION IN SERA COLLECTED FROM TWO SEPARATE PRIMATE-ASSOCIATED OUTBREAKS

Patient	Specimen	Antibody titer[a]	Subject	Specimen	Antibody titer
1	Acute	<10	Chimp 1	Early convalescent	8000
	Convalescent	≧1000		Late convalescent	2000
2	Acute	<10	Chimp 2	Acute	<10
	Convalescent	≧1000		Convalescent	≧1000
3	Acute	NA	Patient 1	Acute	<10
	Convalescent	≧1000		Convalescent	>1000
			Wooley monkey 1	Acute	<10
				Convalescent	>1000
			Wooley monkey 2	Early	<10
				Late	>1000

[a] Antibody titers expressed as the reciprocal of the titer detected by immune-adherence hemagglutination.

While most of the outbreaks which have been documented were associated with recently captured chimpanzees, other primates (including a gorilla, celebes apes, gibbons, and wooley monkeys) have been involved. While it is not certain whether the animals were infected at the time of capture or became infected during the holding period, most of the epidemiological data suggest the latter. Chimpanzees are often captured by African villagers and held in pens until an exporter makes a visit to purchase them. The animals are held in areas in which hepatitis A is endemic, often under poor conditions, so that contact with contaminated water and food is possible. Most of the animals are young and often affectionate as they have only recently been separated from their mothers. Under these circumstances, there is ample opportunity for them to acquire human pathogens and for these to be transmitted either within the holding facility, during transport to another country, or in the place in which they are held prior to final distribution. Another way in which infection could be acquired is from the (now discontinued) practice of some exporters of injecting the animals with pooled human serum to protect them against human infections.[15]

Once HAV is introduced into a monkey colony or holding facility it tends to spread rapidly from animal to animal and to susceptible handlers. If the colony has a large turnover and fresh susceptible animals are regularly introduced, hepatitis A may become endemic in the colony.

Interestingly, while epidemics of hepatitis were being recognized among American primate handlers, the disease was rarely seen among handlers in Africa, presumably because most were already immune. Despite the absence of controlled studies, physicians caring for the handlers have claimed that the regular injection of human immunoglobulin (HNIG) protected both the handlers and family contacts from acquiring the disease. Although few specimens remain from these outbreaks, there are good epidemiological grounds for believing that most were due to hepatitis A. In recent years Dienstag et al.[17] have been able to examine sera obtained from patients and animals involved in two well-documented outbreaks; on both occasions serological evidence of infection with HAV was obtained. In the first instance, three people who apparently acquired infection from a chimpanzee developed rising levels of anti-HAV; in the second, a handler and four animals in the colony (two chimpanzees and two wooley monkeys) showed serological evidence of infection (see Table 2).

C. Subsequent Attempts to Transmit Infection

1. Old World Primates Other Than Chimpanzees

The occurrence of hepatitis in handlers of nonhuman primates stimulated renewed efforts to transmit the disease to animals. During the 1960s, Bearcroft,[18-22] at the West African Center for Medical Research in Nigeria, attempted to transmit infectious hepatitis to 12 different species of nonhuman primates by injection of tissue or fecal extracts obtained from patients with the disease. The groups of animals which he studied are listed in Table 3.

Bearcroft reported pronounced histological changes in the livers of young patas, mangabey, mona, and putty-nosed monkeys during the third week after inoculation with 10% suspensions of liver, spleen, and feces obtained from patients who had died of hepatitis.[19] These changes occurred less frequently and were less severe in older animals. Although the animals were not monitored regularly for biochemical or histological changes (the only tissue examined was obtained by autopsy at the end of each experiment) and disease could not be regularly transmitted from animal to animal, it seems probable that transmission was occurring. However, as the starting inocula were obtained from patients with fatal hepatitis it is possible that the disease being transmitted was not hepatitis A but some other form of hepatitis such as the enterically transmitted form of non-A, non-B hepatitis (HNANB), which may be clinically indistinguishable, and in certain outbreaks has been associated with high-case fatality rates.[23]

Smetana and his colleagues at the Delta Primate Center in Louisiana also carried out a number of studies in chimpanzees and patas monkeys, and reported the spontaneous occurrence of hepatitis in both groups.[24-26] In the first of these studies, patas monkeys were injected intravenously and intramuscularly with plasma (WW-SS) from a patient with hepatitis which had previously transmitted disease to about 50% of human volunteers after an incubation period of 5 weeks.[27] Another group of animals were fed or injected intraperitoneally with a suspension of feces collected from an animal handler who had developed disease after contact with chimpanzees. Three uninoculated controls were included in each experiment and all animals were monitored for changes in liver function by weekly biochemical tests and liver biopsy.

In the first experiment, one of nine patas monkeys infected with WW-SS developed typical hepatitis after an incubation period of 6 to 7 weeks. In the second experiment, three of nine animals developed biochemical and histological evidence of hepatitis after incubation periods of 8 to 10 weeks; the other animals showed only minor changes and all six controls remained healthy.[25] The incubation period of this disease is slightly longer than usually found in hepatitis A, and as neutralization tests with paired sera were not carried out and no specimens remain from these studies, it is not possible to be certain that the disease being transmitted was hepatitis A.

2. Marmosets and New World Monkeys

In 1948 to 1949 a few attempts were made to transmit human hepatitis to the common marmoset (*Callithrix jacchus*), a new world monkey, without success.[8] Interest in these animals waned until the early 1960s when colonies of white-lipped (*Saguinus nigricollis*) and cotton-topped marmosets (*S. oedipus*) were established in Fritz Deinhardt's laboratory in Chicago, by the National Cancer Institute of the N.I.H.[28] (see Figure 1). Although this facility was established primarily for the study of cancer viruses it soon became apparent that the animals were susceptible to a range of human viruses including influenza, parainfluenza, adenovirus, herpes simplex, measles, mumps, and rubella (reviewed by Deinhardt).[12]

The animals were examined and tested regularly, normal biochemical and hematological values defined,[29,30] and their microbial flora,[31] common diseases,[32] and dietary habits described. As a result, conditions were established which enabled the animals to breed, thus removing many of the variables associated with the use of animals obtained from importers.

Table 3
SPECIES OF NONHUMAN PRIMATES INJECTED WITH MATERIAL FROM PATIENTS WITH INFECTIOUS HEPATITIS IN STUDIES IN NIGERIA FROM 1963 - 1969[17-21]

Patas monkeys	*Erythrocebus patas patas* or *Cercopithecus patas patas*
Mangabeys	*Cercocebus torquatus torquatus*
Mona guenons	*Cercopithecus mona mona*
Putty-nosed monkeys	*Cercopithecus nictitans nictitans*
Tantalus guenons	*Cercopithecus aethiops tantalus*
Sclater's guenons	*Cercopithecus erythrotis sclateri*
White-throated guenons	*Cercopithecus nictitans erythrogaster*
Drills	*Mandrillos leucophaeas* or *Papio leucophaeas*
Dog-faced baboons	*Papio ambis*
Demidov's galagos	*Galago demidoxii demidoxii*
Bosman's monkey	*Perodicticus potto potto*
Mamfe guenons	*Cercopithecus erythrotis camerunensis*

FIGURE 1. Dr. Fritz Deinhardt, leader of the group that demonstrated that marmosets were susceptible to infection with human hepatitis virus, pictured in his laboratory in Chicago in the 1960s. (Photo kindly provided by Dr. Jean Deinhardt.)

Table 4
RESULTS OF CROSS-CHALLENGE STUDIES WITH THREE INOCULA KNOWN TO PRODUCE HEPATITIS IN MARMOSETS

Inoculum 1	Inoculum 2	Outcome
GB	GB	No disease
GB	Berlin	No disease
Berlin	GB	No disease
MS-1	MS-1	No disease
GB	MS-1	Hepatitis
MS-1	GB	Hepatitis

During a study of the susceptibility of marmosets to human viruses, several animals were inoculated with serum or plasma which had been collected from patients with hepatitis during the acute phase of their illness. Some of the animals developed biochemical and histological evidence of hepatitis.[9,33] This led to an extensive series of experiments in which white-lipped marmosets *(S. nigricollis)* were inoculated with five sera from patients who were judged, on epidemiological grounds, to have had hepatitis A.

One of these sera (GB), which had been obtained from a surgeon with acute hepatitis, was shown to be highly infectious, and the disease it produced was readily transmitted from animal to animal. This strain was studied in several laboratories and many of its biochemical and biophysical features defined. Evidence gradually accumulated that the GB agent was not responsible for hepatitis A, as infection could not be neutralized with convalescent human serum or normal immunoglobulin (HNIG), nor could convalescent sera from GB prevent hepatitis A infection. The GB agent is now believed to be either a virus of marmosets[34,35] or one of the HNANB agents (see Chapter 2).

About the same time as the GB agent was recognized, Kohler et al.[36] in Berlin, isolated a similar agent (the Berlin agent) in cotton-topped marmosets *(S. oedipus)*. This agent was recovered after sera from patients with acute hepatitis was passed several times in primary pig kidney-tissue culture and then inoculated into marmosets. Comparative studies of the Berlin and GB agents in an independent laboratory suggested that they were antigenically related[37] (see Chapter 2).

Because of doubts about the specificity of the GB and Berlin agents, further attempts were made to transmit hepatitis to marmosets using pre-inoculation and acute-phase plasma collected from three adults who had been experimentally infected with the MS-1 strain of HAV. In 1969 and 1971, Holmes and his colleagues reported the results of a number of experiments carried out under code. In these studies hepatitis was induced in *S. nigricollis* and *S. oedipus* marmosets by inoculation of plasma collected from experimentally infected volunteers during the acute phase of their illness, but not by plasma collected prior to infection.[1,38] The incubation period from infection to the onset of abnormal liver function tests ranged from 26 to 51 days and seemed to be inversely related to the volume of plasma injected. They further showed that marmosets first infected with the MS-1 strain of HAV were resistant to rechallenge with this inoculum, but developed hepatitis after inoculation with the GB agent. In addition, animals inoculated with GB were refractory to challenge with GB but remained susceptible to MS-1, while animals infected with either GB or the Berlin agent failed to develop hepatitis after inoculation with the other agent (see Table 4).

Similar results were reported shortly afterward by Lorenz et al.,[2] who induced hepatitis in white-mustached marmosets *(S. mystax)* with serum from patients with MS-1 hepatitis

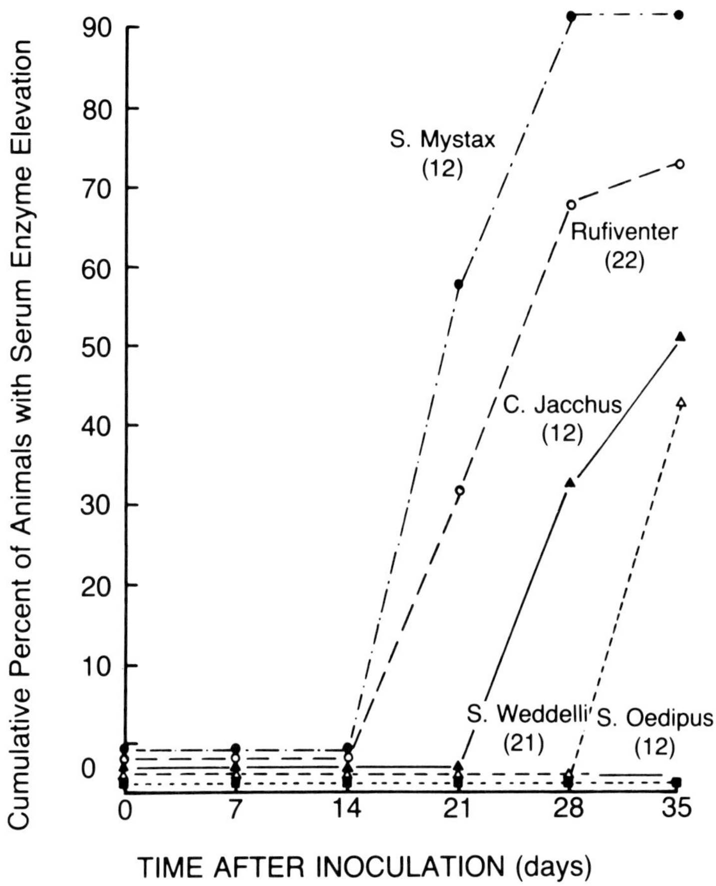

FIGURE 2. Time to onset of biochemical hepatitis and proportion of animals who developed disease when five different species of marmoset were inoculated with CR326. Number of animals used indicated in brackets. (From Provost, P. J., Villarejos, V. M., and Hilleman, M. R., *Proc. Soc. Exp. Biol. Med.*, 155, 283, 1977. With permission.)

and subsequently by Mascoli and her colleagues[39] in marmosets *(S. mystax* and *S. nigricollis)* inoculated with blood obtained from Costa Rican patients with naturally acquired infections. One of the strains isolated in the latter study (CR326) has been worked on extensively. This strain induced biochemical hepatitis in marmosets within 25 to 55 days of inoculation and biopsies performed during the height of the illness revealed acute hepatitis of variable severity characterized by local or diffuse degeneration of hepatocytes, distention of sinusoids, and infiltration of mononuclear cells.

CR326 and other strains such as MS-1 have been used to evaluate the sensitivity of different species of marmosets to HAV[40] (see Figure 2). The availability of this system enabled workers at the Merck Institute for Therapeutic Research to develop a marmoset neutralization test for detecting antibodies to CR326 and to show that Costa Rican patients with naturally acquired infections and volunteers infected with the MS-1 strain of HAV developed neutralizing antibodies to this agent.[41] Subsequently CR326 has been adapted to cell culture and used to produce experimental batches of killed and live hepatitis A vaccines.[42,43]

Although marmosets are relatively cheap and readily available they are not routine laboratory animals. The species which are most susceptible to infection breed poorly in captivity and wild-caught animals frequently die during transportation or adaptation to their new environment. Considerable experience is needed before these animals can be maintained in captivity and attempts to establish a colony should not be approached lightly.

Despite these difficulties marmosets remain important models of hepatitis A and the enterically transmitted form of HNANB as well as for some other infectious diseases of man.

3. Chimpanzees

Outbreaks of hepatitis among handlers of recently imported chimpanzees provided evidence of the susceptibility of these animals to HAV and suggested that the failure of earlier attempts to infect them may have been due to use of animals that were already immune. After the successful identification of the etiological agent and the development of techniques such as IEM for detection of anti-HAV,[44] Dienstag et al. at the N.I.H. were able to test a number of colony-raised and wild-caught chimpanzees and select seronegative animals for further study.[4] Interestingly, of the animals they tested all the jungle-caught chimpanzees had detectable levels of anti-HAV while none of the colony-raised animals showed evidence of previous infection.

Two young seronegative chimpanzees were selected and administered either an i.v. injection or oral dose of a filtered extract of feces, diluted in veal infusion broth. This specimen had been collected from an adult volunteer who had received the MS-1 strain of HAV 33 days earlier (and had developed symptoms of infection 29 days later). Two other animals were administered extracts of feces collected from the same volunteer a week prior to exposure to the virus.

Both animals that received the acute-phase fecal extract developed biochemical and histological hepatitis 3 to 4 weeks after inoculation, shed HAAg in the feces during the acute phase of their illness, and developed anti-HAV in their serum during convalescence (see Figure 3). The control animals remained free of illness during this period but developed acute hepatitis 3 to 5 weeks after exposure to the infected animals, indicating that the disease could be naturally transmitted from animal to animal, probably by contact with feces containing HAV.

At the Hepatitis Laboratories of the Centers for Disease Control in Phoenix, Arizona, Maynard and his colleagues[3] were also able to transmit hepatitis to chimpanzees (which had been shown by Feinstone to be seronegative by the recently developed IEM test) by i.v. injection of 1 $m\ell$ of an extract of a pool of feces collected from seven people involved in a food-borne outbreak of the disease. Three of these animals developed biochemical and histological hepatitis and shed HAAg in their feces. An extract of feces collected from one animal during the acute phase of its illness was injected i.v. into another seronegative chimpanzee and induced an attack of hepatitis.

Following these pioneering studies, chimpanzees have been widely used for studying the pathogenesis of infection, the presence of HAV and HAAg in various tissues and secretions, the timing and pattern of antibody development, and assessing the degree of attenuation of strains which have been repeatedly passaged in cell culture.

D. Detection of Anti-HAV in Nonhuman Primates

Because of the availability of specific assays for anti-HAV, it became possible to examine a wide range of nonhuman primates in zoos, as well as recently captured animals, and animals held in breeding colonies for evidence of past infection with HAV. Apparently, specific antibody has been detected in a number of species both at the time of capture and after the animals have been held for varying periods[45] (see Table 5). Theoretically all these species might be useful models of HAV infection.

Despite their susceptibility of HAV in the laboratory, anti-HAV has rarely been detected in wild-caught and colony-raised marmosets.[44] Neither in their natural habitat nor in captivity

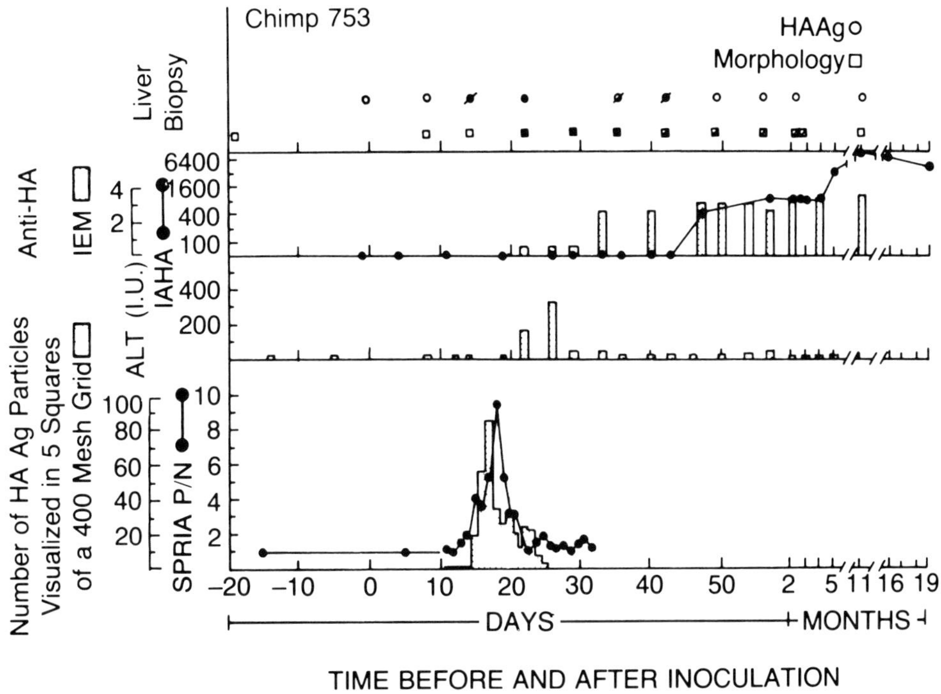

FIGURE 3. Shedding of HAAg in the feces and development of anti-HAV in an experimentally infected chimpanzee correlated to liver enzymes, the appearance of HAAg in hepatocytes, and histological evidence of hepatitis. (□) no hepatitis; (■) acute hepatitis; (◫) resolving hepatitis; (○) no fluorescence; (●) bright fluorescence; (⌀) weak fluorescence. (From Mathieson, L. R., Feinstone, S. M., Purcell, R. H., and Wagner, J. A., *J. Infect. Immun.*, 18, 524, 1977. With permission.)

Table 5
NONHUMAN PRIMATES HELD IN ZOOS OR CAPTURED IN THE WILD IN WHICH ANTIBODIES TO HAV HAVE BEEN DETECTED

Chimpanzee	Macaque
Gorilla	Cebid
Orangutan	Marmoset
Gibbon	Spider monkey
Mangabey	Wooley monkey
Baboon	Owl monkey
Mandrill	Patas
Vervet/grivet	

do these animals become closely associated with man, so that opportunities for infection must be rare.

By contrast, anti-HAV has been frequently detected in wild-caught chimpanzees and animals that have been housed in zoos or colonies for some years. In three separate studies involving 148, 82, and 23 chimpanzees, antibodies were detected in 50%, 36.6%, and 61%, respectively.[46-48] By contrast, antibodies were not detected in colony-raised animals.[49]

Similar data have been provided by Lemon and his colleagues in Panamanian owl monkeys

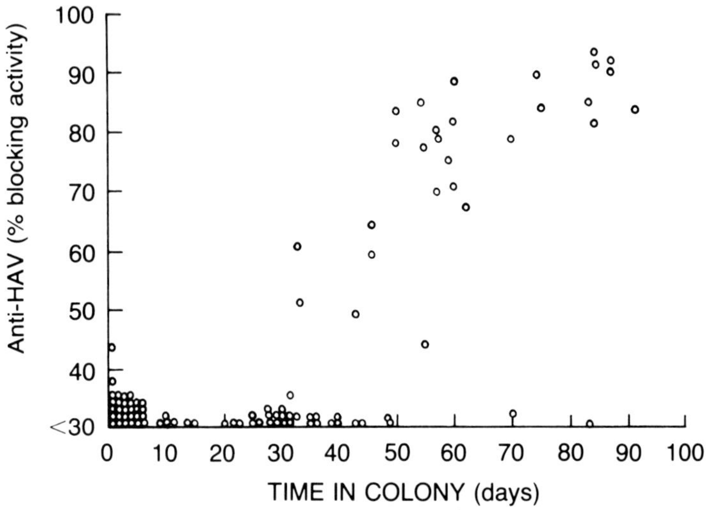

FIGURE 4. Presence of anti-HAV in newly captured *Aotus trivirgatus* monkeys held at the Gorgas Memorial Laboratories, related to time. (From Lemon, S. M., Le Duc, J. W., Binn, L. N., Escajadillo, A., and Ishak, K. G., *J. Med. Virol.*, 10, 25, 1982.)

(Aotus trivirgatus).[50] An initial survey of owl monkeys held at the Walter Reed Army Institute of Research in Washington showed that 55 of 91 (60%) wild-caught animals had detectable levels of anti-HAV, compared with 2 of 79 (3%) colony-bred animals. Studies at the Gorgas Memorial Laboratory in Panama, where the monkeys were held prior to shipment to the U.S., confirmed a high prevalence of antibody (76%), with the highest prevalence being detected among animals which had been held for long periods (see Figure 4), suggesting that exposure had occurred after capture. A study of newly caught animals revealed anti-HAV in only 2 of 145 (1.4%). When 41 seronegative newly caught animals were studied prospectively, serological evidence of HAV infection was detected in 25; only one animal which survived for more than 60 days from admission was not infected. HAV was demonstrated in the feces of infected animals and was presumably being transmitted by the fecal-oral route. Subsequently these authors have been able to demonstrate that owl monkeys are useful models for the study of HAV infection[51] (see Figure 5).

Various species of macaque monkeys are reported to have naturally occurring antibodies to HAV[46] and infection has apparently been transmitted and serially passaged in stump-tailed monkeys *(Macaca speciosa)*[52] which developed histological and biochemical evidence of infection, shed virus in their stools, and seroconverted to anti-HAV. These animals are currently being used as models of HAV infections in studies being carried out in China. Antibody to HAV has also been demonstrated in 20% of 98 recently captured cynomolgus monkeys *(M. fescicularis)*. In a second group of 64 animals, 80% of those which had been in captivity for more than 4 weeks were antibody positive. Of 26 seronegative animals followed for 6 to 8 months after capture, 25 (96%) seroconverted.[53]

In other surveys of nonhuman primate colonies, Purcell and Dienstag[46] detected anti-HAV in 60% of pig-tailed macaques, 50% of stump-tailed macaques and patas monkeys, and 38% of rhesus monkeys. Eichberg and Kalter[47] reported antibody prevalences of 35 to 50% in colonies of baboons, African green monkeys, and celeid monkeys, while Peterson obtained similar data in gorillas and orangutans.[54]

Antibodies have also been detected in rhesus monkeys. Of ten seronegative animals inoculated orally or intravenously with fecal extracts containing HAV, about half seroconverted;[48] however, as none showed clinical or laboratory evidence of hepatitis, rhesus mon-

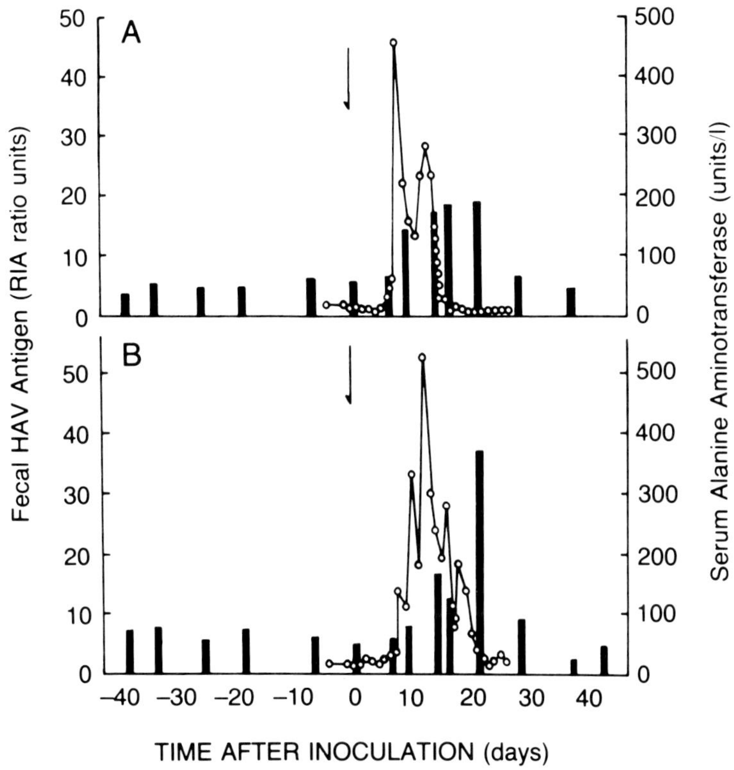

FIGURE 5. The course of infection in two susceptible colony-bred *Aotus trivingatus* monkeys. (—) ALT estimations; (○) detection of HAAg in feces. (From Le Duc, J. W., Lemon, S. M., Kennan, C. M., Graham, R. R., Marchwicki, R. H., and Binn, L. N., *Infect. Immun.*, 766; 40, 1983.)

keys are of limited value as models of the disease in man. Some indication that lesser bushbabies *(Galago sergalensis* and *G. sergalensis moholi)* are susceptible to infection has also been obtained.[55,56]

While apparently specific anti-HAV has been detected in numerous species of nonhuman primates, only chimpanzees, marmosets, owl and stump-tailed monkeys are currently used widely for laboratory studies (see Figure 6). Some species have proved to be difficult to infect or fail to develop evidence of liver damage which limits their value as models of the disease in man, while others are in limited supply or difficult to breed in captivity.

The presence of pre-existing antibodies in some species at the time of capture has been interpreted to indicate that there is a reservoir of infection in nature. However, as these antibodies are usually present in low titer, they more probably represent cross-reactions with naturally occurring animal piconaviruses.[57] Similar cross-reactions have been observed in some dog sera collected in isolated Pacific islands.

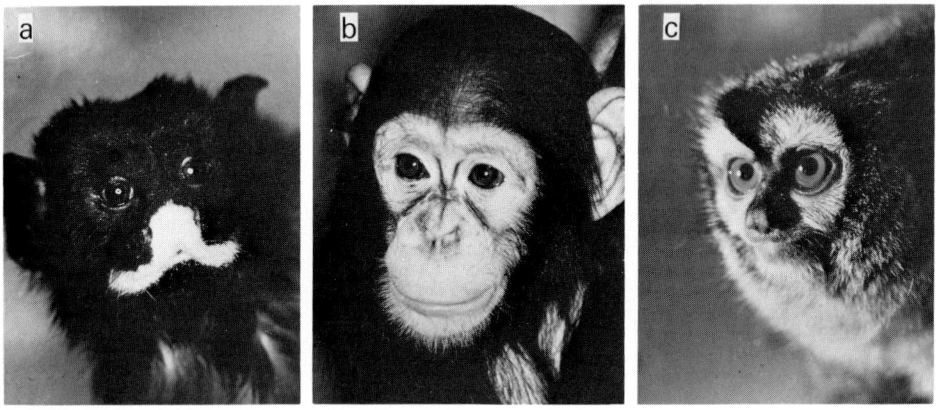

FIGURE 6. Animals which have provided useful models of hepatitis A infection. (a) White-mustached marmoset, *Saguinus mystax;* (b) juvenile chimpanzees, *Pan troglodytes;* (c) owl monkey, *Aotus trivingatus.*

REFERENCES

1. **Holmes, A. W., Wolfe, L. G., Rosenblate, H., and Deinhardt, F.,** Hepatitis in marmosets: induction of disease with coded specimens from a human volunteer study, *Science,* 165, 816, 1969.
2. **Lorenz, D., Barker, L., Stevens, D., Peterson, M., and Kirschstein, R.,** Hepatitis in the marmoset: Saguinus mystax, *Proc. Soc. Exp. Biol. Med.,* 135, 348, 1970.
3. **Maynard, J. E., Bradley, D. W., Gravelle, C. R., Ebert, J. W., and Krushak, D. H.,** Preliminary studies of hepatitis A in chimpanzees, *J. Infect. Dis.,* 131, 194, 1975.
4. **Dienstag, J. L., Feinstone, S. M., Purcell, R. H., Hoofnagle, J. H., Barker, L. F., London, W. T., Popper, H., Peterson, J. M., and Kapikian, A. Z.,** Experimental infection of chimpanzees with hepatitis A virus, *J. Infect. Dis.,* 132, 532, 1975.
5. **Piazza, M.,** *Experimental Viral Hepatitis,* Charles C Thomas, Springfield, Ill., 1969, chap. 2 to 7, 9 and 10.
6. **Colbert, J. W.,** Review of animal experimentation in infectious hepatitis and in serum hepatitis, *Yale J. Biol. Med.,* 21, 335, 1949.
7. **MacCallum, F. O.,** Review of attempts to transmit hepatitis to other animal hosts and review of recent European attempts of propagation of the virus, in *Symposium on Laboratory Propagation and Detection of the Agent of Hepatitis,* Paul, J. R., Ed., Pub 322, Natl. Acad. Sci./Natl. Res. Council, Washington, D.C., 1954, 68.
8. **Evans, A. S.,** Attempts to transmit the virus of human hepatitis to primates other than man, in *Symposium on Laboratory Propagation and Detection of the Agent of Hepatitis,* Paul, J. R., Ed., Pub. 322, Natl. Acad. Sci./Natl. Res. Council, Washington, D.C., 1954, 58.
9. **Deinhardt, F. and Holmes, A. W.,** Epidemiology and etiology of viral hepatitis, in *Progress in Liver Diseases,* Vol 2, Popper, H. and Schaffer, F., Eds., Grune & Stratton, New York, 373, 1965.
10. **Peebles, T. C., McCarthy, K., Enders, J. F., and Holloway, A. J.,** Behavior of monkeys after inoculation of virus derived from patients with measles and propagated in tissue culture together with observations on spontaneous infections of these animals by an agent exhibiting similar antigenic properties, *J. Immunol.,* 78, 63, 1957.
11. **Deinhardt, F., Courtois, G., Dherte, P., Osterrieth, P., Ninawe, G., Henle, G., and Henle, W.,** Studies of liver function tests in chimpanzees after inoculation with human infectious hepatitis virus, *Am. J. Hyg.,* 75, 311, 1962.
12. **Deinhardt, F.,** Hepatitis in primates, in *Advances in Virus Research,* Academic Press, New York, 20, 1976.
13. **Hillis, W. D.,** An outbreak of infectious hepatitis among chimpanzee handlers at a United States Air Force Base, *Am. J. Hyg.,* 73, 316, 1961.
14. **Held, J. R.,** Subhuman primates and the transmission of human hepatitis, Paper read at the 99th meeting of the American Veterinary Medical Association, 1962.

15. **Hillis, W. D.,** Viral hepatitis associated with subhuman primates, *Transfusion,* 3, 445, 1963.
16. Communicable Diseases Center: Veterinary Public Health Notes, May 1962.
17. **Dienstag, J. L., Davenport, F. M., McCollum, R. W., Hennessy, A. V., Klatskin, G., and Purcell, R. H.,** Nonhuman primate-associated viral hepatitis type A: serologic evidence of hepatitis A virus infection, *JAMA,* 236, 462, 1976.
18. **Bearcroft, W. G. C.,** Hepatitis in African monkeys, *Nature (London),* 197, 806, 1963.
19. **Bearcroft, W. G. C.,** Histological studies on the livers of West African primates inoculated with infectious hepatitis, *J. Pathol. Bacteriol.,* 88, 511, 1964.
20. **Bearcroft, W. G. C.,** Studies on the transmission of the agent of infectious hepatitis to patas monkeys, *J. Med. Microbiol.,* 1, 1, 1968.
21. **Bearcroft, W. G. C.,** Further studies on the transmission of the agent of infectious hepatitis to patas monkeys, *Br. J. Exp. Pathol.,* 50, 56, 1969.
22. **Bearcroft, W. G. C.,** Susceptibility of patas monkeys to reinfection with infectious hepatitis, *Br. J. Exp. Pathol.,* 50, 327, 1969.
23. **Khuroo, M. S., Teli, M. R., Skidmore, S., Sofi, M. A., and Khuroo, M. S.,** Incidence and severity of viral hepatitis in pregnancy, *Am. J. Med.,* 70, 252, 1981.
24. **Smetana, H. F.,** Experimental and spontaneous viral hepatitis in primates, *Lab. Invest.,* 14, 1366, 1965.
25. **Smetana, H. F. and Felsenfeld, A. D.,** Viral hepatitis in subhuman primates and its relationship to human viral hepatitis, *Virchows Arch.,* 348, 309, 1969.
26. **Smetana, H. F., Felsenfeld, A. D., and Riopelle, A. J.,** Human viral hepatitis and chimpanzees, *Chimpanzee,* 3, 26, 1970.
27. **Rightsel, W. A., Keltsch, R. A., Taylor, A. R., Boggs, J. D., and McLean, J. W., Jr.,** Virology laboratory studies. I, *JAMA,* 177, 671, 1961.
27. **Purcell, R. H., Feinstone, S. M., Ticehurst, J. R., Daemer, R. J., and Baroudy, B. M.,** Hepatitis A virus, in *Viral Hepatitis in Liver Disease,* Vyas, G. N., Dienstag, J. L., and Hoofnagle, J. H., Eds., Grune & Stratton, Orlando, Fla., 9, 1984.
28. **Deinhardt, F. and Deinhardt, J.,** The use of platyrrhine monkeys in medical research, in *Symposia of the Zoological Society of London,* T.-W.-Fiennes, R. N., Ed., Academic Press, New York, 17, 1966.
29. **Anderson, E. T., Lewis, J. P., Passovoy, M., and Trubaugh, F. E.,** Marmosets as laboratory animals. II. The hematology of laboratory-kept marmosets, *Lab. Anim. Care,* 17, 30, 1967.
30. **Holmes, A. W., Passovoy, M., and Capps, R. B.,** Marmosets as laboratory animals. III. Blood chemistry of laboratory-kept marmosets with particular attention to liver function and structure, *Lab. Anim. Care,* 17, 41, 1967.
31. **Deinhardt, F., Holmes, A. W., Devine, J., and Deinhardt, J.,** Marmosets as laboratory animals. IV. The microbiology of laboratory-kept marmosets, *Lab. Anim. Care,* 17, 48, 1967.
32. **Nelson, B., Cosgrove, G. E., and Gengozian, N.,** Diseases of an imported primate Tamarinus nigricollis, *Lab. Anim. Care,* 16, 255, 1966.
33. **Deinhardt, F., Holmes, A. W., Capps, R. B., and Popper, H.,** Studies on the transmission of human hepatitis to marmoset monkeys. I. Transmission of disease, serial passages and description of liver lesions, *J. Exp. Med.,* 125, 673, 1967.
34. **Parks, W. P. and Melnick, J. L.,** Attempted isolation of hepatitis viruses in marmosets, *J. Infect. Dis.,* 120, 539, 1969.
35. **Parks, W. P., Melnick, J. L., Voss, W. R., Singer, D. B., Rosenberg, H. S., Alcott, J., and Casazza, A. M.,** Characterization of marmoset hepatitis, *J. Infect. Dis.,* 120, 548, 1969.
36. **Kohler, H. and Apodaca, J.,** Hepatitis bei Affen Nach Verimpfung von Adenoviren aus Menschen mit Virushepatitis. I. Bestimmung der Enzymaktivitaten in Verlauf der Erkrangung, *Zentralbl. Bakteriol.* 206, 1, 1968.
37. **Appleton, H.,** Hepatitis transmission in marmosets, *J. Infect. Dis.,* 132, 500, 1975.
38. **Holmes, A. W., Wolfe, L., Deinhardt, F., and Conrad, M. E.,** Transmission of human hepatitis to marmosets: further coded studies, *J. Infect. Dis.,* 124, 520, 1971.
39. **Mascoli, C. C., Ittensohn, O. L., Villarejos, V. M., Arguedas, J. A., Provost, P. J., and Hilleman, M. R.,** Recovery of hepatitis agents in the marmoset from human cases occurring in Costa Rica, *Proc. Soc. Exp. Biol. Med.,* 142, 276, 1973.
40. **Provost, P. J., Villarejos, V. M., and Hilleman, M. R.,** Suitability of the rufiventer marmoset as a host animal for human hepatitis A virus, *Proc. Soc. Exp. Biol. Med.,* 155, 283, 1977.
41. **Provost, P. J., Ittensohn, O. L., Villarejos, V. M., Arguedas, J. A., and Hilleman, M. R.,** Etiologic relationship of marmoset propagated CR326 hepatitis A virus to hepatitis in man, *Proc. Soc. Exp. Biol. Med.,* 142, 1275, 1973.
42. **Provost, P. J. and Hilleman, M. R.,** An inactivated hepatitis A vaccine prepared from marmoset liver, *Proc. Soc. Exp. Biol. Med.,* 159, 201, 1978.

43. **Provost, P. J., Buynak, E. G., McLean, A. A., Hilleman, M. R., and Scolnik, E. M.,** in *Viral Hepatitis and Liver Disease,* Vyas, G. N., Dienstag, J. L., and Hoofnagle, J. H., Eds., Grune & Stratton, New York, 467, 1984.
44. **Feinstone, S. M., Kapikian, A. Z., and Purcell, R. H.,** Hepatitis A: detection by immune electron microscopy of a virus-like antigen associated with acute illness, *Science,* 182, 1026, 1973.
45. **Deinhardt, F. and Deinhardt, J. B.,** Animal models, in *Hepatitis A,* Gerety, R. J., Ed., Academic Press, Orlando, Fla., 185, 1984.
46. **Purcell, R. H. and Dienstag, J. L.,** Experimental hepatitis A virus infection, in *Hepatitis Viruses,* Oda, T., Ed., Univ. Tokyo Press, 3, 1978.
47. **Eichberg, J. W. and Kalter, S. S.,** Hepatitis A and B: serologic survey of human and nonhuman primate sera, *Lab. Anim. Sci.,* 30, 541, 1980.
48. **Miller, W. J., Provost, P. J., McAleer, W. J., Ittensohn, O. L., Villarejos, V. M., and Hilleman, M. R.,** Specific immune adherence assay for human hepatitis A antibody: application to diagnostic and epidemiologic investigations, *Proc. Soc. Exp. Biol. Med.,* 149, 254, 1975.
49. **Kessler, H., Tsiquaye, K. N., Smith, H., Jones, D. M., and Zuckerman, A. J.,** Hepatitis A and B at the London Zoo, *J. Infect. Dis.,* 4, 63, 1982.
50. **Lemon, S. M., Le Duc, J. W., Binn, L. N., Escajadillo, A., and Ishak, K. G.,** Transmission of hepatitis A virus among recently captured Panamanian owl monkeys, *J. Med. Virol.,* 10, 25, 1982.
51. **Le Duc, J. W., Lemon, S. M., Keenan, C. M., Graham, R. R., Marchwicki, R. H., and Binn, L. N.,** Experimental infection of the new world owl monkey (Aotus trivirgatus) with hepatitis A virus, *Infect. Immun.,* 40, 766, 1983.
52. **Mao, J. S., Go, Y. Y., Huang, H. Y., Yu, P. H., Huang, B. Z., Ding, Z. S., Chen, N. L., Yu, J. H., and Xie, R. Y.,** Susceptibility of monkeys to human hepatitis A virus, *J. Infect. Dis.,* 144, 55, 1982.
53. **Burke, D. S., Graham, R. R., and Heisey, G. B.,** Hepatitis A virus in primates outside captivity, *Lancet,* 2, 928, 1981.
54. **Peterson, D. A.,** Prevalence of antibodies to HAV in nonhuman primates, *Abstr. Annu. Meet. Am. Soc. Microbiol.,* 267, 1979.
55. **Grabow, W. O. K. and Prozesky, O. W.,** Lesser bushbabies may be susceptible to CR326 hepatitis A virus, *S. Afr. J. Sci.,* 71, 310, 1975.
56. **Grabow, W. O. K., Prozesky, O. W., Bradley, D. W., and Maynard, J. W.,** Response of Galago bushbabies to the hepatitis A virus, *S. Afr. J. Sci.,* 77, 314, 1981.
57. **Coursaget, P., Levesque, B., Gretillat, E., Eyraud, M., Ferrara, L., and Germain, M.,** Hepatitis A virus in primates outside captivity, *Lancet,* 2, 929, 1981.
58. **Gust, I. D.,** The epidemiology of viral hepatitis, in *Viral Hepatitis and Liver Disease,* Vyas, G. N., Dienstag, J. L., and Hoofnagle, J. H., Eds., Grune & Stratton, Orlando, Fla., 415, 1984.
59. **Mathieson, L. R., Feinstone, S. M., Purcell, R. H., and Wagner, J. A.,** Detection of hepatitis A antigen by immunofluorescence, *Infect. Immun.,* 18, 524, 1977.

Chapter 5

FEATURES OF HEPATITIS A IN EXPERIMENTALLY INFECTED NONHUMAN PRIMATES

I. INTRODUCTION

Transmission of HAV to marmosets, chimpanzees, owl, and macaque monkeys has enabled laboratory workers to study the pathogenesis of infection and address such questions as the efficacy of different routes of inoculation, the presence of virus in various tissues and body fluids, the duration of infectivity, the relationship of incubation period and severity of illness to the dose of virus administered; assess levels of attenuation of strains of HAV which have been passaged many times in cell culture; and assess the safety and efficacy of candidate vaccines. Some of these aspects are considered in detail in other sections of the book.

In this chapter we will confine our attention to studies on the natural history of infection and what is known about the pathology and pathogenesis of hepatitis A in experimentally infected animals.

II. EXPERIMENTAL INFECTION

A. Methods for Detecting Infection

Clinical illness has proven to be an unreliable indication of infection in nonhuman primates.[1] Infected animals occasionally become listless, lose their appetite, and develop ruffled hair. By contrast, with outbreaks of hepatitis A in animal colonies, diarrhea is rarely a feature in experimentally transmitted infections, suggesting that this symptom is caused by concurrent infection with other enteric pathogens. The most reliable markers of infection are seroconversion to anti-HAV, the detection of virus particles or HAAg in the feces or liver, detection of elevated levels of ICD, aspartate aminotransferase (AST), and alanine aminotransferase (ALT) in the serum, and development of a characteristic histological picture in liver biopsies or tissue obtained at laparotomy. As primates may have other causes of liver pathology, such as parasitic infestation, a high degree of expertise is needed to interpret liver biopsies from experimentally infected animals.

B. Origin of the Major Strains Used to Infect Animals

Many of the studies of hepatitis A infection in nonhuman primates have been undertaken with five well-documented strains of HAV; MS-1, SD11, CR326, HM175, and PA33; which were derived from naturally acquired infections in man (MS-1, SD11, CR326, HM175) and Aotus monkeys (PA33). The origin of each of these strains is detailed below.

1. MS-1

The MS-1 and MS-2 serum pools were derived from a child (Mir) at Willowbrook State School who developed two separate attacks of viral hepatitis.[2]

Mir was one of 11 newly admitted children involved in a volunteer study. On September 23, 1964 each of the children was administered 1 mℓ of a serum pool (designated WSP-5) which had been generated by pooling sera collected from another 27 Willowbrook children 3 to 7 days prior to the onset of jaundice. Ten of the 11 children who received WSP-5 developed hepatitis after incubation periods of 30 to 36 days. On March 26, 1965, a second dose of 0.25 mℓ of WSP-5 was administered to most of the children who participated in the first trial. Four of these children, including Mir, developed a second attack of hepatitis after incubation periods of 66 to 82 days suggesting that the WSP-5 pool contained two distinct hepatitis agents.

During the course of these studies serial samples of sera were collected from each child and stored at −20°C. Samples from subject Mir were chosen for further study. Serum samples collected prior to his first and second attacks were pooled and designated MS-1 and MS-2, respectively (Mir Serum pool 1 and Mir Serum pool 2). Subsequent studies showed that these pools contained two antigenically and epidemiologically distinct hepatitis viruses. Oral or parenteral administration of MS-1 regularly induced short-incubation period hepatitis (hepatitis A) in volunteers while inoculation of MS-2 regularly induced hepatitis with a long incubation period (hepatitis B).

The term MS-1 is now generally used to refer to all strains of HAV derived from this original pool or from animals, human volunteers, or cell cultures infected with the pool or materials derived from it.

2. CR326

CR26 was derived from a pool of sera collected from a 9-year-old child during a prospective study of the epidemiology and mode of transmission of viral hepatitis in semi-urban and rural areas of Alajuela, Costa Rica.[3] As soon as a case of hepatitis occurred in a household, serum samples were collected from all other family members and repeat specimens obtained at weekly intervals for 2 months. After standard biochemical tests, residual samples of serum were stored at −20°C until further use.

Aliquots of sera from each of seven children who developed hepatitis A (collected from 42 days prior to the onset of illness and up until 11 days after onset) were pooled and inoculated into seven groups of marmosets, mainly of the *Saguinus mystax* species. Five groups of marmosets which received pooled sera from five different patients developed hepatitis. Serial marmoset passages were carried out with two pools of sera (206-033-03 and 068-330-08) with positive results in all animals. The serum pool from patient 206-033-03 was designated CR326. This term has subsequently been used to designate all isolates of HAV in marmosets or cell culture which originate from this pool.

3. HM175

The HM175 strain of HAV was derived from a fecal sample collected in November 1976 from a 35-year-old Australian male with serologically confirmed hepatitis A.[4] This man was involved in a family outbreak of hepatitis in which he, his wife, and seven children all developed serological or virological evidence of infection with HAV. The fecal sample was collected 7 days after the onset of symptoms on the day of onset of jaundice. The appelation HM175 was derived from a laboratory code, H signifying hepatitis, M the electron microscoper involved, and 175 the laboratory accession number. The term HM175 has subsequently been used to designate strains of HAV which have been recovered from infected nonhuman primates and cell cultures inoculated with the original fecal extract and their progeny.

4. SD11

The SD11 strain of HAV was contained in a fecal sample collected from a naval recruit with serologically confirmed anicteric hepatitis during an epidemic of hepatitis which occurred at the Naval Training Center, San Diego in October 1974.[5] The outbreak, which involved 136 recruits, was traced to an infected food handler who, during the incubation period of his illness, had prepared salads and desserts which were eaten by the recruits. HAV particles were demonstrated in this specimen by IEM and extracts have subsequently been used to transmit infection to nonhuman primates and to infect cell cultures. The letters SD stand for San Diego, while 11 indicates that the specimen came from patient number 11.

Table 1
ANIMAL MODELS MOST COMMONLY USED FOR STUDYING HEPATITIS A INFECTION

Chimpanzee	*(Pan troglodytes)*
Marmoset	*(Saguinus mystax, S. labiatus)*
Owl monkey	*(Aotus trivirgatus)*

5. PA33

The PA33 strain of HAV was derived from a filtered suspension of feces obtained from an owl monkey *(Aotus trivirgatus)* in late 1980, during an outbreak of disease in a monkey colony housed at the Gorgas Memorial Laboratory, Panama City.[6] This material has been demonstrated to contain HAV particles by EM and to be infectious for owl monkeys and for cell culture. The designation PA33 stands for Panamanian Aotus monkey 33. While it is assumed that these animals were infected with human HAV, it is possible that PA33 is a monkey virus closely related to HAV.

C. Shedding of Virus in the Feces

In the three species of nonhuman primates in which HAV infections have been extensively studied (see Table 1), marmosets,[7] chimpanzees,[8,9] and owl monkeys,[10] the period of virus shedding is relatively brief, being detectable only during the late incubation period of the disease and for a few days after the onset of biochemical hepatitis. The appearance of virus particles or specific viral antigen (HAAg) in the feces usually coincides with the first detection of these markers in the liver, presumably because the liver is the major site of virus replication and the particles present in the feces enter in the bile rather than as a result of replication in the lining of the gut.[11,12]

1. Marmosets

When wild-type virus is administered to marmosets intravenously[11] or by nasogastric tube,[13] HAAg or HAV particles usually become detectable in the feces several days to a week prior to the onset of biochemical hepatitis (see Figure 1). The titer of virus increases rapidly, usually reaching a peak about the time that the animal develops histological evidence of liver damage, and declines rapidly over the next few days. No evidence of chronic shedding has been obtained. Virus usually becomes detectable in the feces several days prior to the development of circulating anti-HAV, and may continue to be detected for several more days. When examined by EM, virus particles detected in fecal specimens collected during the incubation period of the disease appear to be free of antibody and are readily aggregated by the addition of specific anti-HAV (see Figure 2). Specimens collected later in the illness occasionally show evidence of the presence of antibody.

While there is no evidence that the dose of virus administered to marmosets affects the amount of virus which is shed in the feces, there is some data to suggest that certain strains cause more prolonged shedding than others. On the other hand, the incubation period does appear to be dose related. In a study with the HM175 strain of HAV,[14] serial tenfold dilutions of a 10% extract of human feces were inoculated into pairs of *Saguinus mystax* marmosets. Animals receiving the undiluted inoculum developed biochemical hepatitis within 5 days of inoculation, whereas those receiving lower doses developed disease after progressively longer incubation periods. The relationship between starting dose and duration of the incubation period appears to be linear, with the onset of disease being delayed by approximately 1 week for each tenfold dilution of inoculum — 5 weeks at 10^{-6}, 6 weeks at 10^{-7}, and 7 weeks at 10^{-8} (see Figure 3).

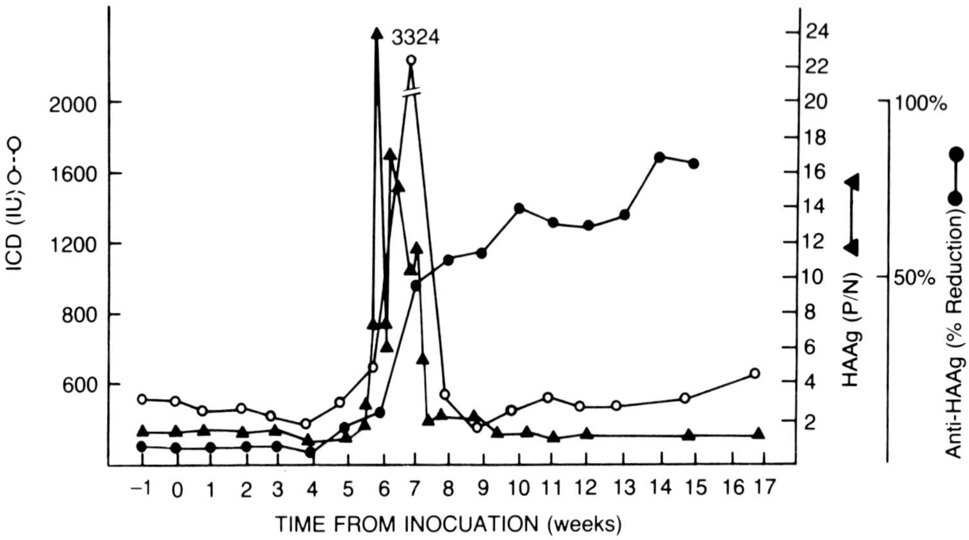

FIGURE 1. Pattern of shedding of HAAg in the feces and development of anti-HAV in a marmoset *(S. mystax)* infected by administration of 1 mℓ of a 10^{-9} of the SD11 inoculum i.v. (From unpublished experiments from the authors' laboratory.)

In some animals which have received a large inoculum of virus, a biphasic illness has been observed (see Figure 4). In a series of experiments by the authors and their colleagues, 1 mℓ of a 10% extract of HM175 was administered intravenously to several marmosets *(S. mystax* or *S. labiatus).* These animals began to shed virus in their feces within 5 to 7 days of inoculation. Shedding of virus was associated with a modest elevation of liver enzymes, the detection of HAAg in the cytoplasm of hepatocytes, and histological evidence of hepatitis suggesting intrahepatic replication of HAV. Shedding of virus continued as evidence of liver damage declined. However, about 3 weeks after inoculation, just at the time that circulating antibodies became detectable, these animals developed a second more pronounced attack of hepatitis. This phenomenon has been reproduced in the authors' laboratory and elsewhere[34] and appears to be dose dependent. The implications of these observations will be discussed further in the section on pathogenesis.

2. Chimpanzees

Shedding of HAV in the feces of intravenously infected chimpanzees was first reported by Maynard et al.[7] and Dienstag and his colleagues.[9] In the first study, three animals were inoculated intravenously with a filtrate of a pool of feces collected from seven patients who had been involved in a common-source epidemic of hepatitis A. All of the animals developed biochemical hepatitis within 17 days of inoculation and examination of pooled fecal specimens collected just prior to and during the acute phase of the illness revealed the presence of HAV particles in two of three animals. A fecal filtrate from one of these animals transmitted hepatitis to another chimpanzee and virus was also recovered from the feces of this animal. These findings were extended by Dienstag et al.[9] who infected chimpanzees by inoculating them intravenously with 1 mℓ of a 0.2% (w/v) filtrate of a human stool in which HAV had been detected by IEM. These animals developed biochemical and histological evidence of hepatitis 19 and 20 days after inoculation. HAV particles were detected, not only in the feces of these animals, but in aliquots of liver and bile collected during the acute phase of the illness.

The shedding of HAV in experimentally infected chimpanzees has been studied in several

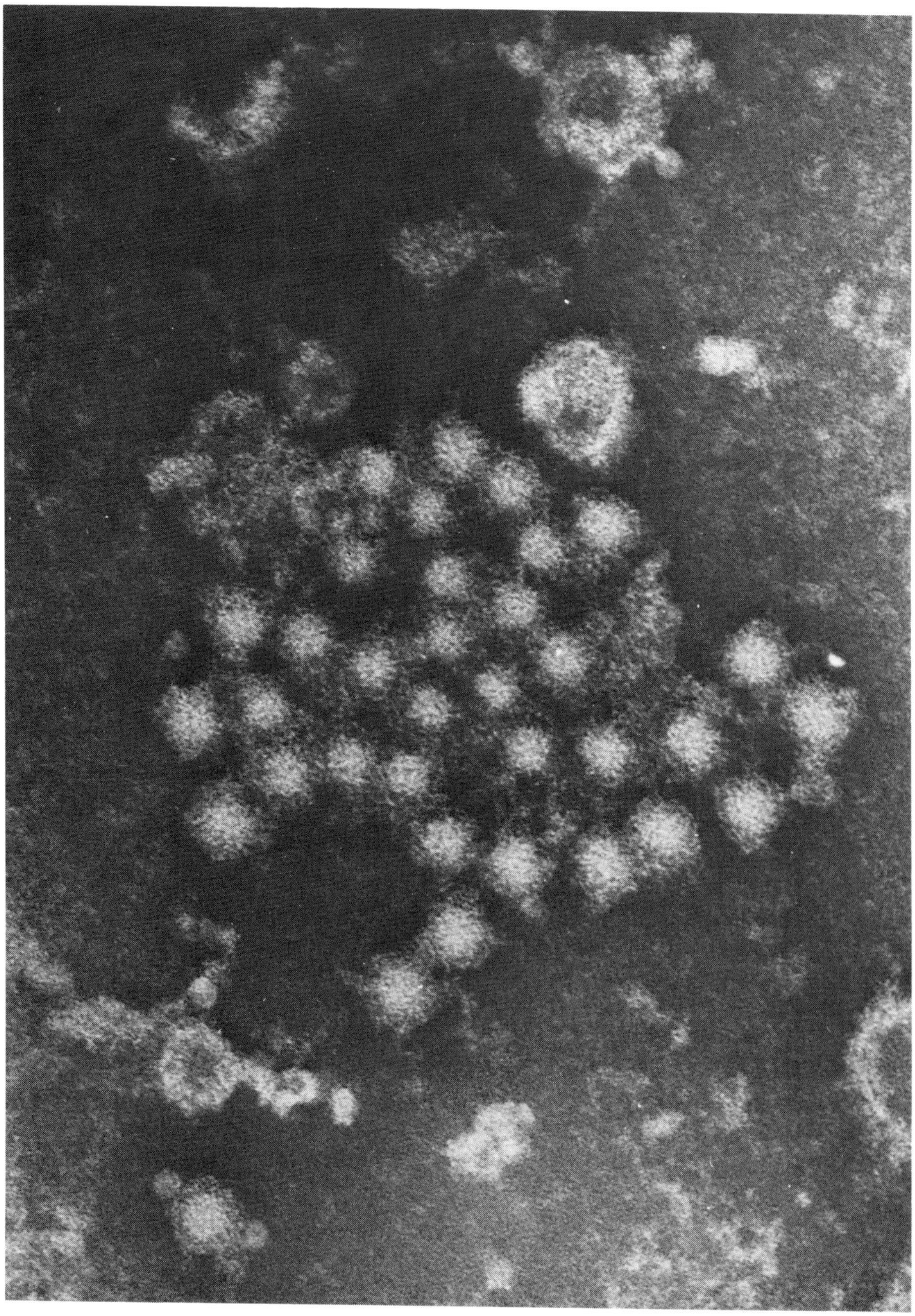

FIGURE 2. HAV particles detected in marmoset feces during the acute phase of infection. (From the authors' collection.)

laboratories using a variety of strains. The results which have been obtained are remarkably similar even in animals which were acutely or chronically infected with other hepatitis viruses.[15] In general, shedding of HAV in chimpanzees is neither as prolonged nor as intense as in marmosets so that despite their size and the volume of feces they produce, chimpanzees have not been a major source of viral antigen for other studies. Fecal shedding has been

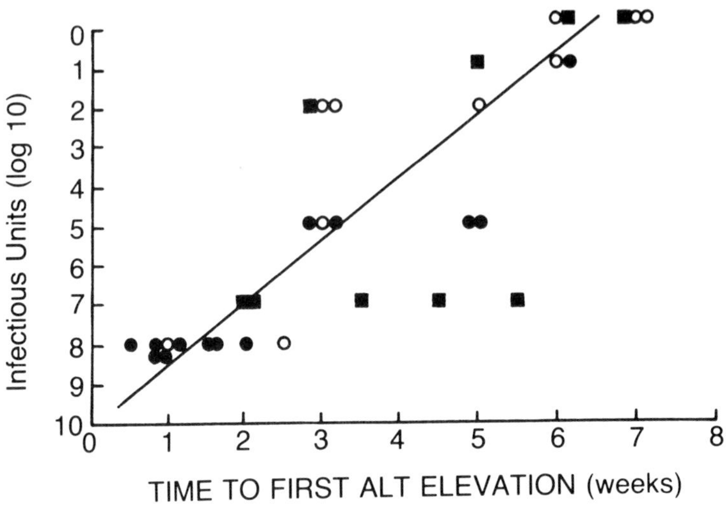

FIGURE 3. Inverse relationship between infecting dose of HAV and incubation period measured to the first elevation of alanine transferase levels. Circles represent data from marmosets; squares data from chimpanzees; open circles represent animals which received the HM175 strain of HAV; solid circles or squares represent animals which received MS-1. (From Purcell, R. H., Feinstone, S. M., Ticehurst, J. R., Daemer, J. R., and Baroudy, B. M., *Viral Hepatitis in Liver Disease*, Vyas, G. N., Dienstag, J. L., and Hoofnagle, J. H., Eds., Grune & Stratton, Orlando, Fla., 11, 1984. With permission.)

observed following both intragastric and intravenous inoculation of HAV and following infections with fecal extracts or aliquots or serum collected during the acute phase of the illness.[16]

3. Owl Monkeys

Workers at the Walter Reed Army Institute of Research in the U.S. have performed an extensive series of studies of hepatitis A infection in new world owl monkeys *(Aotus trivirgatus)*.[6] Initially 12 seronegative, colony-raised owl monkeys were divided into two groups and inoculated intravenously with 1 mℓ of a 0.2% filtered suspension of two different strains of HAV — PA33 (originally isolated from an infected owl monkey) and HM175. All six animals which received the PA33 inoculum developed hepatitis. Abnormal LFTs were usually detected about 14 days after inoculation and persisted for another week. In five of the six animals, HAV became detectable in the feces 3 to 8 days prior to the onset of biochemical hepatitis, persisting for 5 to 11 days (mean 7.5 days). Five of the six animals inoculated with HM175 developed hepatitis; however, the incubation period was longer than for PA33, with abnormal LFTs first becoming detectable 25 to 39 days after inoculation. Fecal shedding was more prolonged with this strain of HAV, with HAAg usually appearing early in the second week after inoculation and persisting for 7 to 29 days (mean 17.6 days) (see Table 2).

D. Detection of Virus in the Liver

While it had always been assumed that HAV replicated in the liver, early attempts to transmit infection to nonhuman primates concentrated exclusively upon acute-phase sera and fecal samples as sources of infectious virus.[3,17,18] The first objective evidence that HAV replicated in the liver was provided by Provost and his colleagues[19] who injected a group of *Saguinus mystax* marmosets intravenously with the CR326 strain of HAV and then

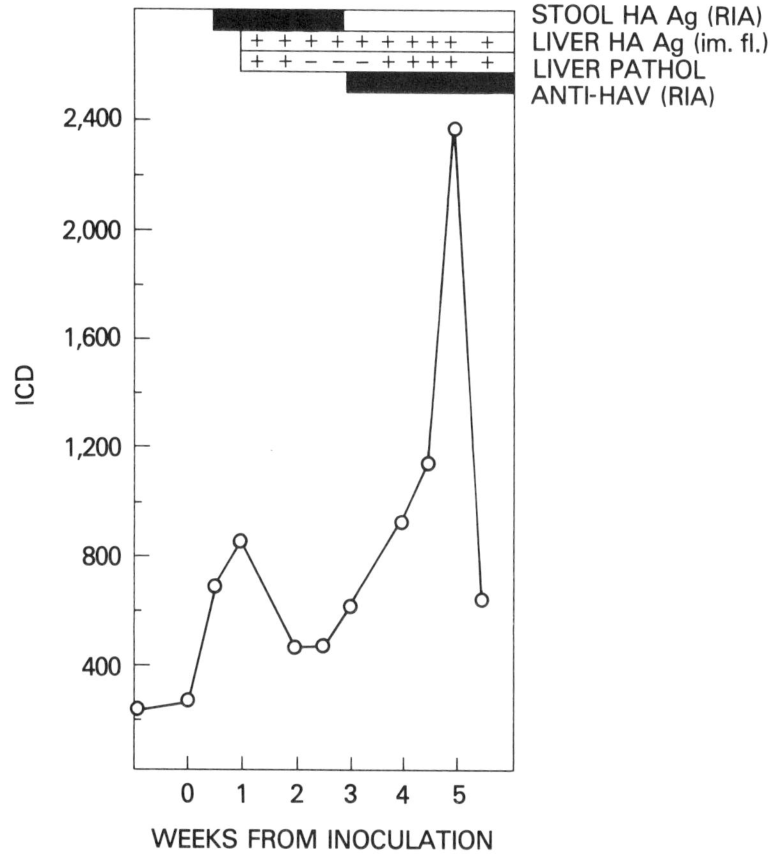

FIGURE 4. Biphasic illness observed in marmoset *(S. mystax)* inoculated with a large dose (1 mℓ) of HAV strain HM175 i.v. (o——o) ICD levels. Note the appearance of HAAg in the liver and shedding of virus into the feces at the time of the first enzyme peak. Cessation of fecal shedding was followed by the development of anti-HAV and a second attack of hepatitis. Interestingly, HAAg could be detected in the liver throughout the period of observation. (From studies in the authors' laboratory.)

Table 2
INFECTION OF NEW WORLD OWL MONKEYS WITH TWO STRAINS OF HAV

	Strain PA33				Strain HM175			
Animal	Preinfection	Max.	Days elevated	HAAg in feces	Preinfection	Max.	Days elevated	HAAg in feces
				ALT U/ℓ				
1	27.3	210	14—21	6—14	20.4	480	25—35	9—21
2	40.9	316	14—21	10—13	46.3	124	32—42	27—34
3	68.5	372	14—21	8—19	37.0	367	28—43	8—32
4	35.4	233	21—28	17—22	47.0	303	39—43	8—37
5	73.3	272	14—28	7—15	35.8	252	28—49	8—24
6	49.3	189	9—21	6—16				
Mean	49.1	265	14—23	9—16	39.1	305	30—42	12—30

From Le Duc, J. W., Lemon, S. M., Keenan, C. N., Graham, R. R., Marchwicki, R. H., and Binn, L. N., *Infect. Immun.*, 40, 766, 1983.

sacrificed the animals at 3 to 4 day intervals. After death the animals were perfused and a variety of organs removed, minced, washed, and frozen. Extracts of these organs were then evaluated for the presence of infectious virus by inoculation into further groups of susceptible animals. Infectious virus was detected in the liver of infected marmosets during the first week after infection, 2 weeks or more prior to the onset of hepatitis. Virus was not detected in specimens of spleen or bone marrow, suggesting that virus was replicating in the liver and not simply being trapped in the reticuloendothelial system. In fact, such large amounts of antigen could be harvested from the livers of infected animals that Provost and his colleagues were able to use this material as a source of antigen and to develop complement fixation[20] and immune-adherence hemagglutination assays[21] for detection of anti-HAV.

HAAg has been identified in the cytoplasm of infected hepatocytes by immunofluorescence and characteristic HAV particles detected in liver extracts by EM (see Figures 5 and 6). These techniques have been useful for studying aspects of the replication and morphogenesis of the virus in infected cells. After inoculation of *Saguinus mystax* or *S. labiatus* marmosets with HAV, the incubation period to the onset of biochemical hepatitis is usually about 4 weeks.[22,23] HAAg usually becomes detectable in the cytoplasm of infected hepatocytes about 1 week after inoculation, and increases in concentration until the onset of liver damage, by which time HAAg or virus particles can be detected in virtually every cell. Although the amount of detectable viral antigen declines during the convalescent phase of the illness, immunofluorescent (IF) tests may remain weakly positive for several months after the animals' liver function tests have returned to normal and all histological evidence of liver damage has disappeared (see Figure 7).

1. Immunofluorescence (IF)

In studies of liver tissue obtained from marmosets by serial biopsy the first objective evidence of virus replication is usually obtained by IF manifested as a fine granular pattern in the cytoplasm of hepatocytes and Kuppfer cells.[12,13] Initially only some areas of the biopsy are positive but the proportion of infected cells increases with time. Fluorescence appears first in hepatocytes adjacent to the portal tracts and has not been detected in sinus lining cells or inflammatory cells. As the disease progresses the presence of HAAg correlates well with pathological changes in the liver parenchyma.

There are relatively few reports on the detection of HAAg by IF in chimpanzees.[24] Extensive experience in one of our laboratories indicates that following either nasogastric or i.v. inoculation, HAAg becomes detectable in the liver about 2 weeks later, coinciding with the time that virus is first detected in the feces. In contrast to marmosets, prolonged expression of HAAg is uncommon and liver biopsies taken during the early convalescent period are usually negative by IF.

2. Electron Microscopy (EM)

HAAg has also been localized in the liver of infected nonhuman primates by the use of immunoperoxidase or ferritin-labeled antibodies. Shimizu et al.[25] used an immunoperoxidase technique to localize HAAg in liver tissue obtained from *Saguinus mystax* marmosets inoculated with stool or liver suspensions of two different strains of HAV and sacrificed 2 weeks after the onset of abnormal liver functions tests. HAAg containing particles measuring 24 to 27 nm were detected in the cytoplasm of infected hepatocytes, almost always located within cytoplasmic vesicles the membranes of which also appeared to contain HAAg. Antigen was also detected in sinusoidal cells, probably of the Kuppfer type. It was not clear whether the virus was synthesized in these vesicles or whether the structures represent the end result of a phagocytic process (see Figures 8 and 9).

In another study, Huang et al.[26] performed IF and indirect immunoferritin staining on fresh and formalin-fixed liver tissue from *S. mystax* marmosets which had been injected i.v.

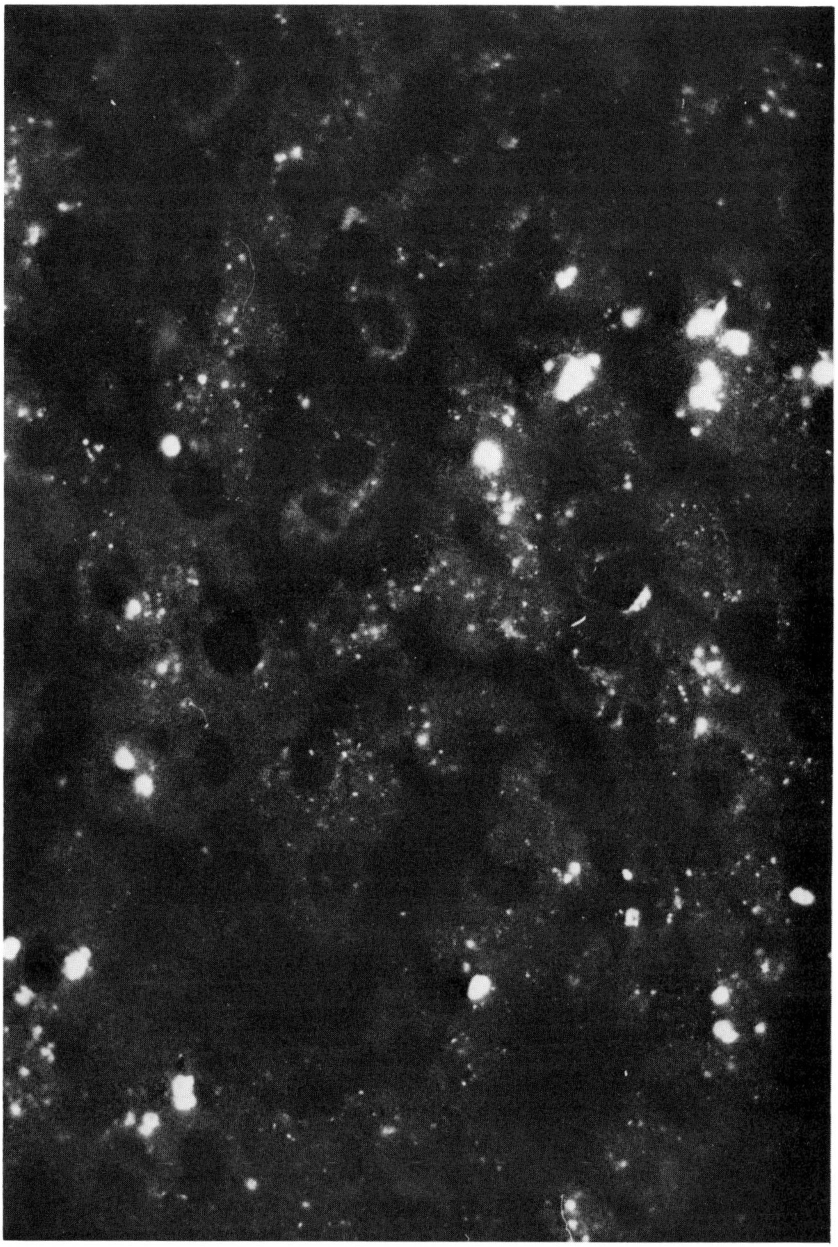

FIGURE 5. IF of liver biopsy tissue obtained from a marmoset during the acute phase of illness. This animal was infected with HAV strain HM175. Note the fine granular intracytoplasmic fluorescence. (From studies in the authors' laboratory.)

with the MS-1 strain of HAV and sacrificed after they developed biochemical evidence of hepatitis. HAAg was detected in the cytoplasm of infected hepatocytes and in sinusoidal mononuclear cells. Structural changes were noted in the endoplasmic reticulum of infected cells which appeared to undergo distention and vesicle formation. Fragments of endoplasmic reticulum were seen within the vesicles often in association with HAV particles, leading the authors to suggest that virus was produced within these vesicles. These workers were unable to confirm the presence of HAAg on the membranes which surround vesicles containing

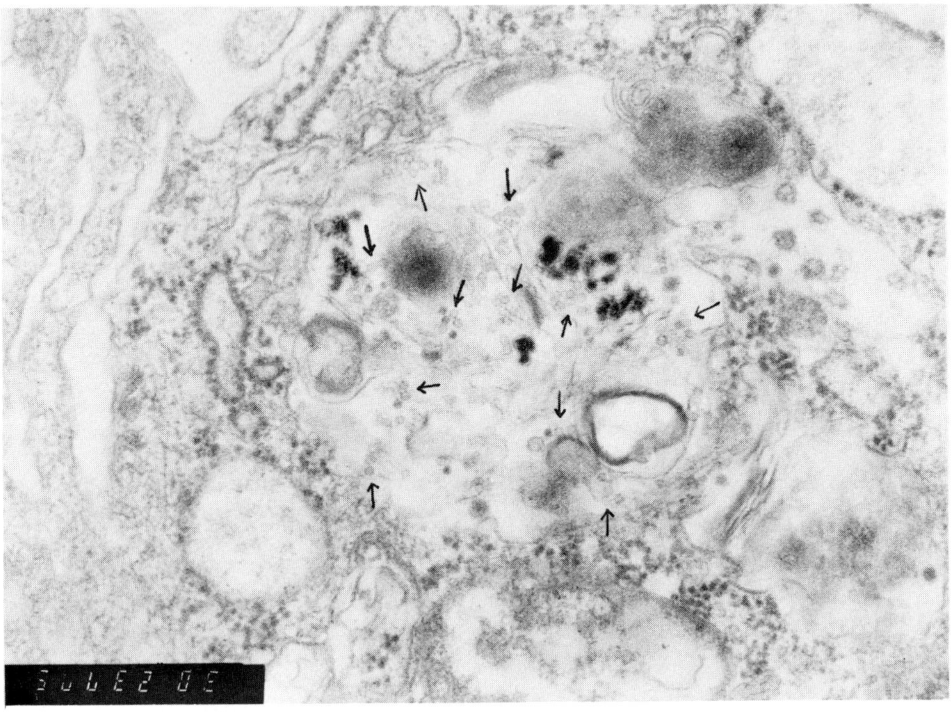

FIGURE 6. Thin-section EM of tissue infected with HAV strain HM175. HAV particles are seen in intracytoplasmic vesicles (see arrows). (Picture kindly provided by Dr. Yohko Shimizu.)

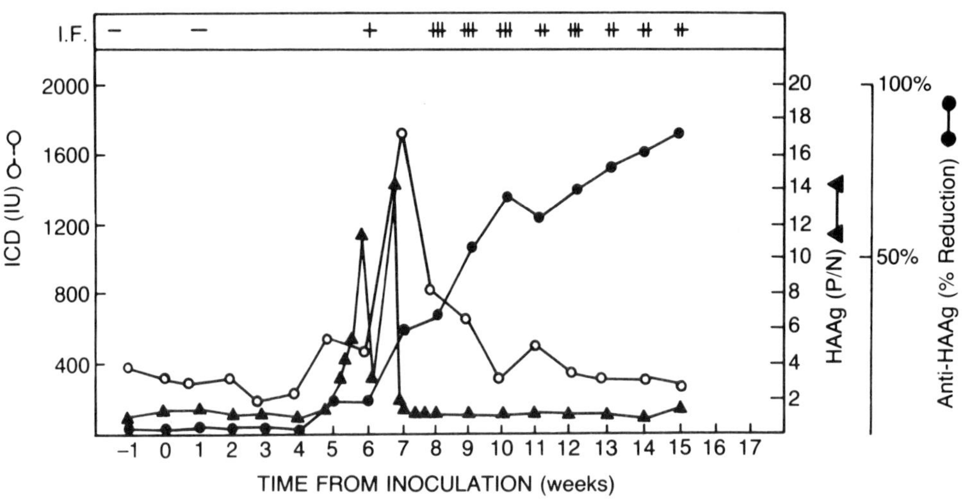

FIGURE 7. Prolonged detection of HAAg in liver biopsy specimens obtained from a marmoset *(S. mystax)* infected with HAV strain SD11. (From unpublished studies in the authors's laboratory.)

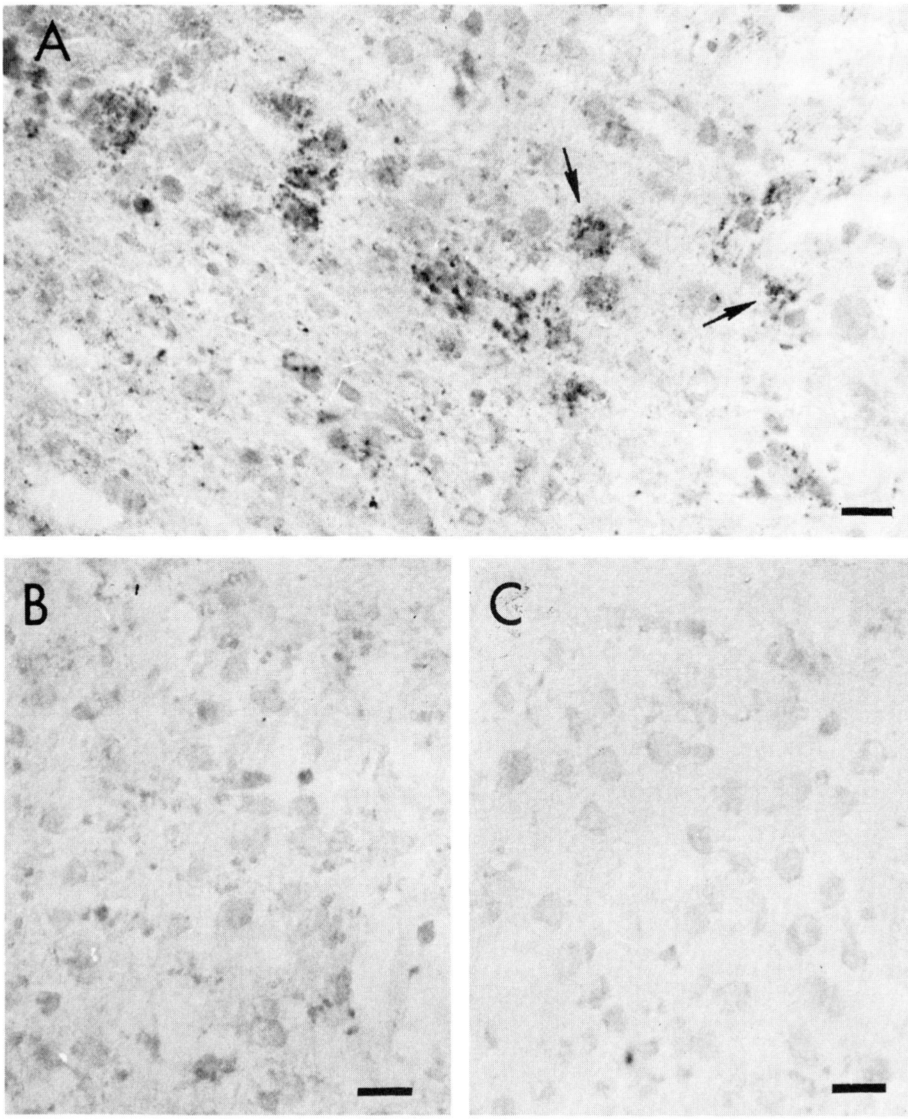

FIGURE 8. Localization of HAAg in the hepatocytes of infected marmosets by immunoperoxidase staining. Bar = 10 μm. (A) Arrows point to granular reaction products scattered through the cytoplasm; (B) control reaction in which peroxidase staining is blocked by unlabeled anti-HAV; (C) control reaction, incubation with 3.3′ diaminobenzidine — H_2O_2 alone. (From Shimizu, Y. K., Mathieson, L. R., Lorenz, D., Drucker, J., Feinstone, S. M., Wagner, J. A., and Purcell, R. H., *J. Immunol.*, 121, 1671, 1978. With permission.)

HAV particles. Virus particles were regularly detected in the bile canaliculi, confirming that the liver is the major source of virus detected in the feces.

HAAg and HAV particles have also been demonstrated in cytoplasmic vesicles in the livers of experimentally infected chimpanzees[27,28] and owl monkeys.[35]

3. *Structural Changes*

The histological changes in the livers of infected animals have been well documented,[9,28] with the most complete account being that of Dienstag et al.[9] in which serial liver biopsies from chimpanzees infected (either orally or i.v.) with the MS-1 strain of HAV, were examined

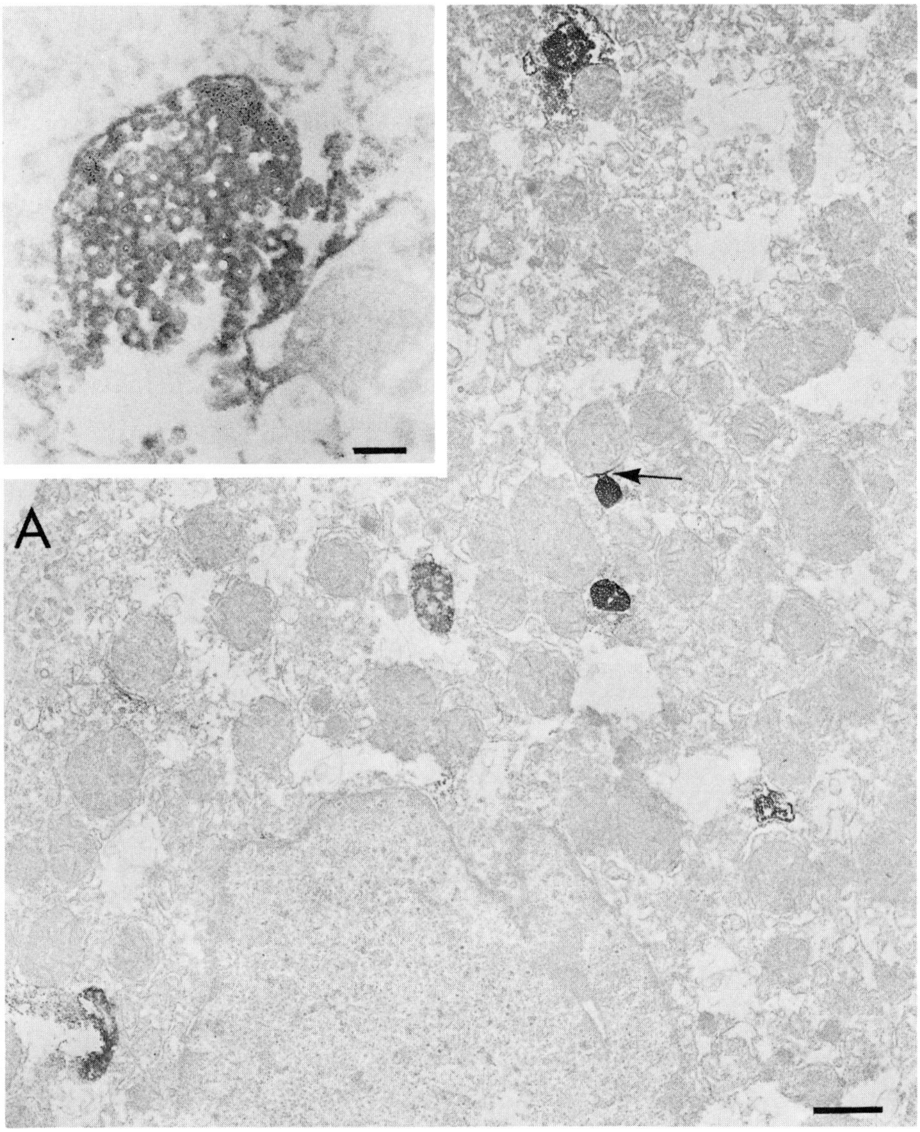

FIGURE 9. Localization of HAAg in the hepatocytes of an infected marmoset by immunoperoxidase staining. (A) HAAg is localized on particles packed within vesicles in the cytoplasm. The surrounding membrane of the vesicle is positively stained and appears to be connected to a mitachondria (see arrow). Bar = 1 μm. Inset shows HAAg particles at higher magnification. Bar = 100 nm; (B) control reaction after blocking with unlabeled serum containing anti-HAV. (From Shimizu, Y. K., Mathieson, L. R., Lorenz, D., Drucker, J., Feinstone, S. M., Wagner, J. A., and Purcell, R. H., *J. Immunol.*, 121, 1671, 1978. With permission.)

under code. The changes observed were similar but milder than those seen in humans with naturally acquired infections. The earliest signs of liver damage which coincided with the first elevation of liver enzymes were focal necrosis of parenchymal cells in the periphery of the lobules and an increase in mononuclear inflammatory cells. As focal necrosis spread throughout the parenchyma, infiltration and enlargement of the portal tracts with mononuclear cells was noted together with erosion of the parenchymal limiting plate and proliferation of bile ducts. Cellular proliferation was most marked around the ductules. These changes resolved rapidly (see Figure 10).

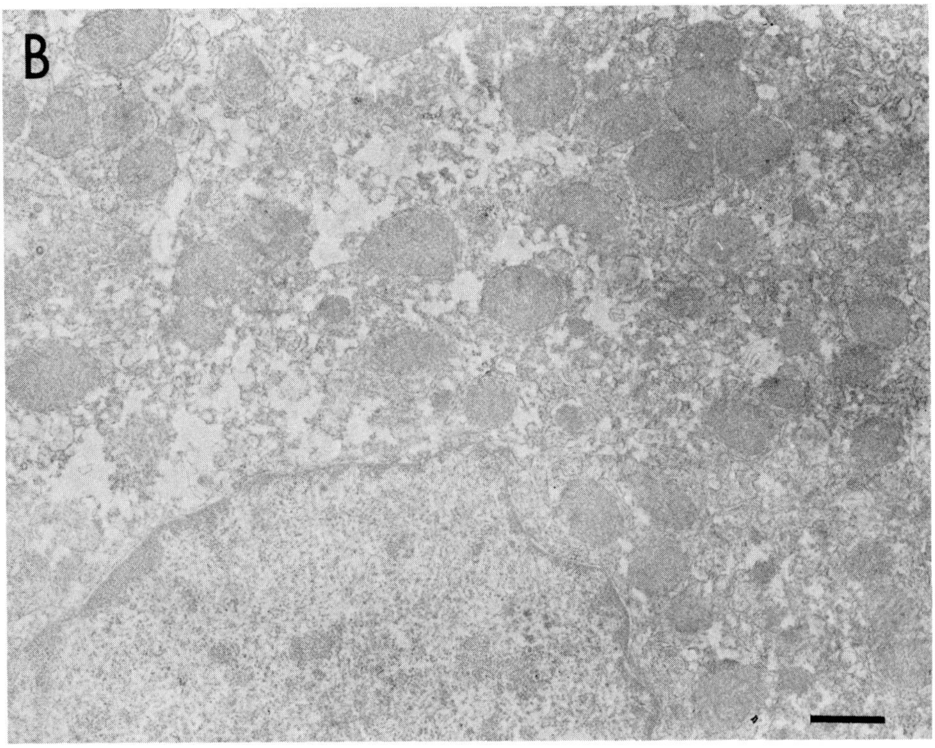

FIGURE 9B.

When examined by EM, hepatocytes from infected animals showed characteristic ultrastructural changes which enable experienced observers to recognize specimens from animals infected with HAV, under code. The nuclei of infected cells showed large clumps of dense heterochromatin-like granules, measuring 35 to 40 nm in diameter, which were not seen in animals infected with HBV or HNANB. Within the cytoplasm of infected cells the smooth endoplasmic reticulum had a clumped vesicular appearance and there was a marked reduction of rough endoplasmic reticulum. In contrast to animals infected with other forms of hepatitis, the mitochondria of chimpanzees infected with hepatitis A had thin cristae and a coarser than usual matrix; no abnormalities in lysosome structure were noted.

E. Detection of Virus in Other Sites

While HAV (or HAAg) has been detected in the liver, bile, and feces of experimentally infected nonhuman primates, it has rarely been detected in other sites. Mathieson et al.[11,13] and Krawczynski et al.[12] sacrificed marmosets infected by both the i.v. and oral routes at various times after inoculation, but consistently failed to detect virus particles or specific viral antigens in the gut, mesenteric lymph nodes, tonsils, larynx, spleen, or kidney of these animals. It would appear that the liver is either the major or the only site of virus replication.

While it is easy to understand how HAV reaches the liver after parenteral inoculation, it is difficult to explain how virus gets to the liver after being swallowed unless there is an intitial cycle of replication in the gut or the virus is engulfed by scavenger cells of the reticuloendothelial system and sequestrated in the liver. The first (indirect) evidence that virus may replicate in the gut was the detection of coproantibodies in the feces of humans with naturally acquired infections.[29] Recently Karayiannis et al.[30] have obtained some confirmatory evidence by demonstrating HAAg by IF in the cytoplasm of duodenal lining cells

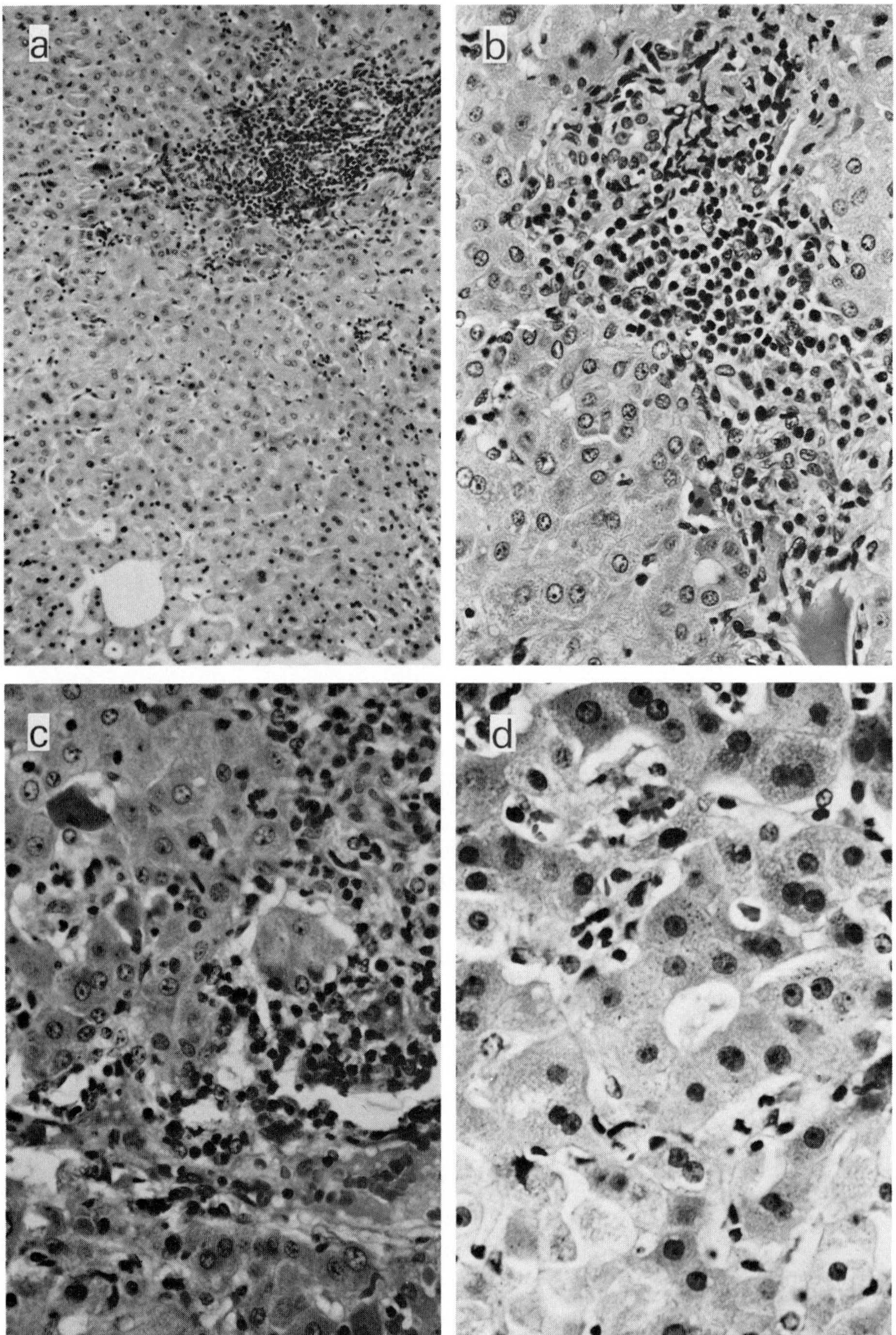

FIGURE 10. Typical changes revealed by light microscopy in chimpanzees experimentally infected with HAV (see text). (From material kindly provided by Dr. Hans Popper.)

in two of three immunosuppressed marmosets. Whether this represents the primary site of replication or whether these cells were secondarily infected by virus shed in the bile, remains to be determined.

F. Pathogenesis

Several groups have looked for the presence of HAAg, different subclasses of anti-HAV, and various complement components in tissues obtained from experimentally infected marmosets. Mathieson et al.[11] detected HAAg by IF in the germinal centers of the spleen and mesenteric lymph nodes and in the glomeruli of some animals. In the spleen and lymph nodes, antigen was located predominantly in the cytoplasm of lymphoid cells located in the center of the germinal center although the quantity of antigen was much lower than that found in the hepatocytes of these animals. HAAg was detected in the glomeruli of a single animal and appeared to be located adjacent to the basement membrane, suggesting the deposition of immune complexes.

Krawczynski et al.[12] were unable to detect deposits of immunoglobulin (IgG, IgM, or IgA) or complement components (C1q or C3) in the livers of orally infected marmosets, although granular masses of IgG were frequently detected in the central veins, sinuses, and blood vessels of the portal tracts. IgG and IgM antibodies were detected in the germinal centers of the spleen and mesenteric, axillary, and cervical lymph nodes of infected animals, but not in greater amounts than in uninfected controls. Interestingly, in view of the debate about whether HAV replicates in the gut, IgA-containing plasma cells were occasionally detected in the lamina propria of the small intestine of infected animals. Although renal glomeruli contained deposits of IgM and C3, located mainly in the glomerular capillary wall, similar deposits were seen in uninfected animals. HAAg was not detected in the glomeruli of infected animals suggesting that its presence in the kidney may be an artifact resulting from i.v. inoculation.

Although liver damage usually occurs at the time that circulating antibodies become detectable, these studies failed to provide evidence that the damage is antibody dependent. Immunoglobulin and complement deposits were not found at the sites of liver cell damage, and resolution of disease occurred at a time when antibody levels were rising and HAAg could still be detected in the liver. Similarly the presence of large quantities of virus in infected hepatocytes prior to the onset of hepatitis and during the convalescent phase of the disease argues against the idea that liver cell necrosis is a direct result of changes in the cell produced by viral replication. The only evidence for a direct cytopathic effect of HAV comes from the studies of Provost et al.[20] and Gust et al.[14] In the latter study a large dose of HAV strain HM175 was inoculated into marmosets resulting in a biphasic illness. The simplest explanation of this phenomenon is that the initial peak of enzymes was a direct result of replication of the virus, while the second peak was a consequence of some immunological response. The first peak was not seen when the animals were inoculated by the oral route, or when diluted suspensions of the original inoculum were inoculated i.v. By a process of elimination it seems most probably that liver cell destruction is either a product of the effects of the virus plus a cellular immune response or entirely due to the cellular immune response, perhaps mediated by cytotoxic T lymphocytes.

Recently, two groups have provided some data to support this hypothesis. Fleming et al.[33] have demonstrated that epidermal cell lines established from humans infected with HAV showed a pronounced cytotoxic effect when washed lymphocytes from these patients (collected during the convalescent phase of the disease) were added to the culture.[30] Furthermore, Kurane and his colleagues[31] have demonstrated that peripheral blood mononuclear cells (PBLs) from nonimmune donors will lyse BS-C-1 cells infected with HAV. The lymphocytes involved were characterized with monoclonal antibodies and found to belong to the Leu-11$^+$ and M1$^+$ subsets, similar to those of natural killer (NK) cells. These workers also found that high titers of IFN-α were produced by PBL in contact with HAV infected cells and have suggested that production of interferon may play a role in limiting spread of infection as well as augmenting lysis by NK cells.

While some aspects of the pathogenesis of HAV infection are still unclear, it seems probable that following oral administration of the virus, replication occurs in the alimentary gastrointestinal tract, after which the virus is transported to the liver which is the major site of replication. Virus is shed from infected liver cells into the hepatic sinusoids and the bile canaliculi and passes into the intestine after which it is excreted in the feces. Liver cell damage is probably mediated by cytotoxic lymphocytes. The presence of circulating antibodies is probably important in limiting the spread of virus to other organs but has not been shown to be a critical factor in liver cell damage.

REFERENCES

1. **Deinhardt, F. and Deinhardt, J. B.,** Animals models, in *Hepatitis A,* Gerety, R. J., Ed., Academic Press, Orlando, Fla., 1984, 197.
2. **Krugman, S., Giles, J. P., and Hammond, J.,** Infectious hepatitis: evidence for two distinct clinical, epidemiological and immunological types of infection, *JAMA,* 200, 365, 1967.
3. **Mascoli, C. C., Ittensohn, O. L., Villarejos, V. M., Arguedas, J. A., Provost, P. J., and Hilleman, M. R.,** Recovery of hepatitis agents in the marmoset from human cases occurring in Costa Rica, *Proc. Soc. Exp. Biol. Med.,* 142, 276, 1973.
4. **Gust, I. D., Lehmann, N. I., Crowe, S., McCrorie, M., Locarnini, S. A., and Lucas, C. R.,** The origin of HAV strain HM175, *J. Infect. Dis.,* 151, 365, 1985.
5. **Dienstag, J. L., Routenberg, J. A., Purcell, R. H., Hooper, R. R., and Harrison, W. O.,** Food handler-associated outbreak of hepatitis type A, *Ann. Intern. Med.,* 83, 647, 1975.
6. **Lemon, S. M., Le Duc, J. W., Binn, L. N., Escajadillo, A., and Ishak, K. G.,** Transmission of hepatitis A virus among recently captured Panamanian owl monkeys, *J. Med. Virol.,* 10, 25, 1982.
7. **Maynard, J. E., Lorenz, D., Bradley, D. W., Feinstone, S. M., Krushak, D. H., Barker, L. F., and Purcell, R. H.,** Review of infectivity studies in nonhuman primates with virus-like particles associated with MS-1 hepatitis, *Am. J. Med. Sci.,* 270, 81, 1975.
8. **Maynard, J. E., Bradley, D. W., Gravelle, C. R., Ebert, J. W., and Krushak, D. H.,** Preliminary studies on hepatitis A in chimpanzees, *J. Infect. Dis.,* 131, 194, 1975.
9. **Dienstag, J. L., Feinstone, S. M., Purcell, R. H., Hoofnagle, J. H., Barker, L. F., London, W. T., Popper, H., Peterson, J. M., and Kapikian, A. Z.,** Experimental infection of chimpanzees with hepatitis A virus, *J. Infect. Dis.,* 132, 532, 1975.
10. **Le Duc, J. W., Lemon, S. M., Keenan, C. M., Graham, R. R., Marchwicki, R. H., and Binn, L. N.,** Experimental infection of the new world owl monkey (Aotus trivirgatus) with hepatitis A virus, *Infect. Immun.,* 40, 766, 1983.
11. **Mathieson, L. R., Drucker, J., Lorenz, D., Wagner, J., Gerety, R. J., and Purcell, R. H.,** Localization of hepatitis A antigen in marmoset organs during acute infection with hepatitis A virus, *J. Infect. Dis.,* 138, 369, 1978.
12. **Krawczynski, K. S., Bradley, D. W., Murphy, B. L., Ebert, J. W., Anderson, T. E., Doto, I. L., Nowoslawski, A., Duermeyer, W., and Maynard, J. E.,** Pathogenic aspects of hepatitis A virus infection in enterally inoculated marmosets, *Am. J. Clin. Pathol.,* 76, 698, 1981.
13. **Mathieson, L. R., Moller, A. M., Purcell, R. H., London, and W. T., and Feinstone, S. M.,** Hepatitis A virus in the liver and intestine of marmosets after oral inoculation, *Infect. Immun.,* 28, 45, 1980.
14. **Gust, I. D., Feinstone, S. M., and Purcell, R. H.,** unpublished observations.
15. **Drucker, J., Tabor, E., Gerety, R. J., Jackson, D., and Barker, L. F.,** Simultaneous acute infection with hepatitis A and hepatitis B viruses in a chimpanzee, *J. Infect. Dis.,* 139, 338, 1979.
16. **Tabor, E., Purcell, R. H., and Gerety, R. J.,** Primate animal models and titered inocula for the study of human hepatitis A, hepatitis B, and non-A, non-B hepatitis, *J. Med. Primatol.,* 12, 305, 1983.
17. **Deinhardt, F., Holmes, A. W., Capps, R. B., and Popper, H. J.,** Studies on the transmission of human viral hepatitis to marmoset monkeys. I. Transmission of disease, serial passages and description of liver lesions, *J. Exp. Med.,* 125, 673, 1967.
18. **Lorenz, D., Barker, L., Stevens, D., Peterson, M., and Kirchstein, R.,** Hepatitis in the marmoset, Saguinus mystax, *Proc. Soc. Exp. Biol. Med.,* 135, 348, 1970.

19. **Provost, P. J., Miller, W. J., Wolanski, B. S., Ittensohn, O. L., Villarejos, V. M., McAleer, W. J., and Hilleman, M. R.,** Studies on human HAV strain CR326, in *Viral Hepatitis,* Vyas, G. N., Cogen, S. N., and Schmid, R., Eds., Franklin Institute Press, Philadelphia, 1978, 49.
20. **Provost, P. J., Ittensohn, O. L., Villarejos, V. M., and Hilleman, M. R.,** A specific complement fixation test for human hepatitis A employing CR326 virus antigen, *Proc. Soc. Exp. Biol. Med.,* 148, 962, 1975.
21. **Miller, W. J., Provost, P. J., McAleer, W. J., Ittensohn, O. L., Villarejos, V. M., and Hilleman, M. R.,** Specific immune adherence assay for human hepatitis A antibody: application to diagnostic and epidemiological investigations, *Proc. Soc. Exp. Biol. Med.,* 149, 254, 1975.
22. **Provost, P. J., Ittensohn, O. M., Villarejos, V. M., Arquedos, J. A., and Hilleman, M. R.,** Etiologic relationship of marmoset-propagated CR326 hepatitis A virus to hepatitis in man, *Proc. Soc. Exp. Biol. Med.,* 142, 1257, 1973.
23. **Provost, P. J., Villarejos, V. M., and Hilleman, M. R.,** Suitability of the Rufiventer marmoset as a host animal for human hepatitis A virus, *Proc. Soc. Exp. Biol. Med.,* 155, 283, 1977.
24. **Murphy, B. L., Maynard, J. E., Bradley, D. W., Ebert, J. W., Mathieson, L. R., and Purcell, R. H.,** Immunofluorescence of hepatitis A virus antigen in the chimpanzee, *Infect. Immun.,* 21, 663, 1978.
25. **Shimizu, Y., Mathieson, L. R., Lorenz, D., Drucker, J., Feinstone, S. M., Wagner, J. A., and Purcell, R. H.,** Localization of hepatitis A antigen in liver tissue by peroxidase-conjugated antibody method: light and electron microscopic studies, *J. Immunol.,* 121, 1671, 1978.
26. **Huang, S. N., Lorenz, D., and Gerety, R. J.,** Electron and immunoelectron microscopic study of liver tissues of marmosets infected with hepatitis A virus, *Lab. Invest.,* 41, 63, 1979.
27. **Schulman, A. N., Dienstag, J. L., Jackson, D. R., Hoofnagle, J. H., Gerety, R. J., Purcell, R. H., and Barker, L. F.,** Hepatitis A particles in liver, bile and stool of chimpanzees, *J. Infect. Dis.,* 134, 80, 1976.
28. **Schaffner, F., Dienstag, J. L., Purcell, R. H., and Popper, H.,** Chimpanzee livers after infection with human hepatitis viruses A and B, *Arch. Pathol. Lab. Med.,* 101, 113, 1977.
29. **Locarnini, S. A., Coulepis, A. G., Kaldor, J., and Gust, I. D.,** Coproantibodies in hepatitis A: detection by enzyme-linked immunoabsorbent assay and immune electron microscopy, *J. Clin. Microsc.,* 11, 710, 1980.
30. **Karayiannis, P., Jowett, T., Enticott, M. J., Moore, D., Pignatelli, M., Brenes, F., Scheuer, P. J., and Thomas, H. C.,** Hepatitis A virus (HAV) replication in tamarins and host immune response in relation to pathogenesis of liver cell damage in acute HAV infection, in press, 1987.
31. **Kurane, I., Binn, L. N., Bancroft, W. H., and Ennis, F. A.,** Human lymphocyte responses to hepatitis A virus-infected cells: interferon production and lysis of infected cells, *J. Immunol.,* 135, 2140, 1985.
32. **Purcell, R. H., Feinstone, S. M., Ticehurst, J. R., Daemer, R. J., and Baroudy, B. M.,** Hepatitis A Virus, in *Viral Hepatitis in Liver Disease,* Vyas, G. N., Dienstag, J. L., and Hoofnagle, J. H., Eds., Grune & Stratton, Orlando, Fla., 9, 1984.
33. **Fleming, B., et al.,** personal communication.
34. **Provost, P. J. and Maynard, J. E.,** personal communication.
35. **Asher, L. S.,** personal communication.

Chapter 6

BIOLOGY OF HAV IN CELL CULTURE

I. INTRODUCTION

Starting in the 1940s numerous attempts were made to isolate the causative agent of hepatitis A in cell culture. Sera and fecal specimens known to be infectious for human volunteers were inoculated into a variety of primary cell cultures and continuous cell lines and evidence of a virus-induced cytopathic effect (CPE) or interference with the growth of other CPE-producing viruses was sought. These efforts were unsuccessful because HAV is not normally a cytolytic virus nor does it stimulate the production of interferon in most cells. As a result, interest in virus isolation began to wane and by the late 1960s workers in most laboratories had turned their attention to other means of detecting the etiologic agent (see Chapter 2).

It was not until HAV had been identified by EM and reliable serological tests had been established that efforts to isolate the virus were again given priority. The most important factor in the eventual success of this approach was the availability of an IF assay for detection of HAAg within infected cells.[1] This assay was originally developed to assist studies of the pathogenesis of HAV infection by localizing HAAg in specimens collected from experimentally infected marmosets and chimpanzees. Its potential for detecting HAAg in cell culture was recognized and the technique was rapidly introduced as a means of screening cultures for evidence of HAV replication. These studies led rapidly to isolation of HAV in liver explant tissue and in a cloned line of fetal rhesus monkey kidney cells.[2]

II. ISOLATION OF HAV IN CELL CULTURE

A. After Adaptation to Marmosets

1. In Liver Explants

In 1979 Provost and Hilleman (see Figure 1) reported the first isolations of HAV using the CR326 strain of HAV which had been passaged five times in *Saguinus mystax* marmosets and a further 26 times in *S. labiatus* marmosets.[2] Their starting inoculum, a clarified, filtered 4% suspension of infected marmoset liver, was inoculated into explant cultures of liver tissue obtained from susceptible marmosets *(S. labiatus)* and a cloned line of fetal rhesus monkey kidney cells (FRhK-6). Both types of cultures were performed in small plastic flasks containing glass cover slips. Passages were made with the contents of the flask after subjecting them to two freeze-thaw cycles, 15 sec of sonication, and a clarifying spin.

HAV was isolated in the explant tissue and successfully passaged five times. Minute fluorescent granules were detected in the cytoplasm of some hepatocytes within 8 days of inoculation and these granules increased in size and number with time, so that by the first passage (day 23), fluorescence could be demonstrated in more than 75% of hepatocytes (see Table 1). The specificity of the IF reaction was confirmed by demonstrating that it could be blocked by preincubation of the cells with convalescent (but not preinfection) sera from patients with serologically confirmed hepatitis A.

2. In FRhK-6 Cells

Having isolated HAV in explant tissue and passaged it several times, Provost and Hilleman[2] attempted to adapt the CR326 strain to primate cell cultures which could be readily obtained and were more easily manipulated in the laboratory. In preliminary experiments, evidence of virus replication was obtained in primary *Cercopithecus aethiops* (African green monkey)

FIGURE 1. Phillip Provost and Maurice Hilleman of the Merck Institute for Therapeutic Research, who first demonstrated that HAV could be adapted to growth in cell culture. (Photo kindly provided by Dr. Provost.)

Table 1
ISOLATION OF HUMAN HEPATITIS A STRAIN CR326 IN LIVER EXPLANT TISSUE FROM SERONEGATIVE MARMOSETS

Passage	Inoculum	1	2	3	4	5
Day of harvest	—	23	27	27	20	17
HAAg — IF	—	4+	4+	4+	4+	4+[a]
RIA	3.1	5.7	5.3	4.4	4.0	6.5[b]

[a] Intensity of IF expressed on a 0 to 4+ scale; 0 = no fluorescence; 1+ = up to 25% of cells with fluorescent granules; 2+ = 25 to 50% of cells involved; 3+ = 50 to 75% of cells involved; 4+ = 75 to 100% of cells involved.

[b] RIA results expressed as a P/N ratio, i.e., the ratio of the radioactivity of the specimen compared with the radioactivity of negative controls.

kidney cells; however, as primary monkey kidney tissue varies in sensitivity from batch to batch and some batches are infected with endogenous viruses, a continuous cell line which would support the growth of HAV was sought.

Numerous cell lines were studied and the virus eventually propagated in FRhK-6 cells, a cloned line of epithelial-like cells derived from fetal rhesus monkey kidney. When monolayers of FRhK-6 were inoculated with CR326, minute intracytoplasmic fluorescent particles developed in some cells within 3 days, with both the size and number of particles detected increasing rapidly over the next few weeks. CR326 was serially propagated in FRhK-6 eight times and as it became adapted to these cells, the period from inoculation to peak IF decreased and the titer of HAAg detectable in the cell lysate increased (see Table 2).

The identity of the agent which was being transmitted was confirmed by demonstrating that a lysate of infected cells prepared after five passages of CR326 in FRhK-6 was infectious when inoculated into seronegative *S. labiatus* marmosets. Further studies in which serial dilutions of the fifth and sixth passage harvests were inoculated into FRhK-6 cells and evidence of virus replication sought by IF revealed that titers of infectious virus of 10^6 ID/mℓ or greater were being obtained.

While these studies clearly established that it was possible to propagate HAV in cell

Table 2
PROPAGATION AND PASSAGE OF CR326 IN FRhK-6 CELLS

Passage	Inoculum	1	2	3	4	5	6	7	8
Day of harvest	—	23	11	12	12	12	10	11	9
HAAg — IF	—	4+	4+	4+	4+	4+	4+	4+	4+[a]
RIA	3.1	31.6	10.9	11.3	15.5	22.4	46.5	23.2	25.8[b]

[a] Expressed on a 0 to 4+ scale.
[b] Expressed as a P/N ratio (see legend to Table 1.)

culture, it was not clear whether prior adaptation to marmosets was an essential prerequisite. A number of groups attempted to confirm these findings and to determine whether HAV could be isolated directly from clinical material.

B. Direct Isolation From Clinical Material

It was soon demonstrated that although some strains of HAV are more readily isolated in cell culture after prior adaptation to marmosets, many strains can be isolated directly from clinical material. In retrospect, it appears that the key to successful isolation is access to inocula with sufficiently high titers of infectious virus.

1. In PLC/PRF5 Cells

The first strain of HAV obtained directly from clinical material (MBB) was isolated by Frosner and his colleagues at the Pettenkoffer Institute in Munich, from a fecal specimen which had been collected from a patient 5 days prior to the onset of jaundice.[3] An extract of this specimen, which contained large numbers of HAV particles was inoculated into the PLC/PRF 5 (''Alexander'') cell line, a continuous line of human liver cells (derived from a hepatoma removed from a chronic carrier of HBV) which contains copies of the hepatitis B genome and secretes HBsAg. After prolonged incubation, increasing titers of HAAg were detected in the cells and in cell lysates but never in the supernatant fluid. HAAg was first detected (by IF) about 3 weeks after inoculation then increased dramatically with peak levels being attained 5 to 7 weeks post inoculation; no cytopathic effect was observed on primary isolation or on subsequent passage. The replication of the MBB strain of HAV in PLC/PRF5 cells was considerably slower than that of the CR326 strain in FRhK-6 cells, perhaps because the titer of the starting inoculum was lower or because the temperature at which the cells were maintained (32 to 35°C) was suboptimal.

About the same time as the German studies were published, Balayan and his colleagues at the Institute of Poliomyelitis and Viral Encephalitis in Moscow reported the successful propagation of a local strain of HAV in HeLa, Hep2, and PEO (a continuous line obtained from the liver of a green guenon monkey) cells. On primary isolation evidence of virus replication was obtained by IF, however this effect appeared to be lost on further passage.[4]

2. In FRhK-4 Cells

Subsequently, Flehmig in Germany was able to isolate three strains of HAV (GBG, GBM, GJA) by inoculation of another cloned line of fetal rhesus monkey kidney cells, FRhK-4. This line was derived from the same source as the FRhK-6 line[5] but appears to retain its sensitivity for a longer period. As with the previous studies, primary isolation of HAV was very time consuming, with incubation periods of 8 to 10 weeks being required before significant amounts of HAAg could be detected in the cells. On serial passage the time needed for the production of antigen was considerably reduced and some virus could be detected in the supernatant fluid. After ten passages it was possible to passage the infection with virus released into the supernatant fluid and to detect low levels of HAAg in this fluid by radioimmunoassay (RIA).

Flehmig was also able to establish a persistently infected line of FRhK-4 cells from which HAV could be harvested for a period of 12 months. The persistently infected cells appeared to be healthy and produced HAV particles which were morphologically and immunologically indistinguishable from virus purified from human feces but had a slightly lower buoyant density (1.32 to 1.34 g/cm³). Cell culture-derived virus proved to be indistinguishable from fecal-derived virus in assays for detection of HAV-specific IgM.[6]

3. In AGMK Cells

The first report of the isolation of HAV in a cell substrate suitable for vaccine production was made in 1981 by Daemer and his colleagues at the NIH.[7] These workers inoculated serum samples, fecal extracts, and extracts of marmoset liver containing three well characterized strains of HAV, HM175, MS-1, and SD11 (see Chapter 5) into primary African green monkey *(Cercopithecus aethiops)* kidney cell cultures (AGMK). The three strains which were used were selected because they had been obtained from different parts of the world and from different epidemiological situations (a family outbreak of hepatitis A, a strain which was endemic in an institution for the mentally retarded, and a strain obtained from a food-borne outbreak, respectively) and all had been shown to contain infectious virus by inoculation of susceptible chimpanzees and/or marmosets.

Initially extracts of liver tissue obtained from animals infected with each of the three strains were inoculated into AGMK cells and evidence of virus replication sought by IF. In this experiment maximum fluorescence was observed in AGMK cells inoculated with the sixth marmoset passage of HM175, with virtually all the cells containing fluorescent foci within 3 weeks of inoculation. Replication was also observed in AGMK cells inoculated with liver extracts containing the MS-1 strain of HAV which had been passaged 6 to 8 times in marmosets; by contrast, cells infected with marmoset-passaged SD11 showed transient fluorescence only (see Table 3).

The results of attempts to isolate HAV directly from stool and serum samples obtained from patients infected with these three strains are shown in Table 4. Although HAAg was not detected in cell cultures inoculated with either serum or fecal specimens containing the MS-1 strain of HAV, after 14 to 21 days incubation some fluorescence was detected in cells inoculated with specimens containing the SD11 strain. By far the most rapid replication was observed in AGMK cells inoculated with the HM175 strain, fluorescent foci becoming detectable in some cells by day 7, with the number of particles and number of cells affected increasing rapidly.

Because of these results further studies were limited to the HM175 strain. Additional passages were made with isolates which had been obtained directly from feces or after several (six) passages in marmosets (see Table 5). In both systems further passage of the virus resulted in increased yields of HAAg and a decrease in the time taken to produce it. When AGMK cells were inoculated with the stool extract containing HM175 and held for prolonged periods, IF was observed in a few cells within a week. The intensity of the fluorescence tended to decline during the second and third week of incubation but increased from week 4 so that by week 8 about 50% of the cells were affected. By the fourth passage, 10% of cells showed strong fluorescence within 2 weeks of inoculation, with the proportion rising to 60% by week 3, and 100% by week 4.

C. Effect of Mutation Pressure

Many characteristics of HAV can be altered by applying mutation pressure; for example, it has been possible to select variants of the HM175 strain which grow rapidly in cell culture.[27] To achieve this, a strain of HM175 which had been passaged 17 times in primary

Table 3
ISOLATION OF THREE STRAINS OF HAV IN AGMK USING EXTRACTS OF INFECTED MARMOSET LIVER AS INOCULA

Strain	Marmoset passage	IF results (week) 1	2	3
HM175	2	—	2+	3+[a]
	5	2+	3+	2+
	6	2+	3+	4+
MS-1	6	—	1+	1+
	7	1+	2+	—
	8	2+	2+	2+
SD11	2	1+	—	—
	3	—	—	—
	4	—	—	—

[a] Intensity of IF rated on a 0 to 4+ scale (see legend to Table 1).

Table 4
ISOLATION OF HAV IN AGMK USING CLINICAL SPECIMENS OBTAINED FROM PATIENTS INFECTED WITH THREE DIFFERENT STRAINS OF THE VIRUS

Strain	Inoculum	IF results (week) 1	2	3
HM175	Feces	1+	1+	3+[a]
MS-1	Feces	—	—	—
	Serum	—	—	—
SD11	Feces	±	—	±
	Serum	—	1+	2+

[a] Intensity of IF rated on a 0 to 4+ scale (see legend to Table 1).

or secondary AGMK cells was used to infect a continuous cell line derived from AGMK known as Buffalo green monkey kidney cells (BGMK) and blind passages undertaken every 3 or 4 days. During the first few passages only tiny amounts of HAAg could be detected by IF or RIA at the time of harvest. Detectable virus virtually disappeared between passages 4 and 10 after which small quantities of HAAg could again be detected; however, by passage 18 large quantities of HAAg were detected by both IF and RIA after only 3 days of culture. This "rapid passage" virus has some unusual characteristics in that it does not spread readily from cell to cell and has a low infectivity titer relative to the amount of antigen produced. Because it produces very small plaques, this strain is not suitable for incorporation in a radioimmunofocus assay (RIFA), however it can be used to establish a rapid test for neutralizing antibodies based on reduction of fluorescent foci. This assay which does not require

Table 5
RESULTS OF PASSAGE OF HAV STRAIN HM175 IN AGMK CELLS

Inoculum	AGMK pass no.	IF (week) 1	2	4	6	8	11
Stool extract	1	1+	—	1+	1+	2+	4+[a]
	2	NT[b]	NT	NT	NT	4+	
	3	—	±	3+	4+		
	4	2+	2+	4+	4+		
	5	1+	2+	3+	4+		
	6	1+	1+	2+	3+		
	7	1+	2+	2+	2+		
Marmoset	2 (3w)	—	4+	4+	4+[a]		
Passage	(8w)	1+	1+	4+	4+		
	3 (3w)	—	1+	—	—		
	(8w)	1+	1+	1+	1+		
	4 (3w)	—	—	1+	2+		
	(8w)	—	—	1+	2+		
	5 (3w)	2+	3+	3+	4+		
	(8w)	2+	3+	3+	4+		
	6 (3w)	1+	2+	3+			
	(8w)	1+	2+	3+			
	7 (3w)	2+	2+	2+	2+		
	(8w)	±	1+	3+	3+		

[a] Intensity of IF estimated on a 0 to 4+ scale (see legend to Table 1).
[b] NT = not tested.

an overlay (because there is essentially no secondary spread of the virus) can be performed in 3 days using eight-chamber tissue-culture slides so that a dilution series can be read on a single slide.

III. SENSITIVITY OF VARIOUS TYPES OF CELL CULTURE

Numerous other strains of HAV have been isolated in cell cultures of primate origin with cells of rhesus and AGMK appearing to be most sensitive (see Table 6). In an attempt to evaluate the relative sensitivity of these cells Binn and his colleagues tested seven different lines (FRhK-4, FRhK-6, B-SC-1, FRhL-2, OMK-210, CV-1, and MA104) for their ability to support replication of wild-type HAV and strains which had been passaged in nonhuman primates.[11] Nine fecal samples (containing HAV from seven epidemiologically distinct sources) were inoculated into these seven cell lines, each of which was monitored for virus replication by IF, RIA, and RIFA.

Low-passage AGMK cells proved to be most sensitive to HAV, resulting in the recovery of seven of nine strains of virus. Virus was less often recovered in FRhK-4 and FRhK-6 cells, occasionally in the other cell lines, and never in OMK-120. Repeated passage of HAV strain HM175 in AGMK cells resulted in significant quantities of virus being detected in both the supernatant fluid and cell lysate without evidence of a cytopathic effect. Because primary AGMK cells are in limited supply and often contaminated with simian viruses, attempts were made to adapt the various strains to continuous cell lines of AGM origin. After several passages most strains of HAV could be adapted to growth in B-SC-1 cells (a continuous line derived from AGMK).

The effect of certain parameters on cell growth was also studied. The PA-21 strain of

Table 6
RANGE OF PRIMATE CELLS WHICH WILL SUPPORT THE GROWTH OF HAV

Human	Cell type
PLC/PRF/5	Human hepatoma cells secreting HBsAg
FL	Continuous line of human amnion cells
HEF	Continuous line of human diploid fibroblasts
HEK	Human embryonic kidney cells
A549	Human lung carcinoma
	Primate
Rhesus monkey	
FRhK-4	Cloned line of fetal rhesus monkey kidney cells
FRhK-6	Cloned line of fetal rhesus monkey kidney cells
MA104	Continuous line of fetal rhesus monkey kidney cells
FRhL-2	Fetal rhesus monkey lung cells
African green monkey	
AGMK	Primary or secondary African green monkey kidney cells
B-SC-1	Continuous line of AGMK
Vero	Continuous line of AGMK
CV-1	Continuous line of AGMK

Table 7
PROPERTIES OF STRAINS OF HAV ADAPTED TO GROWTH IN CELL CULTURE

Relatively slow growing
Grow best in cells of primate origin
Virus largely cell associated
No cytopathic effect
Relatively low titers of virus produced
Replication detected by IF or RIA

HAV was found to grow better in B-SC-1 cells at 35°C than at 32°C or 39°C while higher titers of virus were obtained when the cells were maintained in serum-free medium than in medium supplemented with 20% bovine calf serum.

Following these studies workers in a number of laboratories have attempted to select mutants which grow more rapidly in cell culture and produce higher yields of virus, either by passaging at short intervals or selecting appropriate mutants in RIF or plaque assays. Despite these efforts and the use of cell factories, pleated surfaces, and continuously infected cell lines, the yields of HAV which can currently be obtained in cell culture are lower than those obtained with many other picornaviruses and pose difficulties for large-scale production of low-cost vaccines.

The major properties of strains of HAV which have been isolated in cell culture to date are shown in Table 7.

IV. METHODS OF QUANTITATING HAV IN CELL CULTURE

Although a number of strains of HAV have been adapted to cell culture, quantitation of the amount of infectious virus in an inoculum has proved to be difficult. Most strains of HAV grow slowly and show no evidence of a cytopathic effect so that detection of viral replication is dependent upon demonstrating the development of virus antigens in infected

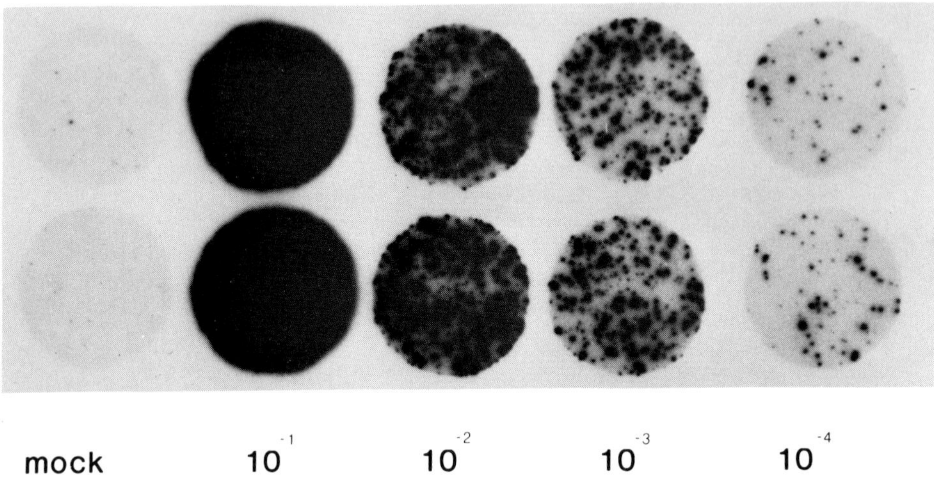

FIGURE 2. An RIFA showing a titration of HAV strain HM175 in log dilutions B-SC-1 cells. Each focus represented the results of replication of a single particle or cluster of virus particles. (From an experiment in the authors' laboratory. Photo kindly provided by David Anderson.)

cells by EIA immunofluorescence, or RIA. Because HAAg may take several weeks to develop, titration of pools of HAV and measurement of neutralizing antibodies are difficult and time-consuming procedures.

A. Radioimmunofocus Assay (RIFA)

A major advance was the development of a RIFA which made it possible to measure the amount of virus in a given inoculum accurately, to develop a sensitive test for neutralizing antibodies, and to undertake a number of manipulations usually performed with plaque assays. In this assay, which was developed at the Walter Reed Army Institute for Research in Washington, D.C.,[12] cultures of susceptible cells (usually AGMK or B-SC-1) are inoculated with HAV and overlayed with agarose. After incubation at 35°C for about 14 days in an atmosphere of 5% CO_2, the overlay is gently removed, the cells dried and fixed, and ^{125}I-labeled anti-HAV added. After incubation to allow the antibody to bind to HAAg, the fixed cells are washed and dried, and the bottom of each dish removed and subjected to autoradiography. After development of the films, foci of infected cells are revealed as black spots on a clear or gray background analogous to stained plaques in a conventional plaque assay. At high concentrations of virus, the foci are confluent but at lower dilutions discrete foci (measuring from 1 to 6 mm in diameter) appear and can be readily counted. The titer of infectious virus in the inoculum can be calculated from the number of foci observed. When serial titrations are performed, the number of foci detected bears a linear relationship to the dilution used (see Figure 2).

The RIFA was originally performed by inoculating 80 radioimmunofocus units (RIFU) of HAV strain HM175 into B-SC-1 cells. The time at which the test was read was critical, as only under these conditions a few scattered foci were apparent after 7 days incubation; discrete, easily counted foci could be detected by day 14; however, by day 21 these had overlapped to such an extent that quantitation was impossible.

The sensitivity of the RIFA was compared with conventional titrations in cell culture in which the presence or absence of HAAg was detected by IF after 28 days incubation. Using three different inocula the sensitivity, reproducibility, and accuracy of the two techniques were found to be comparable.

1. Applications

This modified plaque assay is widely used and has become the standard means of quantitating HAV and testing sera for the presence of neutralizing antibodies. In addition, it provides a means of selecting mutants with different growth characteristics (as revealed by the size of the foci produced) and comparing the antigenic structure of different isolates.

Lemon and his colleagues have used the RIFA to examine the extent of antigenic similarity between strains of virus isolated in different parts of the world.[13] Cross-neutralization studies performed with convalescent sera obtained from primates infected with HM175 and PA-21 strains of HAV (see Chapter 5) produced virtually identical results, suggesting that these strains are closely related. They have also used the RIFA to study the development of neutralizing antibodies to HM175 and PA21 in colony-bred owl monkeys.[14] In both instances neutralizing antibody became detectable early in the course of infection while HAV was being shed into the feces, and then rose rapidly. Although the appearance of neutralizing antibody coincided with a decline in the shedding of virus, HAAg could be detected in fecal samples collected from the animals for several days despite the presence of high-titered neutralizing antibody in the serum. The pattern of the neutralizing antibody response was found to parallel the development of anti-HAV detectable by RIA (see Figure 3). When serial sera from these animals were fractionated and tested, neutralizing activity was demonstrated in both the IgM and IgG fractions of specimens collected early in the disease, but could only be demonstrated in the IgG fraction of sera collected during convalescence.

The RIFA has a number of practical applications and is likely to be of particular value in assessing the immune response in primates and human volunteers immunized with candidate hepatitis A vaccines and for standardizing the titer of anti-HAV in preparations of HNIG (immune serum globulin).

B. Plaque Assay

It has become apparent that after prolonged passage in cell culture, certain strains of HAV release infectious virus and viral antigens into the supernatant fluid. Whether this represents extrusion of virus from the cell or lysis of a subpopulation of infected cells is not known. Several groups have also reported the existence of strains which appear to produce a CPE in tissue culture. The first account from Venuti and his colleagues was associated with a fast-growing strain of HAV which had been derived by repeated passage at short intervals.[15] Although cells infected with this virus appear to show degenerative changes, other workers have been unable to consistently reproduce this finding.

Recently Anderson et al.[16] reported a CPE in B-SC-1 cells grown in serum-free medium infected with the HM175 strain of HAV. The cytopathic effect was reproducible, neutralized by convalescent but not preinfection sera from patients and animals infected with HAV, and was detectable after the virus had been purified by limiting dilution (see Figure 4). In B-SC-1 cells (at passage levels 44 to 60) infected with 1 RIFU of plaque-purified HAV strain HM175 per cell, the CPE developed within 72 hr of inoculation and was enterovirus-like. The infected cells initially appear pyknotic and tend to form small clusters; this is followed within a few days by complete destruction of the monolayer. The ability of this strain to produce a CPE appears to be dependent on some cellular factors, as the appearance of a CPE is delayed when higher passages of B-SC-1 (passage levels 75 to 80) are used, and if the cells were infected 7 days after seeding rather than after 3 or 4 days.

Using this strain it has proved possible to develop a conventional plaque assay (see Figure 5) in which plaques of infected cells can be recognized by absence of staining with crystal violet (when quantitation is required) or by their refractile nature under the microscope (when virus is to be reisolated from the culture).[17] It has proved possible to use this system to select mutants with different plaquing characteristics and to plaque purify strains of virus. Plaque-purified HAV isolated in this fashion can be neutralized by postinfection (but not

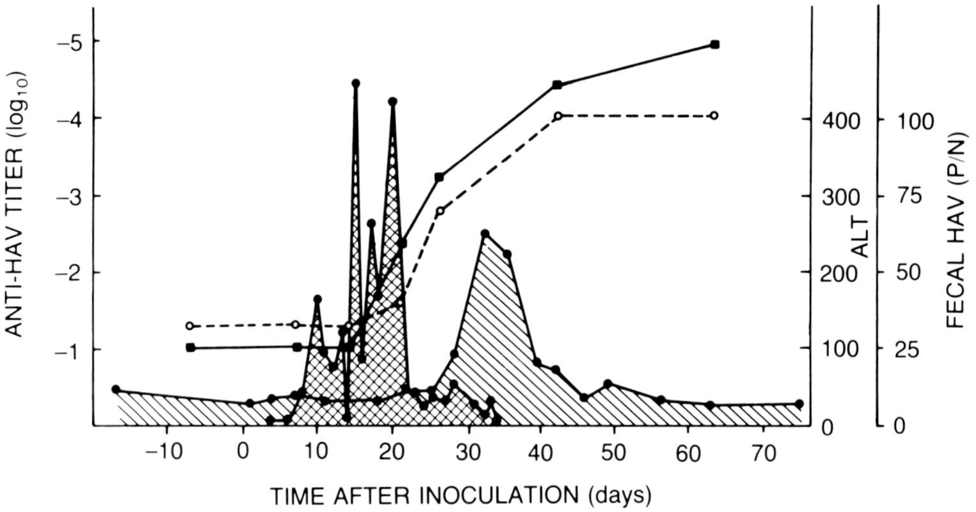

FIGURE 3. The development of neutralizing antibodies (●——●) and antibodies detectable by RIA (○- - -○) in an Aotus monkey infected with the HM175 strain of HAV. (From Lemon, S. M. and Binn, L. N., *J. Infect. Dis.*, 148, 1033, 1983. With permission.)

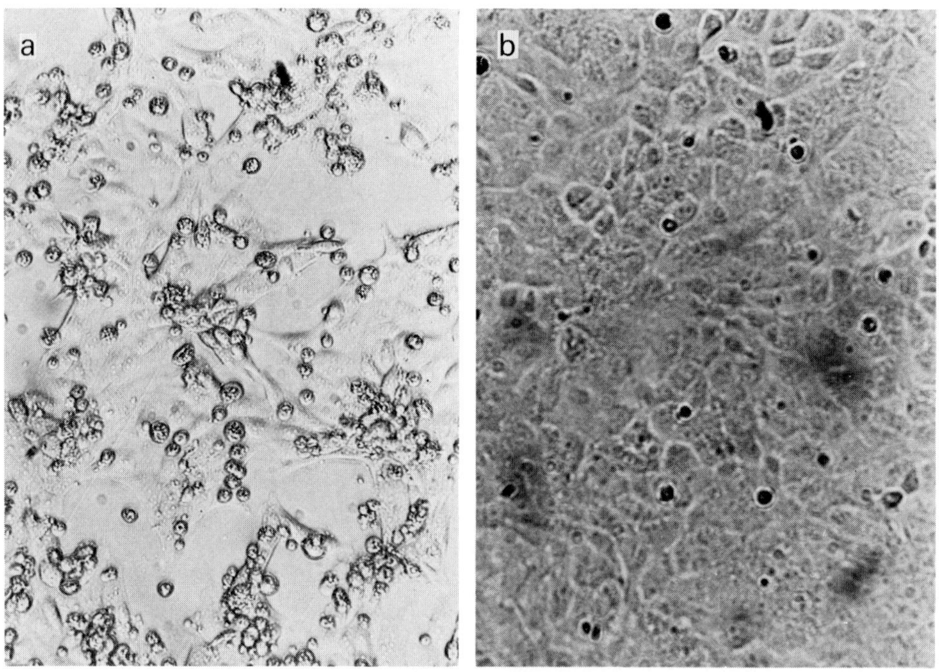

FIGURE 4. CPE produced by the HM175 strain of HAV after 3 days infection of B-SC-1 cells with IRFU virus/cell. (a) Infected cells; (b) after neutralization with monoclonal antibody. (From Anderson, D. S., Locarnini, S. A., Ross, B. C., Coulepis, A. G., Anderson, B. N., Gust, I. D., *Positive Stranded RNA Viruses,* Rueckart, R., Ed., Alan R. Liss, New York, in press. With permission.)

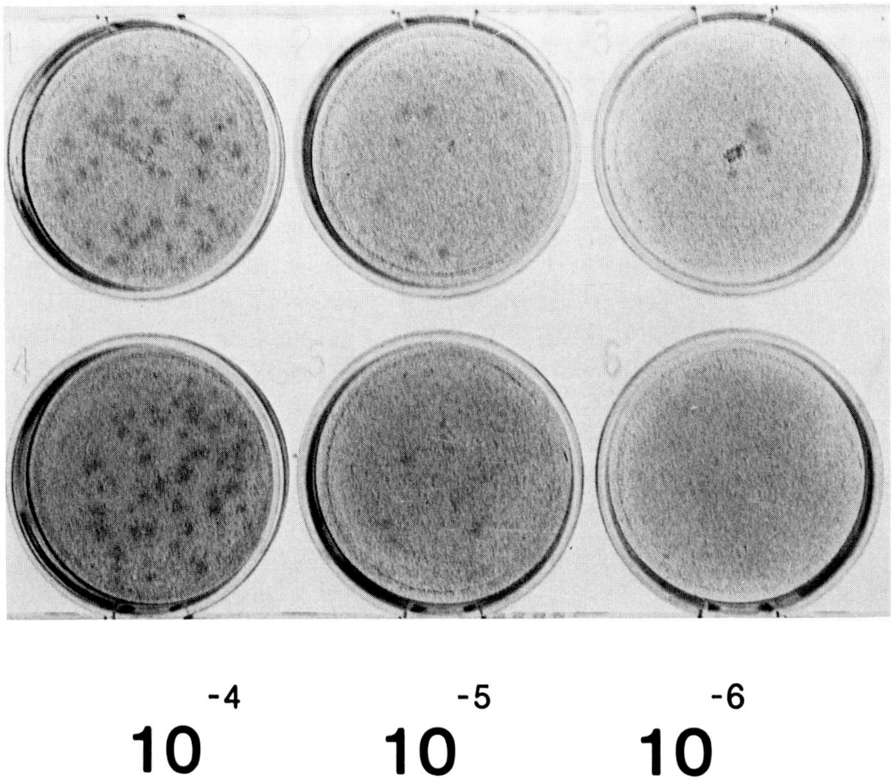

FIGURE 5. Plaque assay of HAV strain HM175 in B-SC-1 cells. Cell monolayers were infected with serial tenfold dilutions of virus and incubated for 9 days under 0.5% agar, fixed with 4% formalin, and stained with 0.2% crystal violet in 20% ethanol. Note that the plaques stain more readily than surrounding uninfected cells. (From Anderson, D. S., Locarnini, S. A., Ross, B. C., Coulepis, A. G., Anderson, B. N., Gust, I. D., *Positive Stranded RNA Viruses,* Rueckart, R., Ed., Alan R. Liss, New York, in press. With permission.)

pre-infection) human and chimpanzee sera and mouse monoclonal antibodies to HAV. CPE-producing strains of HAV have recently been reported from at least two other laboratories using the HM175 strain of HAV[18] and other isolates.[19]

V. KINETICS OF HAV REPLICATION

With the development of assays for detection of viral antigens and infectious virus particles and adaptation of several strains of virus to cell culture, it has become possible to study the mechanism and kinetics of HAV replication. Using several strains of HAV obtained from patients with naturally acquired infections, Locarnini and his colleagues were able to demonstrate the appearance of HAAg in the cytoplasm of infected Vero and FRhK-6 cells within 12 hr of inoculation.[20] The amount of antigen increased steadily for 4 days and then declined over the next few weeks. Using a double labeling coelectrophoresis and subtraction technique they were able to demonstrate that infected cells produced at least 11 apparently virus-coded proteins, with MWs ranging from 14 to 160 kDaltons, within 6 hr of inoculation (see Figure 6). Cell lysates collected 2 to 6 hr after inoculation contained newly synthesized virus-specific, single-stranded, and double-stranded RNA species, the latter presumably representing a replicative intermediate (see Figure 7).

Similar proteins were detected when Vero cells were inoculated with single-stranded RNA

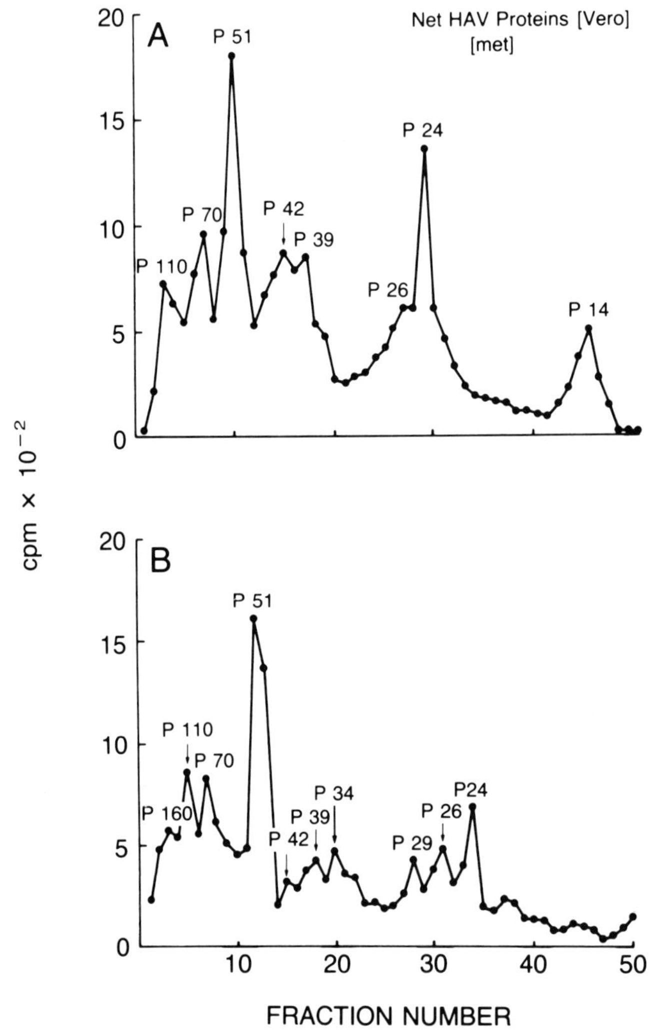

FIGURE 6. Polyacrylamide gel electrophoresis of proteins synthesized at 3 to 6 hr p.i. in infected Vero cells. Cells were treated with 3 μg/mℓ of actinomycin D 1 for 3 hr and then labeled with 10 μCi/mℓ of ^{3}H methionine. Control uninfected cells were labeled with ^{35}S methionine under the same conditions the monolayers dissolved in 2% SDS, the infected and uninfected cytoplasms mixed and electrophoresed. The net profiles of virus-coded proteins were obtained by subtraction. (A) Eight virus-coded proteins, p110 to p14, were detected; (B) shows three additional proteins p160, p34, p29 resolved by longer periods of electrophoresis. In this experiment p14 ran of the gel. (From Locarnini, S. A., Coulepis, A. G., Westaway, A. G., Gust, I. D., *J. Virol.*, 37, 216, 1981. With permission.)

which had been extracted from purified HAV virions, suggesting that the genome of the virus was infectious. In contrast to experience with other picornaviruses, the authors obtained no evidence that infection with HAV resulted in inhibition of host-cell macromolecular synthesis. They suggested that in view of the detection of virus-coded proteins in infected cells but the relatively poor production of infectious virus, some form of cellular restriction must be operating.

These observations were confirmed and extended by Anderson et al.[21] who studied the

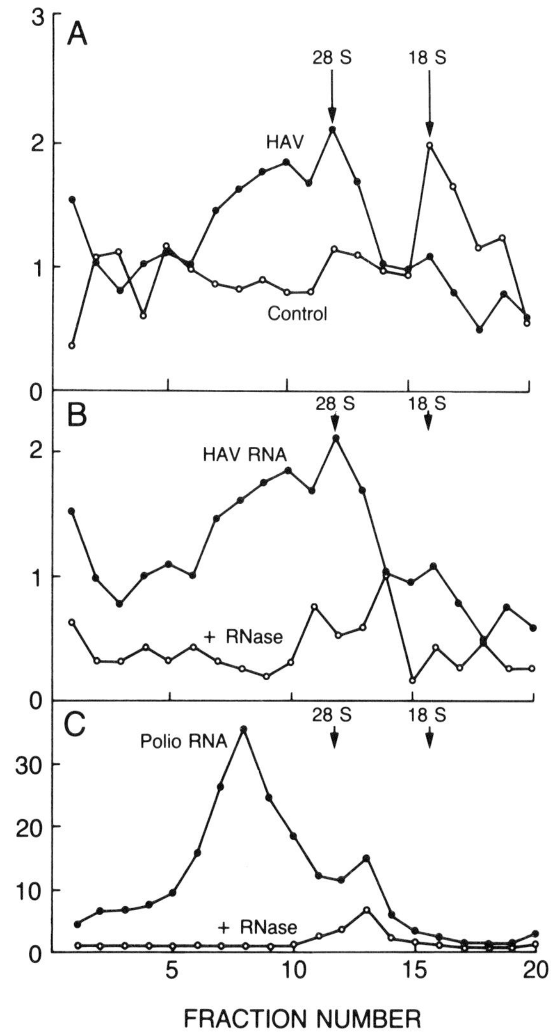

FIGURE 7. Sucrose gradient analysis of cell extracts of PV, HAV, and control RNA, synthesized 2 to 6 hr postinfection. RNA was extracted from the ^{3}H-uridine-labeled cytoplasm by the SDS-phenol chloroform method; carrier RNA was added, and all RNAs were precipitated with ethanol. Each precipitate was dissolved in NET buffer and analyzed by ultracentrifugation in sucrose gradients. (A) Compares the untreated sucrose gradient fractions from the ^{3}H-uridine labeled cytoplasm of the control (○——○) and HAV-infected (●——●) cells; (B) the sensitivity of the ^{3}H-uridine-labeled HAV RNA (●——●) to pancreatic RNase (○——○); (C) the sensitivity of the ^{3}H-uridine-labeled PV type 1 RNA (●——●) synthesized at 2 to 6 hr postinfection to pancreatic RNase (○——○). The relative mobilities of peaks of ^{3}H-labeled acid-insoluble radioactivity were determined by comparison with externally run ^{3}H-labeled 28 S and 18 S rRNA. (From Locarnini, S. A., Coulepis, A. G., Westaway, A. G., Gust, I. D., *J. Virol.*, 37, 216, 1981. With permission.)

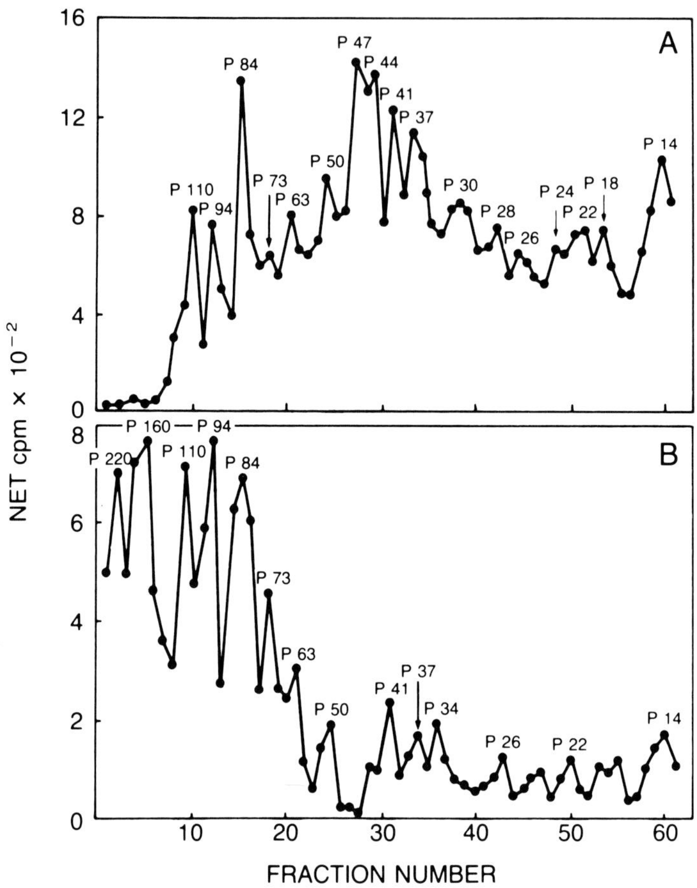

FIGURE 8. Net electrophoretic profile of virus-specific proteins synthesized from 3 to 6 hr postinfection in FRhK-4 cells infected with HM175 or HM790. The net electrophoretic profile of virus-specified proteins was obtained by subtraction after an appropriate correction factor (in these examples 2.5) was applied to the ^{14}C cpm so as to equal the ^{3}H cpm in fractions apparently devoid of virus-specified proteins. Proteins are labeled according to their MWs ($\times$ 10^{-3}) as determined by coelectrophoresis of HAV-infected lysates with ^{14}C-methylated MW markers. (A) HM175-infected FRhK-4 cells; (B) HM790-infected FRhK-4 cells. Note the accumulation of high MW products in FRhK-4 cells infected with HM790 as compared to the products detected in HM175-infected FRhK-4 cells. (From Anderson, D. S., Locarnini, S. A., Coulepis, A. G., Gust, I. D., *Intervirology*, 24, 26, 1985. With permission.)

effects of two strains of HAV (HM175 and HM790) in a permissive cell line (FRhK-4) using the double-label coelectrophoresis and subtraction technique with a modified electrophoresis system which allowed greater resolution of proteins. In FRhK-4 cells infected with HM175, virus-specified proteins were estimated to account for approximately 20% of proteins synthesized in the period 3 to 6 hr after infection. Seventeen virus-coded proteins ranging in MW from 110 to 114 kDaltons were regularly detected (see Figure 8). In cells infected with the HM790 strain, 12 of these proteins were detected plus an additional two (MW of 220 and 160 kDaltons) not recognized in cells infected with HM175. In cells infected with either HM175 or HM790, four of the proteins detected had MWs which were very similar to the MW of the four major structural proteins of HAV.

Despite the fact that both strains of HAV were obtained from patients in the same city

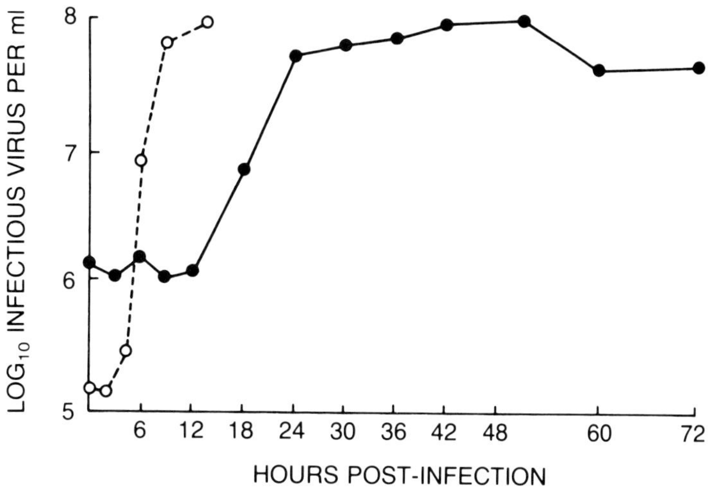

FIGURE 9. Single-cycle growth curve of the HM175 strain of HM175 (●——●) and Mahoney strain of PV 1 (○----○) in B-SC-1 cells. (From Anderson, D. S., Locarnini, S. A., Ross, B. C., Coulepis, A. G., Anderson, B. N., Gust, I. D., *Positive Stranded RNA Viruses,* Rueckart, R., Ed., Alan R. Liss, New York, in press. With permission.)

and similar titers of input virus were used, the protein profiles which were detected showed some differences. In cells infected with HM790 the accumulation of high-MW products was similar to the pattern seen in polio-infected cells grown in the presence of 1 m*M* $ZnCl_2$, suggesting that the polyproteins of this strain are cleaved poorly. The pattern of protein synthesis in cells infected with HM175 was similar to that seen in cells infected with polio virus (PV) with the exception that some accumulation of high-MW viral-coded proteins was observed. From these data it was suggested that the rate-limiting step in replication of HAV in this system was a defect in the processing of the polyprotein.

Subsequently this group inoculated B-SC-1 cells with HM175 (1 RIFU per cell) and compared the single-cycle growth kinetics of HAV with PV[17] (see Figure 9). By contrast with PV, in which a 2-hr eclipse phase was followed by a logarithmic increase in mature virus to 9 hr postinoculation (p.i.), eclipse of HAV was complete with up to 25% of the input virus being recoverable for 9 hr. A logarithmic increase in infectious virus occurred between 12 to 24 hr p.i. with peak yields occurring at 51 hr p.i. HAV-related RNA (detected by dot-blot hybridization) was detectable within 6 hr of inoculation and increased rapidly from 12 to 25 hr (Figure 10). HAAg was first detected (by IF) 18 hr after inoculation and continued to accumulate until 51 hr p.i. Double-stranded RNA (detected by indirect IF) was first detected 24 hr p.i. by which time most of the infectious virus and viral RNA had been produced. The relatively long period between inoculation and detection of virus replication does not appear to be due to defects in penetration or uncoating of the virus, but perhaps due to a restriction in RNA replication.

Similar data on the kinetics of replication were reported by Wheeler et al.[22] using the HAS-15 strain of HAV which had been adapted by weekly passages in FRhK-4 cells. In this system logarithmic production of infectious virus commenced about 18 hr p.i. and continued until day 5 after which production of virus plateaued. From days 5 to 14 p.i., yields of 10^9 to 10^{10} RIFU/mℓ could be obtained from cell lysates and similar, but slightly lower titers of virus were detected in the supernatant fluid. These authors calculated that during the log phase of growth at least 50% of the virus which was produced was released into the supernatant fluid, whereas during the stationary phase about 80% of the virus

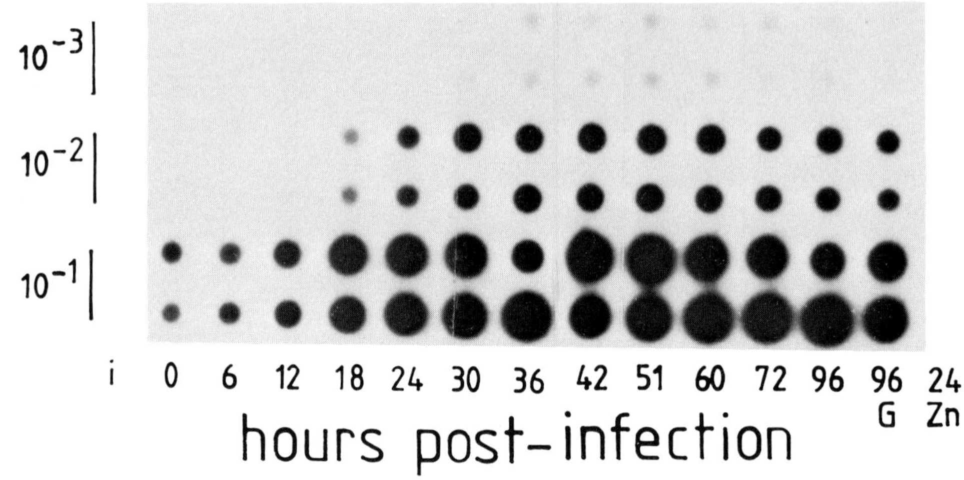

FIGURE 10. Dot-blot hybridization of RNA from HM175 infected B-SC-1 cells under single-cycle growth conditions. (From Anderson, D. S., Locarnini, S. A., Ross, B. C., Coulepis, A. G., Anderson, B. N., Gust, I. D., *Positive Stranded RNA Viruses,* Rueckart, R., Ed., Alan R. Liss, New York, in press. With permission.)

remained cell associated. Surprisingly, while high titers ($\geqq 10^8$ RIFU/mℓ) of infectious virus particles were detected in the supernatant fluid, HAAg was never detected by RIA.

VI. MORPHOLOGICAL CHANGES IN INFECTED CELLS

Relatively little is known about the morphogenesis of HAV or the ultrastructural changes which are associated with infection. Some aspects of the replication have recently been addressed by Kiernan and her colleagues at Fairfield Hospital, Melbourne.[23] B-SC-1 cells persistently infected with the HM175 strain of HAV and uninfected controls were maintained for up to 60 days to allow various aspects of the replication of the virus to be studied.

A. Light Microscopy

After seeding, both persistently infected and uninfected cells showed an exponential growth phase lasting about 14 days after which the cell numbers remained stationary. No major difference in cell density or mitotic index were observed (see Figures 11 and 12). HAAg became detectable (by immunoperoxidase staining) in the cytoplasm of the infected cells midway through the exponential growth phase, with the proportion of antigen-containing cells rising rapidly, so that by early in the stationary phase virtually all cells contained detectable antigen (see Figure 13). Initially the granules were small and scattered through the cytoplasm, but with time they increased in size, being most prominent in the region surrounding the nucleus. Accumulation of intracellular antigen appears to begin toward the end of the exponential growth phase (see Figure 14).

B. Electron Microscopy

When uninfected B-SC-1 cells were examined by thin-section EM no unusual features were observed. The nuclei were round or oval with an undulating nuclear membrane. The cytoplasm contained numerous organelles (rough and smooth endoplasmic reticulum, filamentous and tubular structures, vesicles, lysozyme-like bodies, and tubular structures) bounded by an undulating plasma membrane (see Figure 15).

Persistently infected cells showed a similar appearance with the exception that the cytoplasm was more vacuolated, less dense, and contained more lysozyme-like bodies (see Figure 16). Regular arrays of particles which could be mistaken for HAV were occasionally seen

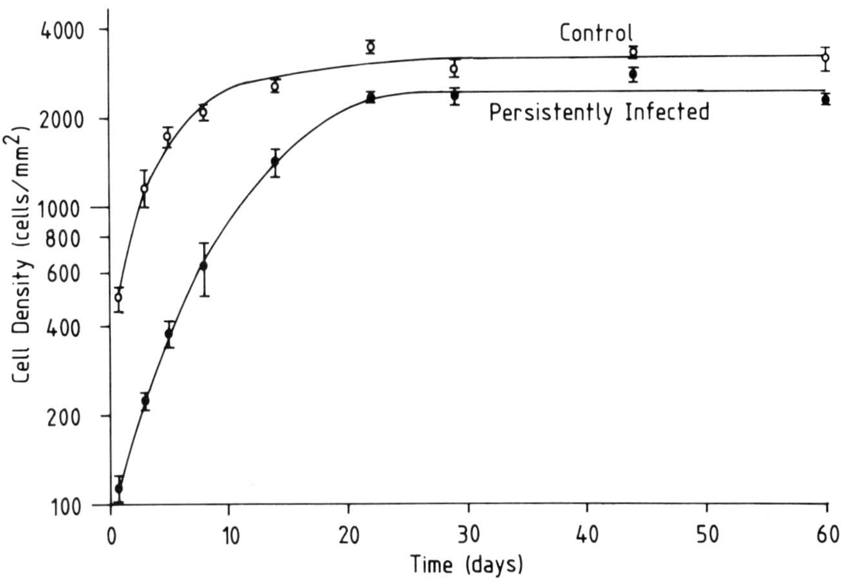

FIGURE 11. Cell density of persistently infected monolayers and uninfected control monolayers over a 60-day interval. Results are expressed as mean ± standard error of the mean (n = 6). (From Kiernan, R., Marshall, J. A., Coulepis, A. G., Anderson, D. A., Gust, I. D., *Arch. Virol.*, in press. With permission.)

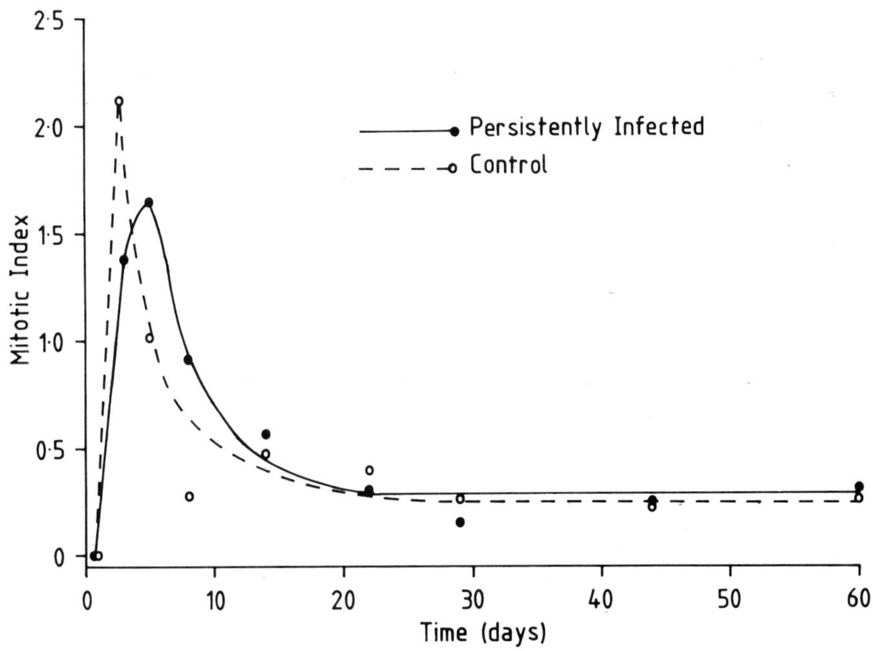

FIGURE 12. The mitotic index of persistently infected cells (●) and uninfected cells (○) over a 60-day growth period. (From Kiernan, R., Marshall, J. A., Coulepis, A. G., Anderson, D. A., Gust, I. D., *Arch. Virol.*, in press. With permission.)

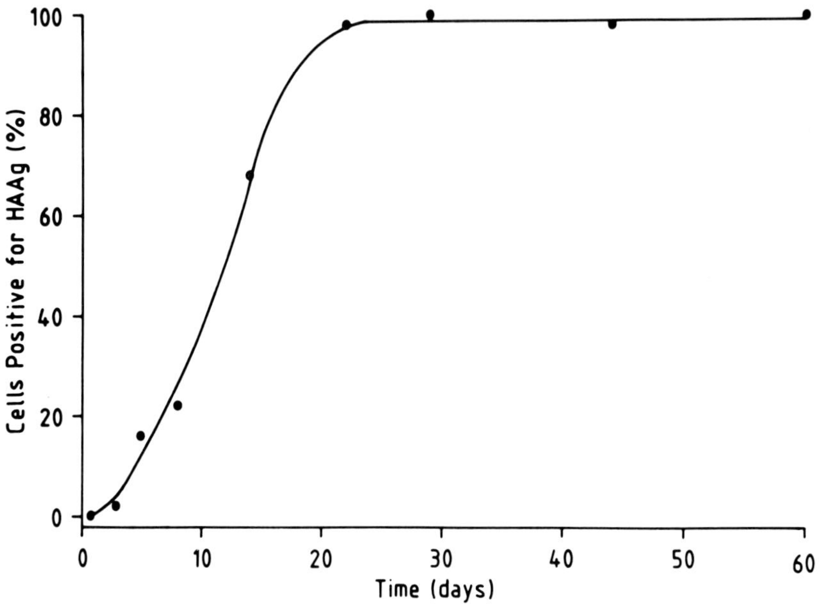

FIGURE 13. Percentage of persistently infected cells found to be positive for intracellular HAAg peroxidase immunoperoxidase staining at various times p.i. (From Kiernan, R., Marshall, J. A., Coulepis, A. G., Anderson, D. A., Gust, I. D., *Arch. Virol.*, in press. With permission.)

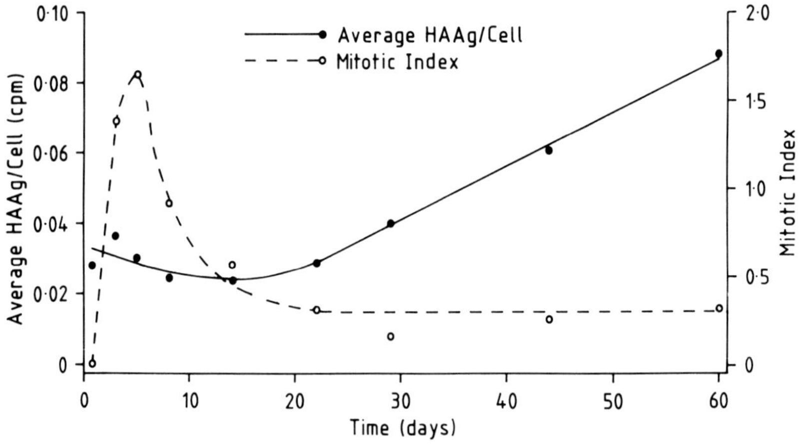

FIGURE 14. Relationship between cell growth and antigen accumulation. The average HAAg-specific cpm per persistently infected cell is calculated for each collection time (solid line). The mitotic index of persistently infected monolayers at each collection times is also shown (dashed line). (From Kiernan, R., Marshall, J. A., Coulepis, A. G., Anderson, D. A., Gust, I. D., *Arch. Virol.*, in press. With permission.)

in cells showing degenerative changes and appear to be fragments of intracellular membranes. The major difference between infected and uninfected cells was the appearance of annulate lamellae which were regularly seen in the cytoplasm of infected cells in both the exponential and stationary phases of growth (see Figure 17). These structures which are composed of parallel arrays of cisternae with regular perforations, were seen in about 10% of cells, usually adjacent to the nuclear membrane or the rough endoplasmic reticulum. No mature HAV particles were demonstrated in the infected cells, which is not surprising in view of the fact

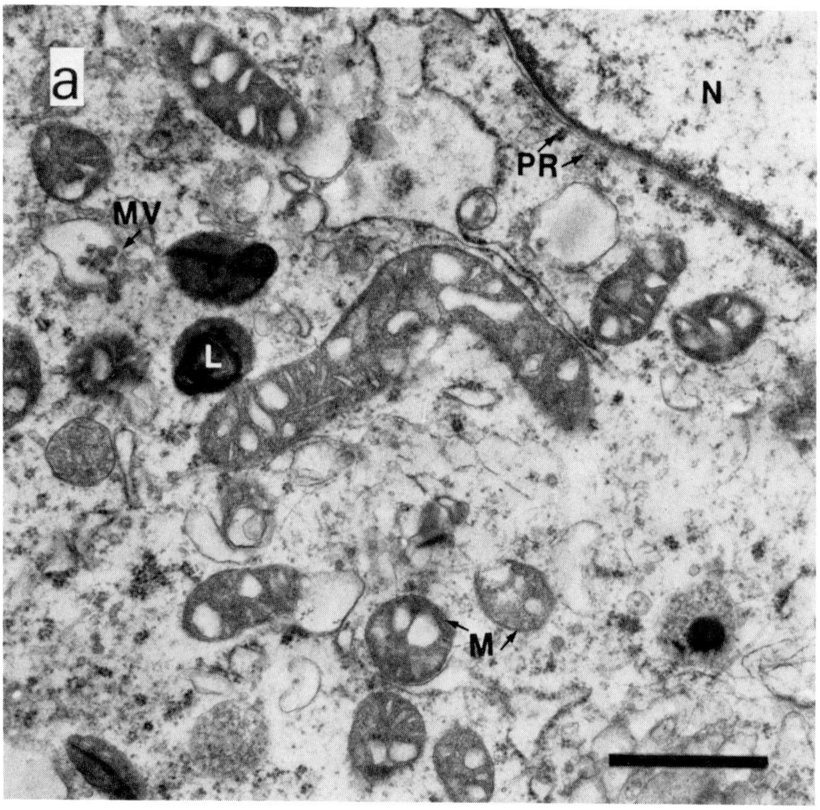

FIGURE 15. (a) An uninfected cell during the exponential phase of growth. The cytoplasm contains numerous tubular mitochondria (M) as well as some lysozyme-like bodies (L) and multivesicular bodies (MV). The nuclear envelope is studded with polyribosomes (PR). (N: nucleus.) Stained with uranyl acetate and lead citrate. Bar represents 1 μm; (b) an uninfected B-SC-1 cell collected during the stationary phase. The cytoplasm contains many vacuoles (V), lysozyme-like bodies (L), and tubular mitochondria (M). (N: nucleus.) Stained with uranyl acetate and lead citrate. Bar represents 2 μm. (From Kiernan, R., unpublished observations.)

that the virus titer in cells lysed after 60 days in culture was only $8.4 + 10^5$/mℓ, equivalent to less than 30 particles per cell.

The lack of major structural changes in infected cells is perhaps not surprising, as in this system infection with HAV is not associated with cell destruction — growth of the virus is relatively slow and unlike other picornaviruses there is no evidence that infection results in a switch off of host-cell macromolecular synthesis. The detection of increasing numbers of membranous structures, "annulate lamellae" in the cytoplasm of infected cells, is of particular interest. In infections with other picornaviruses, replicating forms of proviral RNA and newly formed virions are associated with smooth membranes which appear to proliferate as viral RNA replicates.[24,25] It is possible that these membranes serve as the site of replication of HAV.

In parallel studies at the Walter Reed Army Institute for Research, Asher and her colleagues used conventional thin-section EM and indirect immunoperoxidase staining to localize HAV in B-SC-1 cells infected with HM175 (at an input multiplicity of 2 to 4 infectious particles per cell) which were held for 6 weeks prior to sectioning. Virus-like particles were detected only in the cytoplasm of infected cells, usually in vesicles surrounded by a double membrane

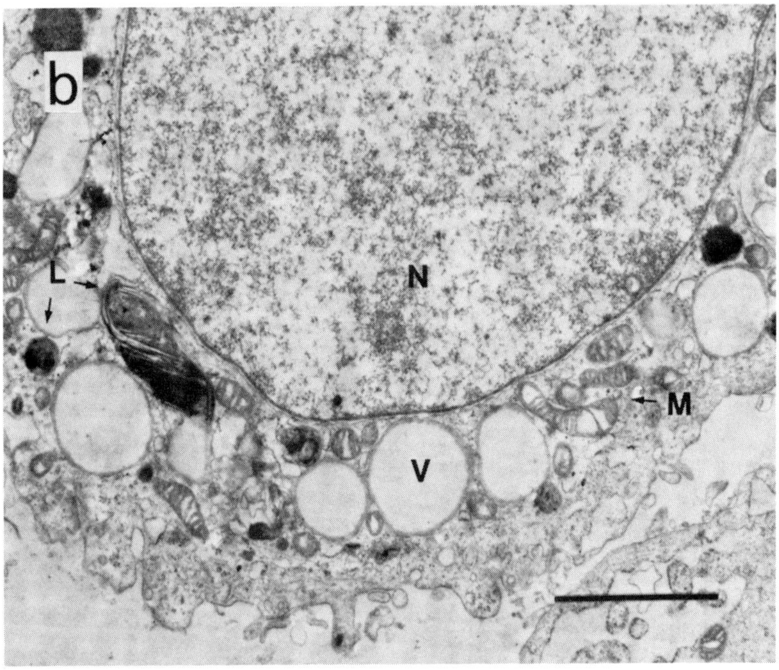

FIGURE 15b.

and containing myelin-like structures, and occasionally free in the cytoplasm[26] (see Figure 18). The particle containing vesicles stained darkly with immunoperoxidase-labeled anti-HAV and similar staining was observed on parts of some mitochondria (see Figure 19). Although no crystalline arrays of virus were observed, occasional clusters of virus-like particles, measuring approximately 24 nm in diamater were detected. HAV particles were also identified in the supernatant fluid of infected cells sometimes contained within vesicles (see Figure 20) similar to those which have been occasionally detected in the feces of infected patients (see Chapter 2). The presence of these structures may explain the presence of a non-neutralizable fraction of HAV in some inocula.

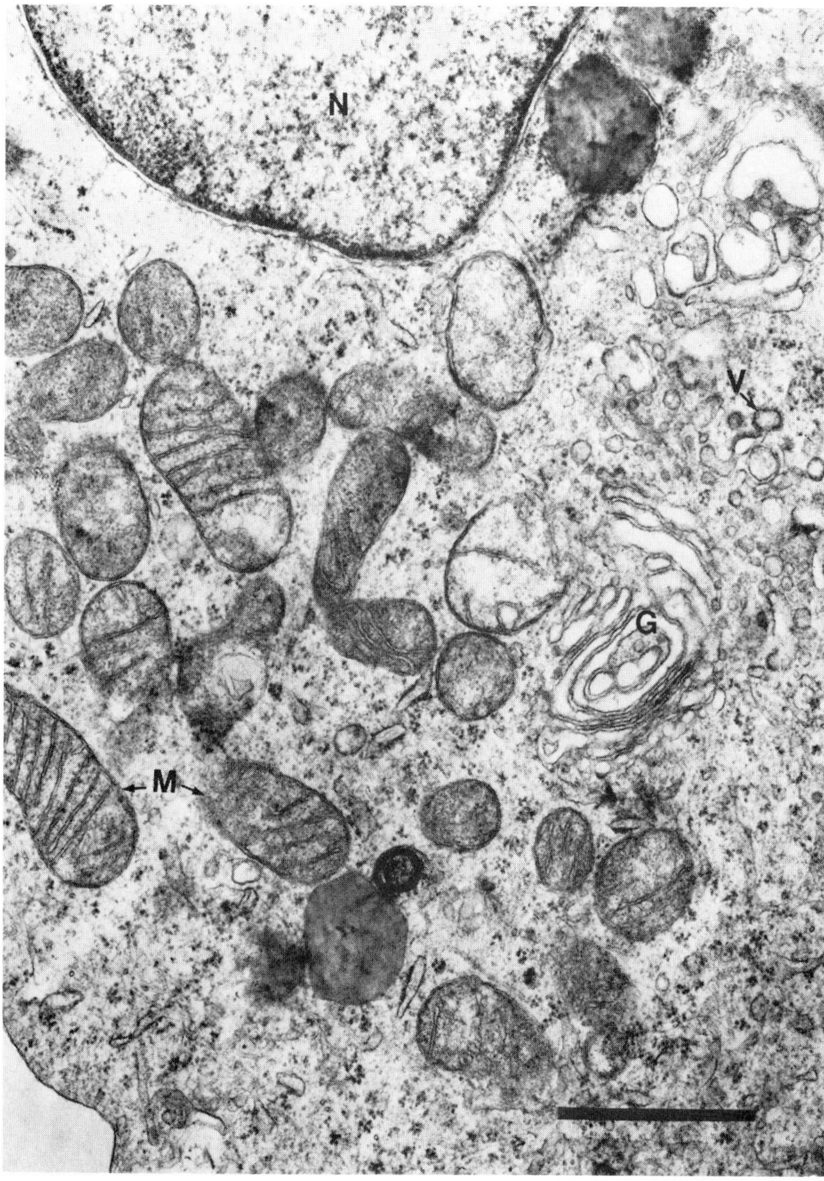

FIGURE 16. A persistently infected B-SC-1 cell from the exponential phase of growth. The cytoplasm contains numerous shelf-like mitochondria and vesicles (V). A well-formed Golgi body (G) can also be seen. (N: nucleus.) Stained with uranyl acetate and lead citrate. Bar represents 1 μm. (From Kiernan, R., unpublished observations.)

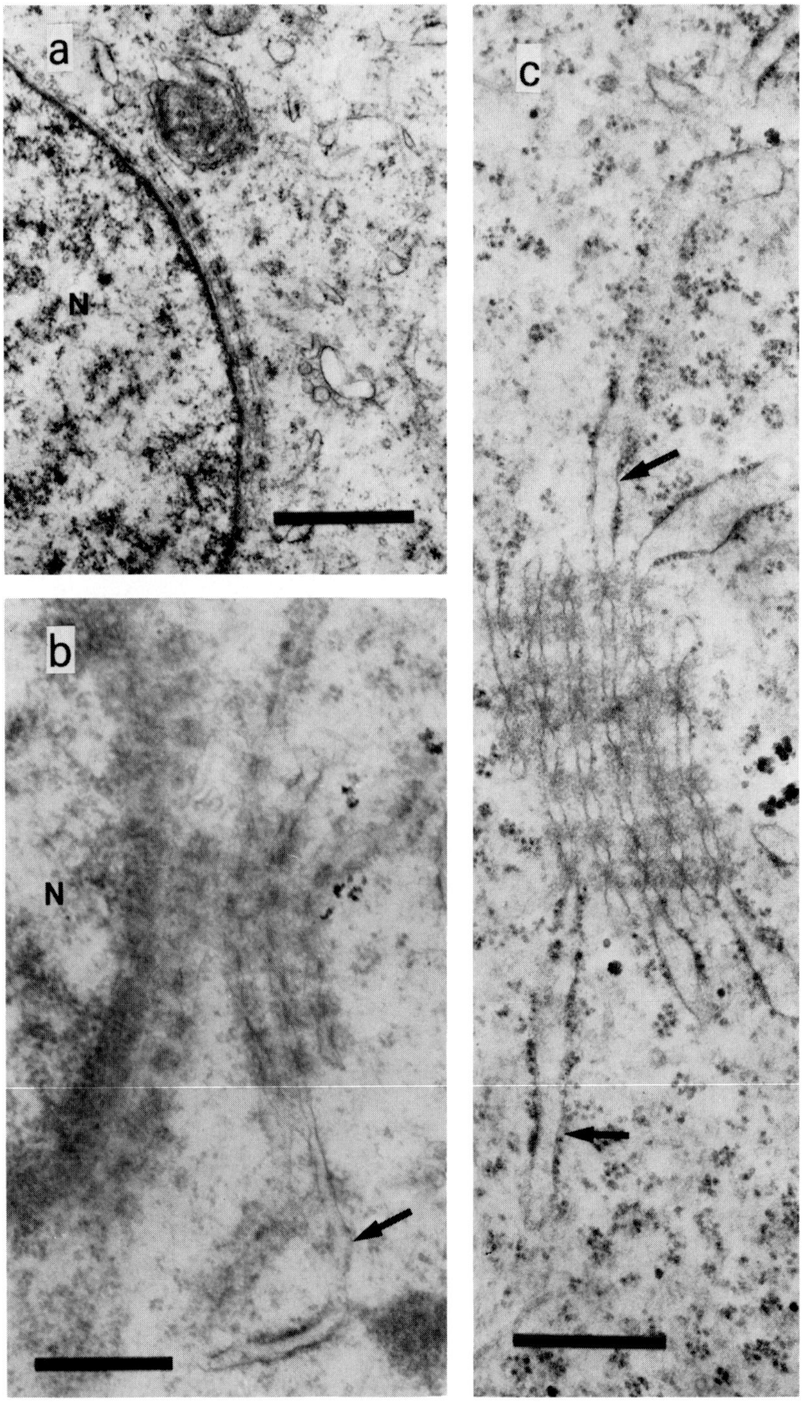

FIGURE 17. Thin-section EMs demonstrating the presence of annulate lamellae in infected cells. (a) Demonstrates a single annulate lamella adjacent to the nucleus (N) from an 8-day culture. Bar equals 1 μm; (b) shows stacks of annulate lamellae apparently peeling off the nucleus (N) in an 8-day culture. The arrow points to apparent continuity between these structures and the endoplasmic reticulum. Bar equals 0.3 μm; (c) demonstrates stacks of annulate lamellae without obvious association with the nucleus in a 29-day culture. The arrow points to apparent continuity with the endoplasmic reticulum. Bar equals 0.5 μm. (From Kiernan, R., Marshall, J. A., Coulepis, A. G., Anderson, D. A., Gust, I. D., *Arch. Virol.*, in press. With permission.)

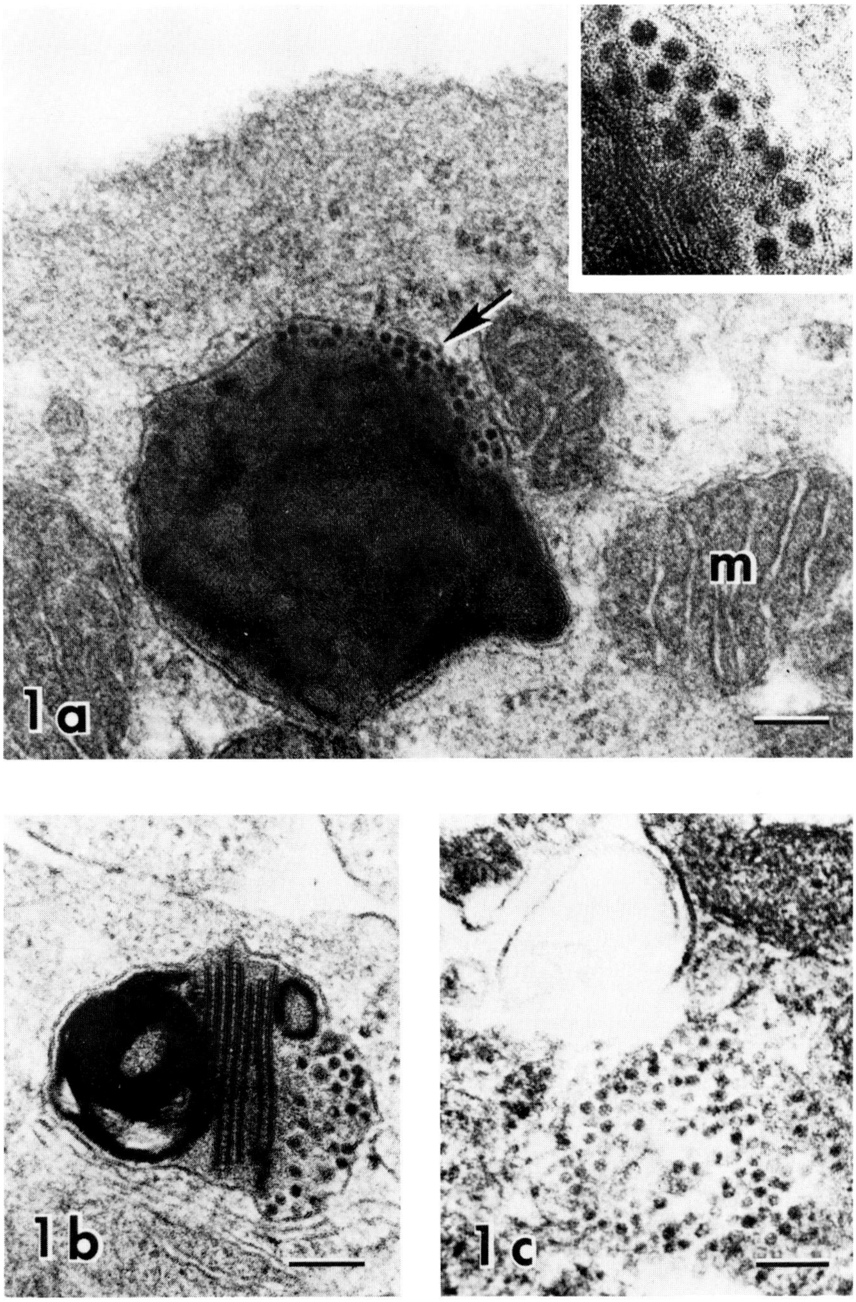

FIGURE 18. Thin-section EMs of HAV strain HM175 in B-SC-1 cells held for 6 weeks in culture. M = mitochondria; N = nucleus, bar = 100 nm. (a) Cytoplasmic vesicle containing virus-like particles (17 to 20 nm) and myelin-like membranes. Insert shows particles at higher magnification); (b) cytoplasmic vesicle containing virus-like particles and trilaminar membranes apparently originating from the myelin-like figures; (c) virus-like particles found free in the cytoplasm near a ruptured vesicle. (From Asher, L. S., Binn, L. N., Marchwicki, R. H., *J. Virol. Methods,* in press. With permission.)

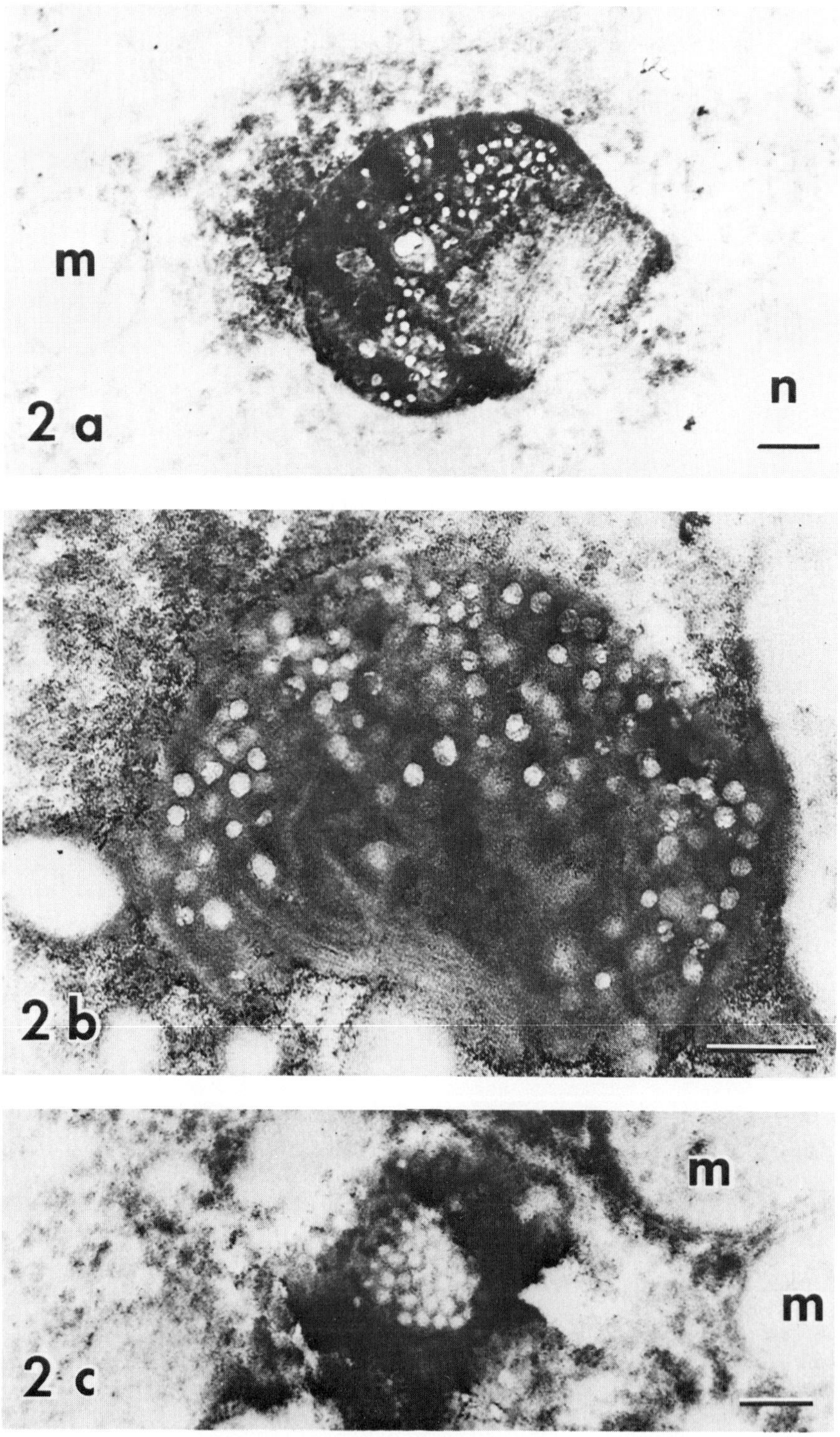

FIGURE 19. Indirect immunoperoxidase staining of B-SC-1 cells infected with HAV strain HM175. (a) A densely strained vesicle in the cytoplasm containing electron-lucent virus-like particles; (b) a higher magnification of one of the stained vesicles, showing virus-like particles measuring 20 to 22 nm in diameter; (c) aggregate of virus-like particles, measuring 24 nm in diameter. Bar = 100 nm. (From Asher, L. S., Binn, L. N., Marchwicki, R. H., *J. Virol. Methods,* in press. With permission.)

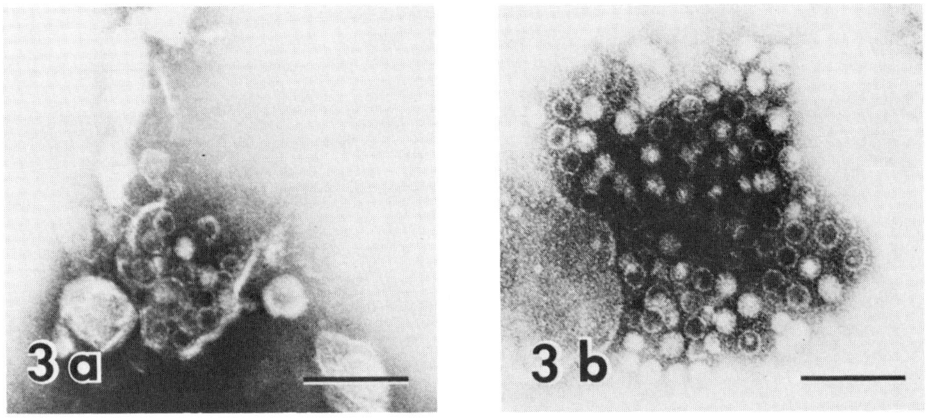

FIGURE 20. Negative-stained EM of (a) clusters of HAV particles in a membrane-enclosed sack detected in the supernatant fluid of infected cell cultures; (b) an immune aggregate of particles detected in the supernatant fluid, showing both full and empty particles measuring 24 to 27 nm in diameter. Bar = 100 nm. (From Asher, L. S., Binn, L. N., Marchwicki, R. H., *J. Virol. Methods*, in press. With permission.)

REFERENCES

1. **Mathieson, L. R., Feinstone, S. M., Purcell, R. H., and Wagner, J. A.,** Detection of hepatitis A antigen by immunofluorescence, *Infect. Immun.*, 18, 524, 1977.
2. **Provost, P. J. and Hilleman, M. R.,** Propagation of human hepatitis A virus in cell culture in vitro, *Proc. Soc. Exp. Biol. Med.*, 160, 213, 1979.
3. **Frosner, G. G., Deinhardt, F., Scheid, R., Gauss-Muller, V., Holmes, N., Messelberger, V., Siegl, G., and Alexander, J. J.,** Propagation of human hepatitis A virus in a hepatoma cell line, *Infection*, 6, 303, 1979.
4. **Balayan, M., Andzhaparisze, A. G., Tol' Skaia, E. A., and Kolesnikova, M. S.,** Possibility of reproducing hepatitis A virus infection in cell systems, *Vopr. Virus.*, 6, 675, 1979.
5. **Flehmig, B.,** Hepatitis A-virus in cell culture. I. Propagation of different hepatitis A-virus isolates in a fetal rhesus monkey kidney cell line (FRhK-4), *Med. Microbiol. Immunol.*, 168, 239, 1980.
6. **Flehmig, B.,** Hepatitis A virus in cell culture. II. Growth characteristics of hepatitis A virus in FRhK-4/R cells, *Med. Microbiol. Immunol.*, 170, 73, 1981.
7. **Daemer, R. J., Feinstone, S. M., Gust, I. D., and Purcell, R. H.,** Propagation of human hepatitis A virus in African green monkey kidney cell culture: primary isolation and serial passage, *Infect. Immunol.*, 32, 388, 1981.
8. **Kojima, H., Shibayama, T., Sato, A., Suzuki, S., Ichida, F., and Hamada, C.,** Propagation of human hepatitis A virus in conventional cell lines, *J. Med. Virol.*, 7, 273, 1981.
9. **Gauss-Muller, V., Frosner, G. G., and Deinhardt, F.,** Propagation of hepatitis A virus in human embryo fibroblasts, *J. Med. Virol.*, 7, 233, 1981.
10. **Fleming, B., Vallbracht, A., and Wurster, G.,** Hepatitis A virus in cell culture. III. Propagation of hepatitis A virus in human embryo kidney and human embryo fibroblasts, *Med. Microbiol. Immunol.*, 170, 83, 1981.
11. **Binn, L. N., Lemon, S. M., Marchwicki, R. H., Redfield, R. R., Gates, N. L., and Bancroft, W. H.,** Primary isolation and serial passage of hepatitis A virus strains in primate cell cultures, *J. Clin. Microsc.*, 20, 28, 1984.
12. **Lemon, S. M., Binn, L. N., and Marchwicki, R. H.,** Radioimmunofocus assay for quantitation of hepatitis A virus in cell culture, *J. Clin. Microsc.*, 17, 823, 1983.
13. **Lemon, S. M. and Binn, L. N.,** Antigenic relatedness of two strains of hepatitis A virus, determined by cross-neutralization, *Infect. Immunol.*, 42, 418, 1983.
14. **Lemon, S. M. and Binn, L. N.,** Serum neutralizing antibody response to hepatitis A virus, *J. Infect. Dis.*, 148, 1033, 1983.

15. **Venuti, A., Di Russo, C., del Grosso, N., Patti, A.-M., Ruggeri, F., De Stasio, P. R., Martiniello, M. G., Pagnotti, P., Degener, A. M., Midulla, M., Pana, A., and Perez-Bercoff, R.,** Isolation and molecular cloning of a fast growing strain of human hepatitis A virus from its double stranded replicative form, *J. Virol.*, 56, 579, 1985.
16. **Anderson, D. S., Locarnini, S. A., Ross, B. C., Coulepis, A. G., Anderson, B. N., and Gust, I. D.,** Single-cycle growth kinetics of hepatitis A virus in B-SC-1 cells, in *Positive Stranded RNA viruses*, Rueckart, R., Ed., Alan R. Liss, New York, in press.
17. **Anderson, D. S.,** Cytopathology, plaque assay and heat inactivation of hepatitis A virus strain HM175, *J. Med. Virol.*, in press.
18. **Cromeans, T., Sobsey, M. D., and Fields, H. A.,** Development of a plaque assay for a cytopathic, rapidly replicating isolate of hepatitis A virus, submitted.
19. **Shen, W., Lian-Ci, J., Hui-Xun, S., Xi-Tan, Z., Qing-Hai, M., and Xue-Guang, L.,** Strain of hepatitis A virus causing cytopathic effects isolated in A549 cell line, *Chin. Med. J.*, 99, 387, 1986.
20. **Locarnini, S. A., Coulepis, A. G., Westaway, A. G., and Gust, I. D.,** Restricted replication of human hepatitis A virus in cell culture: intracellular biochemical studies, *J. Virol.*, 37, 216, 1981.
21. **Anderson, D. S., Locarnini, S. A., Coulepis, A. G., and Gust, I. D.,** Restrictive events in the replication of hepatitis A virus in vitro, *Intervirology*, 24, 26, 1985.
22. **Wheeler, C., Fields, H. A., Schable, C. A., Meinke, W. J., and Maynard, J. E.,** Absorption and growth characteristics of hepatitis A virus strain HAS-15 propagated in fetal rhesus monkey kidney cells (FRhK-4), *J. Clin. Microsc.*, 23, 434, 1986.
23. **Kiernan, R., Marshall, J. A., Coulepis, A. G., Anderson, D. A., and Gust, I. D.,** Cellular changes associated with persistent hepatitis A infection in vitro, *Arch. Virol.*, in press.
24. **Butterworth, B. E., Shimshik, E. J., and Yin, F. H.,** Association of polioviral RNA complex with phospholipid membranes, *J. Virol.*, 19, 457, 1976.
25. **Bienz, K., Egger, D., Rasser, Y., and Bossart, W.,** Kinetics and location of poliovirus macromolecular synthesis in correlation to virus-induced cytopathology, *Virology*, 100, 390, 1980.
26. **Asher, L. S., Binn, L. N., and Marchwicki, R. H.,** Demonstration of hepatitis A virus in cell culture by electron microscopy with immunoperoxidase staining, *J. Virol. Methods*, in press.
27. **Feinstone, S. M.,** unpublished observations.

Chapter 7

MOLECULAR BIOLOGY

I. INTRODUCTION

The picornaviridae comprise four genera — enteroviruses, cardioviruses, rhinoviruses, and aphthoviruses — which have many similarities in structure and genome organization.[1] While hepatitis A has proved difficult to characterize (because of its relatively recent identification, restricted host range, and limited growth in cell culture), it is clearly a picornavirus and has been finally classified as an enterovirus.[2] although it has some distinguishing characteristics from other members of this genus (see Chapter 3).

Although HAV is not as readily manipulated in the laboratory as other picornaviruses, modern molecular biological techniques have provided insight into the structure and organization of the virus and allowed detailed comparisons to be made with other members of the group.

II. STRUCTURE AND ORGANIZATION OF PICORNAVIRUSES

All picornaviruses which have been studied to date have the same general structure and replicative strategy, and HAV appears to share these characteristics.[3] Molecular cloning techniques have provided new information on the organization of the picornavirus genome and increased our understanding of the function of many of the nonstructural proteins.[4] At least one member of each of the picornavirus genera has been molecularly cloned as complementary DNA (cDNA), and the cDNA has been sequenced. Comparisons between the sequences of several picornaviruses reveal an overall similarity of genome structure but major variations in the nucleotide sequence. In addition, new details of picornavirus organization have been provided by solving the X-ray crystallographic structures of poliovirus, rhinovirus, and mengovirus at 2.9 Å resolution.[5-7]

A. General Genome Structure

All picornaviruses have a single-stranded RNA genome of 7000 to 8000 nucleotides in length (see Figure 1) which is plus (or message) sense and infectious in its own right. The viral RNA acts both as a message for translation of new viral protein and as the template for replication of negative-strand RNA, which in turn serves as the template for replication of virion RNA (see Figure 2). Neither the viral RNA nor the viral messenger molecules are capped at their 5′ end as would be typical for eukaryotic messenger RNA (mRNA). Instead, a small virus-coded protein referred to as VPg (virion protein, genome) is covalently attached from the phenolic hydroxyl group of a tyrosine in the VPg to the terminal — pUpUp of the RNA.[8] A noncoding region (of variable length in different picornaviruses) occupies 500 to 800 bases at the 5′ end of the genome, before the site of initiation of translation. Cardioviruses and aphthoviruses have a polycytidylic acid [poly(C)] tract of about 100 to 250 bases within the 5′ noncoding region.[9] Secondary structural characteristics of the virion RNA in this 5′ noncoding region may be important for RNA replication. The coding regions begins with an AUG codon, which initiates translation of the viral polyprotein; translation is terminated by a stop codon or codons located within 100 bases of the 3′ end. There is a polyadenylic acid [poly(A)] tract at the 3′ end of the virion RNA which is viral specified, that is it, is not added to the RNA by a cellular poly(A) addition enzyme.[10]

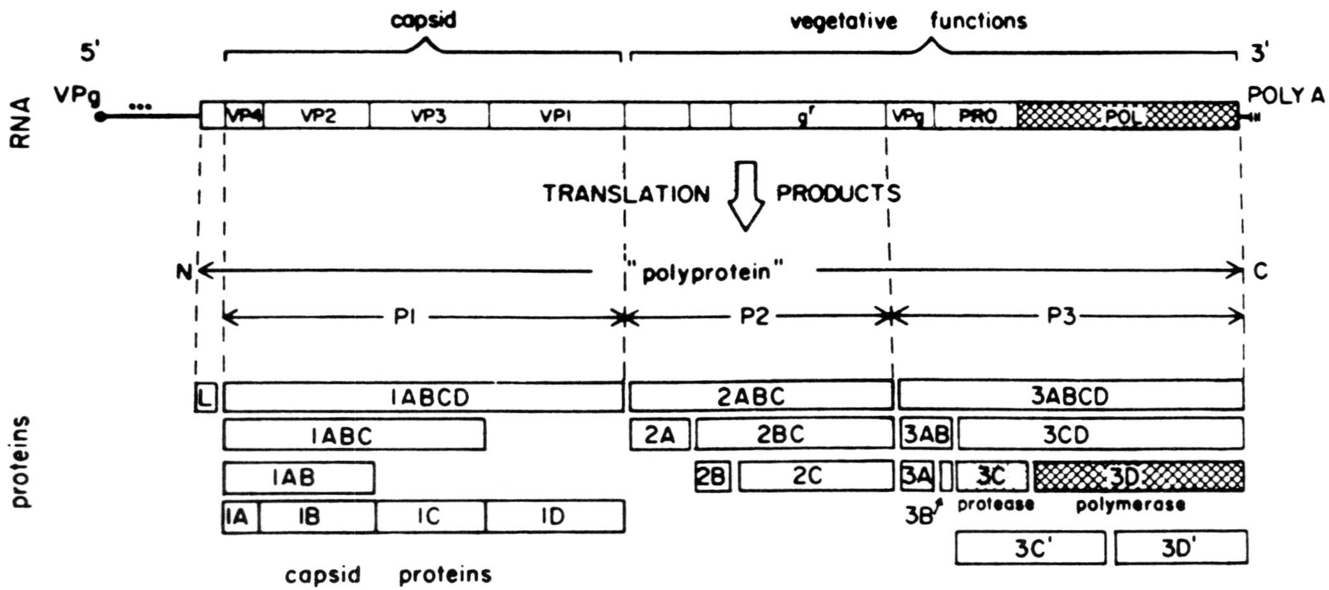

FIGURE 1. Genome organization and translation products of picornaviruses. The three "dots" over the 5′ noncoding region indicate the area of poly(C) found in cardio- and aphthoviruses. "L" designates the leader protein also found in cardio- and aphthoviruses. Protein 2A is now thought to be a second protease. (From Rueckert, R. R., *Virology,* Fields, B. N., Ed., Raven Press, New York, 1985, 705. With permission.)

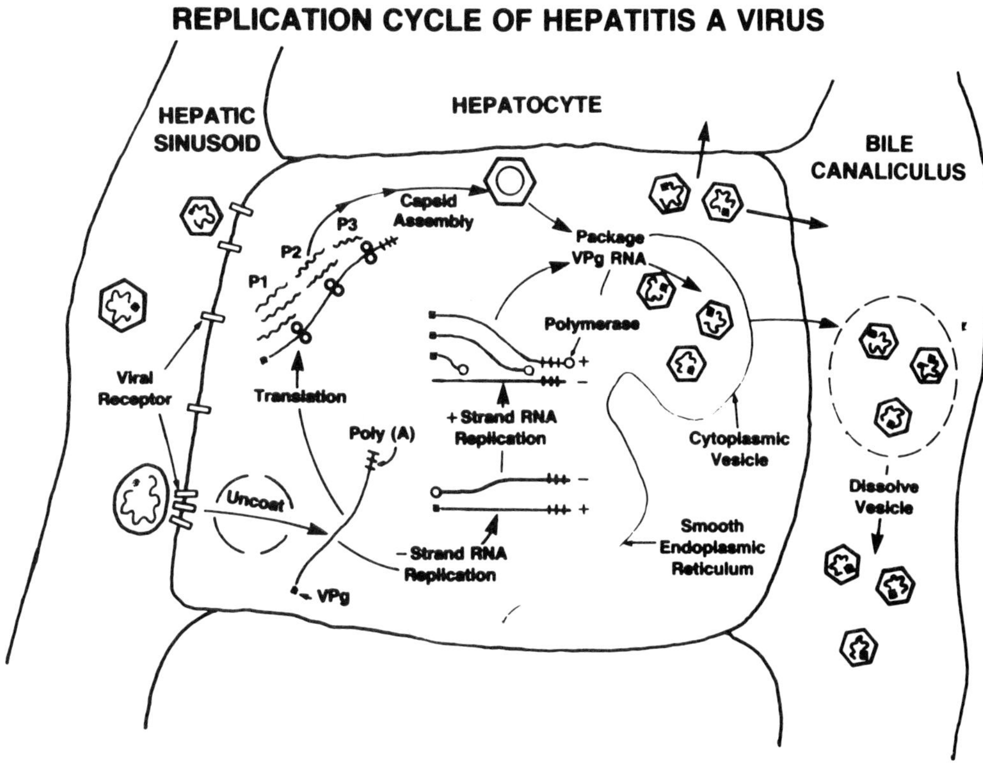

FIGURE 2. Proposed replication cycle of HAV. The virus presumably reaches the hepatocyte in the blood via the hepatic sinusoids and binds to specific receptors on the cell surface. After the virus enters the cell, VP4 is lost from the capsid and uncoating occurs. The viral genome serves as both a messenger molecule for translation of viral peptides and a template for RNA replication. Negative strands of virion RNA are transcribed from the viral RNA and new virion RNA and messenger RNA are made from this template. This process requires the viral polymerase (protein 3D) and probably protein 2C as well as host factors. Viral replication occurs in association with cellular membranes and intracellular HAV particles are usually seen within cytoplasmic vesicles. The virion RNA is packaged which probably also results in the final maturation cleavage of VP0 into VP2 and VP4. The virus may be released from the cell inside a vesicle which would probably be dissolved by the action of the bile. Virus may also spread directly to neighboring hepatocytes or be released back into the blood. In general the hepatocytes are not lysed by infection with HAV. (Adapted from Rueckert, R. R., *Virology,* Fields, B. N., Ed., Raven Press, New York, 1985, 705. With permission.)

B. Genome Structure of HAV

1. Biochemical and Biophysical Studies

HAV has the morphologic characteristics of a picornavirus and its structural proteins have similar MWs to those of other picornaviruses[10-13] (see Figure 9, Chapter 3). Evidence that the genome of HAV consisted of RNA was first provided by Provost[14] who showed that purified HAV stained an orange-red color with acridine orange (a characteristic of single-stranded DNA or RNA) and that the virus could be partially inactivated by treatment with RNase. More definitive data was provided by Siegl and Frosner[15] who showed that the nucleic acid released from purified HAV particles by treatment with 4 *M* urea and 90% formamide, appeared to be single-stranded when examined by EM and could be hydrolyzed by alkali treatment, a characteristic of RNA but not DNA. By measuring the length of the molecules by EM, the MW of the HAV RNA was estimated to be about $1.9 \times 10.^6$ Coulepis et al.[16] showed that the RNA extracted from purified HAV particles sedimented at 35 S, banded at a density of 1.64 g/cm^3 in CsCl and bound to a polyuridylic acid-Sepharose column and suggested that the HAV genome consisted of single-stranded RNA and contained poly(A).

Similar results were obtained by Moritsugu[17] and his colleagues who showed that the genome could be transcribed into DNA using reverse transcriptase and oligo deoxythymidine (oligo dT) as a primer. Subsequently Locarnini et al.[18] produced evidence that naked viral RNA is infectious, which proves that it is plus sense.

2. Molecular Cloning

Since the picornavirus genome resembles eukaryotic messenger RNA, techniques developed for transcribing mRNAs into cDNA have been useful for molecular cloning. Racaniello et al.[4] cloned the poliovirus (PV) genome by synthesizing cDNA with avian myeloblastosis virus reverse transcriptase, using oligo dT to prime the reaction off the poly(A) tract at the 3′ end of the viral RNA. The same technique has also been used to produce cDNA clones of the HAV genome.[19-25] Ticehurst et al.[19] at the NIH in collaboration with Racaniello and Baltimore at MIT, used wild-type HAV strain HM175 amplified by two passages in marmosets as the starting material for cDNA cloning studies. HAV was purified from the liver of acutely infected marmosets, the viral RNA extracted, and cDNA synthesized and cloned into a plasmid vector using an *E. coli* host. In addition to this strain derived from marmoset liver, Von der Helm[20] and colleagues in Munich have cloned cDNA from virus purified directly from human feces while Najarian et al.,[21] Venuti et al.,[22] Ross et al.,[23] Ovchinnikov et al.,[24] and Linemeyer et al.[25] used strains which have been isolated and passaged in cell culture.

As it proved to be very difficult to synthesize and clone cDNA representing the entire HAV genome, Ticehurst et al.[19] studied a series of clones. By restriction endonuclease mapping, overlapping clones representing all but a few bases at the 5′ end of the genome were identified and this region was eventually cloned by primer extension off the virion RNA.[26]

3. Organization

The initial cloning and restriction endonuclease mapping by Ticehurst et al.[19] revealed that the HAV genome was approximately 7500 nucleotides in length. The 3′ end of the genome was sequenced and revealed a single open-reading frame beginning at the start of the sequence and terminating with two closely spaced nonsense codons followed by 63 bases that do not contain an open-reading frame, preceded by an initiation codon (ATG). The sequence ended with the expected poly (A) tract which was presumed to represent the 3′ end of the genome.

More complete sequence data by Baroudy et al.[26] followed by complete sequences by Najarian et al.[21] and Cohen et al.[27] have allowed for extensive analysis of the HAV genome[23-25] (see Figure 3). Several other laboratories have published partial sequences. Both the wild-type HAV strain HM175 obtained from Australia and a tissue-culture adapted strain from California which have been cloned and sequenced contain 7478 bases excluding the 3′ poly (A) tract. There is a 5′ noncoding region of 734 bases, a coding region of 6681 and a short 3′ noncoding region of 63 bases followed by a poly (A) tract.

a. 5′ Noncoding Region

Just as in other picorna viruses, the sequence of the HAV genome begins at the 5′ end with *UU*. The longest potential open-reading frame initiated by an *AUG*, within the first 734 bases begins at base 673 and ends at base 732 (a coding capacity of 20 amino acids) just before the beginning of the translation initiation site at base 735. Cardioviruses and aphthoviruses begin the coding region with leader proteins of more than 100 amino acids whose function is unknown. Enteroviruses and rhinovirus do not have this leader protein but translation begins with the capsid protein-coding region. It is not known if this short open-reading frame found in HAV that is located just before the capsid protein-coding

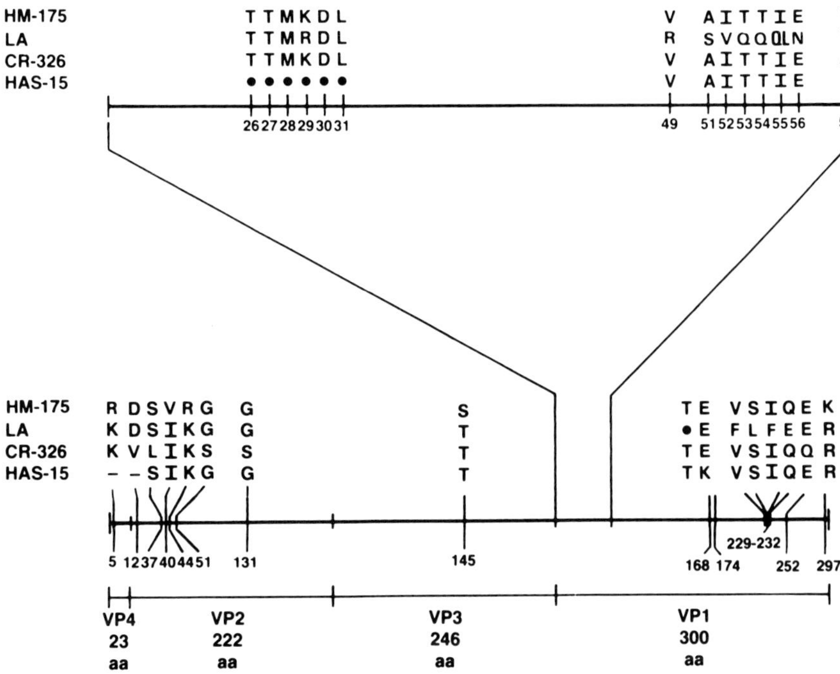

FIGURE 3. Location of amino acid differences between capsid proteins of four strains of HAV — HM175, 2A, CR-326, and HAS-15. Letters indicate amino acids according to the single-letter abbreviation code. Numbers refer to the position of the amino acids within each capsid protein. Upper case letters indicate amino acids of consensus sequence. Amino acid not determined are indicated by — , and deleted amino acids are indicated by (•). (From Cohen, J. I., Ticehurst, J. T., Purcell, R. H., Buckler-White, A., and Baroudy, B. M., *J. Virol.*, 61, 50, 1987. With permission.)

sequences is translated or if it is in any way analogous to the leader protein of aphthoviruses and cardioviruses. The long open-reading frame that codes for the viral polyprotein has a separate initiation codon(s) (see below) and, therefore, this short open-reading frame upstream from the likely initiation site is not precisely analogous to the longer leader proteins.

The 5′ noncoding region of picornaviruses has been demonstrated to have the potential for a high degree of secondary structure with highly stable stems and loops.[28] Cohen et al.[27] have shown that the 5′ end of the HAV genome can be folded into a similar stable stem and loop (see Figure 4) with a free-energy heat calculated to be −32.4 k cal/mol. While it is not certain that these structures exist in nature,[29] the potential for such structures has been found in every picornavirus which has been studied and is assumed to have biological significance.[30]

Both cardioviruses and aphthoviruses have stretches of poly(C) within the 5′ noncoding regions of their genomes. While the function of these poly(C) stretches is not known, the length of these poly(C) regions may be important in determining certain characteristics such as virulence, and affect the efficiency of transcription or even translation. While no poly(C) stretches have been identified in the 5′ region of the HAV genome, two pyrimidine-rich regions (bases 99 to 138, and 712 to 720) exist which could have a similar function. Following the initiation codon at base 735 there is a single open-reading frame stretching 6681 bases to a termination codon. There are 63 noncoding bases at the 3′ end of the genome followed by a poly (A) tract.

b. Coding Region

Following the 5′ noncoding region, a single open-reading frame is initiated by either of

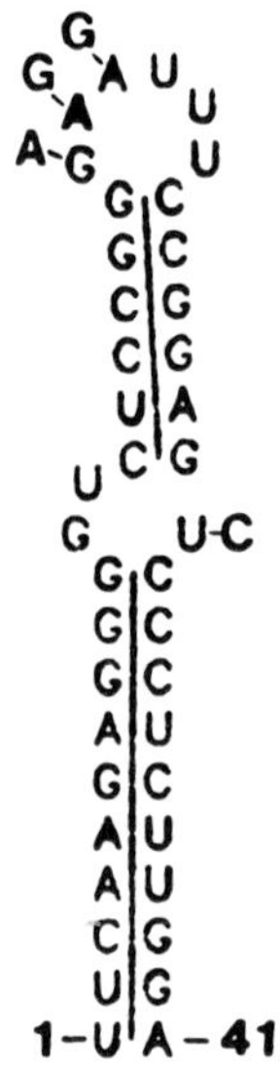

FIGURE 4. Predicted secondary structure of RNA at the 5′ terminus of HAV HM175. The bases surrounding the primary structure are those of the HAV strain sequenced by Najarian. Similar structures have been predicted for the 5′ terminus of other picornaviruses. (From Cohen, J. I., Ticehurst, J. T., Purcell, R. H., Buckler-White, A., and Baroudy, B. M., *J. Virol.*,. With permission.)

two methionine codons (AUG) at position 735-7 and 741-3 (*A*UA AUG *A*AC AUG *T*). Both of these initiation codons are flanked by sequences[31] (italicized) that are recognized preferentially by eukaryotic ribosomes for initiation of translation. It is not known which of the two AUG codons is the active initiation site for HAV protein translation. It is possible that both are active, producing VP4 molecules differing in length by two amino acids, but it will require amino-acid sequencing of VP4 in order to determine the precise translation initiation point.

The coding region of picornaviruses can be divided into three parts based on the initial points of cleavage of the translated polyprotein. The three regions are termed P1, P2, and P3, and the individual peptides that are ultimately cleaved from the translation products of these regions are referred to as 1A, 1B, etc.; 2A, 2B, etc.; 3A, 3B, etc., in order of coding from the 5′ to the 3′ end of the genome.[32] Peptides that have a known function may be further described such as 3D polymerase, which is the fourth peptide coded in the P3 region and in many picornaviruses has been shown to be an RNA polymerase.

The P1 region codes for the four picornavirus capsid proteins and although these peptides are properly referred to as 1A, 1B, 1C, and 1D they are usually referred to as VP1, VP2, VP3, and VP4 — which are the capsid proteins in descending order of molecular size. The actual coding sequence of the capsid proteins in the picornavirus genome is generally 1A = VP4, 1B = VP2, 1C = VP3, and 1D = VP1. Since there is little direct peptide sequence data available for HAV, the precise amino-acid sequence of the structural peptides has been derived by sequencing the genome.

The coding region is terminated by 2-in. frame-stop codons, UGA at position 7417, and UAA at position 7426.

c. 3′ Noncoding Region

There is an untranslated region at the 3′ end of the genome that runs 63 bases from the first termination codon to the start of the poly (A) region. There are no potential open-reading frames in this region. The length of the virion 3′ poly (A) has not been determined because different clones from the 3′ end may have different lengths of poly (A) depending on where in the poly (A) region on the oligo (T) primer initiated the reverse transcription of the virion RNA into cDNA for molecular cloning.

4. Cleavage Sites

It was hoped that typical dipeptide cleavage sites that had been determined for PV and other picornaviruses would be present in appropriate locations in HAV. However, most of these sites were not present in the amino acid sequence predicted from the genomic sequence (see Table 1). However, by aligning the predicted amino acid sequence of HAV with other picornaviruses and utilizing hydrophilicity-hydrophobicity profiles and other computer-assisted comparisons it has been possible to locate probable proteolytic cleavage sites and likely coding regions for the capsid proteins and most of the nonstructural peptides of HAV[27,33,34] (see Figure 5).

The known and predicted dipeptide cleavage sites of HAV are compared with the known cleavage sites of other picornaviruses in Table 1. The uniqueness of HAV is again apparent. The virus shares four of ten dipeptide cleavage sites with encephalomyocarditis virus (EMCV) but no more than two with any other picornavirus. Since it is now believed that all the cleavages are performed by the viral-encoded proteases, proteins 2A and 3C,[35] with the VP2/VP4 cleavage being autocatalytic,[36] it must be assumed that the HAV proteases have different specificities from those of the other picornaviruses.

In the capsid-coding region, analysis of the nucleotide and amino acid sequences leads to some very interesting predictions. In the strain of HAV they studied, Najarian et al.[21] suggested that there was a 100 amino acid shift in the position of VP1 so that the coding position of VP4 was not in the first part of the P1 region (protein 1A) but came after VP1 in what would be position 1D. Alignments made in other laboratories, however, predict the usual picornavirus coding sequence for HAV capsid proteins.[27,33,34] These analyses also predict that VP4 is a very small molecule made up of only 23 amino acids with a MW of approximately 2500 daltons, including the initial methionine. Considerably higher MWs (ranging between 7,000 and 14000 daltons) have been reported for VP4 in studies based on electrophoretic mobility of capsid proteins[12,13] obtained by disrupting preparations of purified virus. Using an antibody prepared against a synthetic peptide with the same sequence of 23 amino acid predicted for VP4, we were unable to identify a 2500-dalton protein by western blots but did detect a protein of about 27,000 daltons which presumably represents VP0 (VP4 plus VP2 prior to cleavage). Antibody directed against a peptide at the carboxyl terminus of VP2 also recognized this presumptive VP0 as well as a slightly smaller peptide, presumably VP2 (Feinstone et al. unpublished observations). These data suggest that VP0 is cleaved to VP2 and VP4; however, it is not known if VP4 remains in the mature capsid of HAV or is lost after cleavage. Using similar methodology, Wheeler et al.[37] demonstrated that VP0 is primarily associated with a slower sedimenting (70 to 80 S) fraction of HAV (presumably empty capsids) whereas VP2 is more often associated with the mature 156 S form of the virion. This suggests that the final maturation cleavage in picornaviruses, the cleavage of VP0 into VP2 and VP4, occurs in HAV. Arnold et al.[36] has shown that this step is autocatalytic and occurs at the time of insertion of the viral RNA which acts as the proton donor for the serine protease (see below).

The cleavage site between VP2 and VP3 (1B and 1C) was located by Linemeyer et al.[25] by direct amino acid sequencing and by Baroudy et al.[26] by sequence alignments. The carboxy terminus of VP2 is believed to be a glutamine coded by nucleotides 1467 to 1469

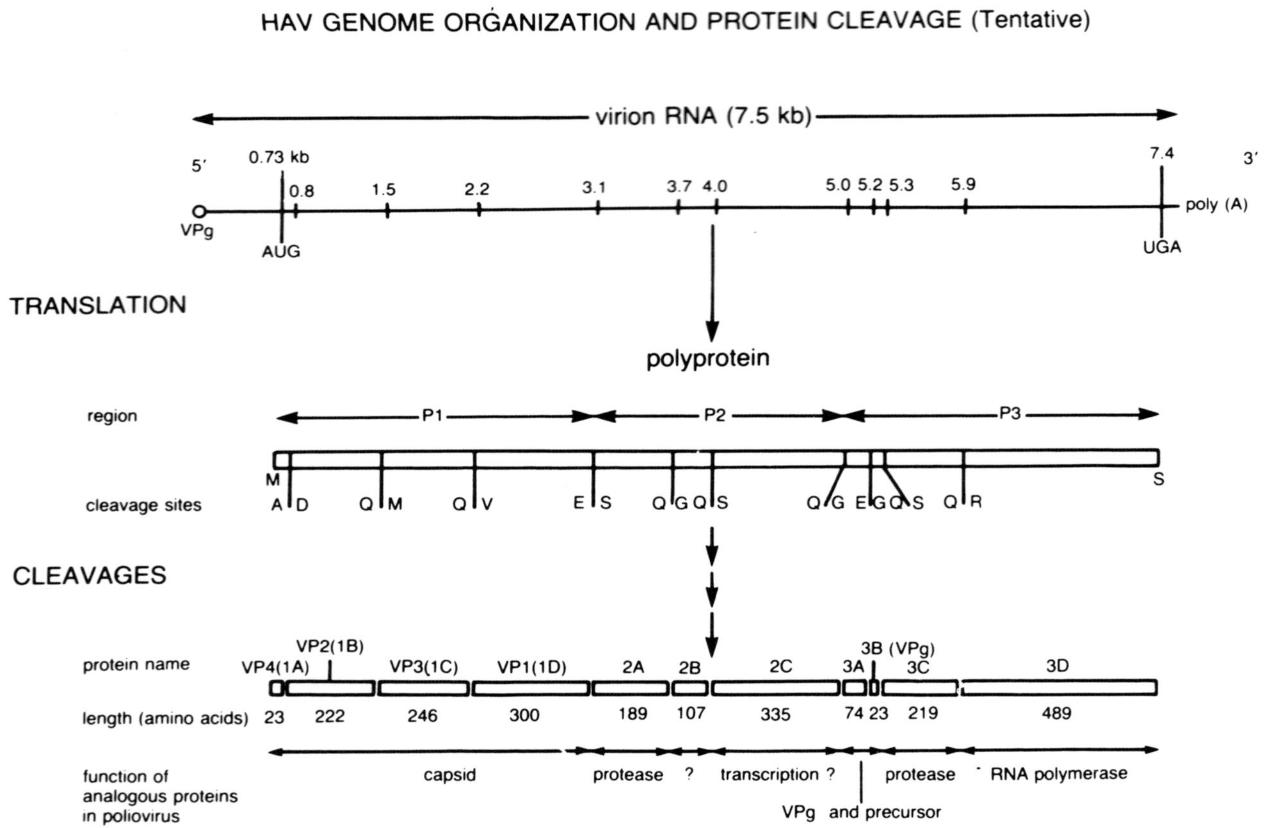

FIGURE 5. Genome organization, probable cleavage sites, and likely translation products of HAV. The dipeptide cleavage sites are designated by the single-letter amino-acid abbreviation code. (From Ticehurst, J. R., *Semin. Liver Dis.*, 6, 46, 1986. With permission.)

Table 1
KNOWN AND PROPOSED MATURATION CLEAVAGE SITES IN PICORNAVIRUS POLYPROTEINS

	Enterovirus		Rhinovirus	Cardiovirus	Aphthovirus
Site	HAV	PV1M	HRV14	EMCV	FMDV A12
L/4	—	—	—	Gln/Gly	Lys/Gly
4/2	Ala/Asp	Asn/Ser	Asn/Ser	Ala/Asp	Ala/Asp
2/3	Gln/Met	Gln/Gly	Gln/Gly	Gln/Ser	Val/Gly
3/1	Gln/Val	Gln/Gly	Glu/Gly	Gln/Gly	Gln/Thr
1/2A	Glu/Ser?	Tyr/Gly	Tyr/Gly	Tyr/Pro	Leu/Asn
2A/2B	Gln/Gly	Gln/Gly	Gln/Gly	Gln/Gly	Glu/Ser
2B/2C	Gln/Ser	Gln/Gly	Gln/Ala	Gln/Ser	Gln/Leu
2C/3A	Gln/Gly	Gln/Gly	Gln/Gly	Gln/Gly	Gln/Ile
3A/3B	Glu/Gly	Gln/Gly	Gln/Gly	Gln/Gly	Glu/Gly
3B/3C	Gln/Ser	Gln/Gly	Gln/Gly	Gln/Gly	Glu/Ser
3C/3D	Gln/Arg	Gln/Gly	Gln/Gly	Gln/Gly	Glu/Gly

Note: Courtesy of J. Ticehurst.

and the amino terminus of VP3, a methionine coded by nucleotides 1470 to 1472. This cleavage results in a VP2 of 222 amino acids and a MW of approximately 24.7 kdaltons.

The amino acid cleavage site between VP3 (1C) and VP1 (1D) has been precisely located by direct amino acid sequencing and alignment to nucleotide sequence. The cleavage occurs between a glutamine coded by nucleotides 2205 to 2207 and a valine coded by nucleotides 2208 to 2210. This results in a VP3 of 246 amino acids with a MW of approximately 27.3 kdaltons.

It has only been possible to locate the cleavage site between the carboxy terminus of VP1 (1D) and the amino terminus of protein 2A by analysis of predicted amino acid sequence. This cleavage is thought to occur between a glutamate and a serine coded by nucleotides 3106 to 3108 and 3109 to 3111. This cleavage results in a VP1 of 300 amino acids with a MW of approximately 33.2 kdaltons.

This analysis suggests that the VP3 peptide (1C) is larger than VP2 (1B), a fact which has recently been verified by western blot studies on HAV capsid proteins using antibody probes made against synthetic oligopeptides located within the putative coding regions.[37] Antibody to a 1D oligopeptide (VP1) recognized a 33-kdalton protein, antibody to a 1C (VP3) oligopeptide recognized a 29-kdalton protein, and antibody to a 1B (VP2) oligopeptide recognized a 27-kdalton peptide. While the MWs of these proteins are not identical to the MWs predicted for the capsid proteins, the relative sizes of VP1, VP3, and VP2 were as expected.

Most of the nonstructural proteins and cleavage sites of HAV have only been located by sequence analysis and comparisons with other better studied picornaviruses. However, VPg (protein 3B) has been identified experimentally.[38] A 23 amino acid protein-coding region similar to the VPg of other picornaviruses has been located in the HAV genome. This was verified experimentally by showing that antibody to a synthetic peptide corresponding to the ten C-terminal amino acids of the predicted 23 amino acid protein could be used to immunoprecipitate HAV viral RNA. In common with the VPgs of other picornaviruses, the predicted HAV VPg has a tyrosine residue at position 3 from the N-terminus which is the likely RNA binding site, and a similar hydrophobicity. However, the calculated isoelectric point for the HAV VPg was 7.15 whereas for PV1 and FMDV A12 the pIs are 11.4 and 10.02 respectively. Thus, again HAV seems to have unique properties compared to other picornaviruses. While this experiment does not precisely locate the termini of the VPg protein it is likely that the 23 amino acids represent the entire molecule because it has characteristics (size and hydrophobicity profile) similar to the VPgs of other picornaviruses.

5. Comparison with the Genomes of Other Picornaviruses

a. Other Strains of HAV

While all HAV isolates analyzed to date are closely related, some phenotypic variability has been noted. For instance, Siegl et al.[39] has found wide range in the ability of various strains to replicate in cell culture. Weitz and Siegl[40] have analyzed a number of HAV isolates by oligonucleotide mapping of the genomic RNAs and demonstrated changes in 10 to 40% of the large oligonucleotides corresponding to an estimated 1 to 4% variation in nucleotide sequence. Only minor differences were detected in strains isolated from human feces even after multiple passages in cell culture. By contrast, strains of HAV that had been passaged in marmoset monkeys prior to adaptation to cell culture showed changes in up to 75% of their large oligonucleotides or an estimated 10% sequence difference.

Cohen et al.[27] have compared the nucleotide sequence of the wild-type HM175 strain of HAV with a tissue-culture adapted strain of HAV sequenced by Najarian et al.[21] The HM175 strain which originated from an Australian patient with hepatitis[41] was passaged three times in marmosets to amplify it prior to cloning, and was shown to maintain virulence for both marmosets and chimpanzees. The strain of HAV cloned and sequenced by Najarian et al.[21] was isolated in Los Angeles (LA) and had been passaged multiple times in cell culture. Both strains had a genome length of 7478 nucleotides but HM175 strain had one more base in the 5′ noncoding region and one less base in the 3′ noncoding region than the LA strain. There was a total of 624 nucleotide differences between the two strains (91.7% identity). Although 589 changes were detected in the coding region, 511 were in the third codon position resulting in only 34 amino acid differences (98.5% identity). In the 5′ noncoding region the two strains were over 96% identical. Only 24 nucleotide and 12 amino acid changes were found between the wild-type HM175 and this strain after a further 35 passages in cell culture. Cohen[27] compared the sequences of four strains of HAV for which capsid region sequence had been reported (see Figure 4). He found the most diversity in the VP1 which could be expected because it is not only the longest capsid peptide, but also the one with the most exposure on the surface of the virion. There was a high degree of identity between the HM175 strain and LA in the region of their genome coding for nonstructural proteins. Changes were found only in proteins 2C, 3A, and 3D. However, the 3′ noncoding region exhibited the greatest diversity with only 88.9% of the nucleotides identical.

Lemon has isolated viruses from a naturally occurring epizootic of HAV in a colony of New World owl monkeys *(Aotus trivirgatus)* in Panama.[42] These viruses, termed PA21 and PA33, cause hepatitis in the monkeys and are morphologically, biophysically, and antigenically indistinguishable from HAV. Considerable differences between the human HAV strain (HM175) and the owl monkey isolates were revealed by molecular hybridization.[43] Several cDNA probes prepared from the genome of wild-type HM175 were reacted with RNA extracted from the PA21 and PA33 strains using nitrocellulose paper and a slot-blot apparatus. Probes representing the 5′ noncoding region of the HM175 genome hybridized to the PA21 virus nearly as well as to HM175. A probe representing the P3 region of HM175 hybridized to PA21 RNA less efficiently than to human isolates under standard stringency conditions (SSC) (5 × SSC, 50% formamide at 42°C). Under these conditions, however, probes representing the P1 and P2 region of HM175 failed to give a hybridization signal with the owl monkey strains. Weak hybridization could be detected under low stringency hybridization. Thus it appears the 5′ noncoding region and the P3 region of the genome is relatively well conserved but that there is significant diversity between the human and owl monkey isolates in the P1 (capsid) and P2 regions of the genome. Despite these differences, cross-neutralization, cross-protection, and monoclonal antibody studies indicated that the immunodominant antigens of the owl monkey and human viruses were closely related. It is not

Table 2
COMPARISON OF PICORNAVIRUS GENOMES

Virus	HAV	PV1	HRV14	EMCV	FMDV
RNA length nucleotides					
Total	7478	7440	7209	7843	8400
5′ Noncoding	733	742	625	844	1200
3′ Noncoding	63	71	47	126	96
Coding region	6681	6627	6537	6873	6996
G + C %	38	47	40	50	53
5′ noncoding poly (C) (polypyrimidine)	—	—	—	+	+
Leader protein	—	—	—	+	+
VPg (number coded)	1	1	1	1	3

known if PA21 and PA33 were derived from a human virus that was introduced to the monkey colony resulting in the epizootic or if they are related monkey-hepatitis viruses. Weitz and Siegl[40] have shown that rapid changes can occur in the genome of HAV when the virus is passaged in marmosets so that the differences between HM175 and PA21 and PA33 could be accounted for by serial passage of human HAV within the Aotus colony. On the other hand, there may be naturally occurring monkey hepatitis A-like viruses which may be related to human HAV. While antibody to HAV has been found in the serum of many primates (see Chapter 4), most of the samples have been obtained after the animals were in contact with humans in captivity. However, Burke[44] has proposed a ''sylvatic cycle'' of HAV based on data obtained from studies of cynomolgus monkeys in Malaysia. He found the prevalence of antibody in newly captured animals was age-dependent, suggesting transmission of HAV among these monkeys in the wild. It will be necessary to isolate, clone, and sequence a virus from jungle-caught animals to determine how closely related they are to human HAV. Such a study could have important implications concerning the ecology of HAV.

b. Other Picornaviruses

Sequencing data has revealed that the genome of HAV has similar characteristics to other picornaviruses. However, at the level of nucleotide sequence and amino acid sequence HAV is more divergent from the picornaviruses that have been sequenced than any of these viruses are from one another. It has been shown that the proportion of guanidine plus cytidine in the HAV genome is only 38%, the lowest among picornaviruses (see Table 2). Cohen et al.[27] have shown that the only areas of significant homology between HAV and other picornaviruses occur in proteins 2C and 3D which are involved in viral RNA transcription (see Figure 6). These proteins not only exhibit a high degree of homology within the picornavirus family but have some homology with proteins from other virus groups which perform a similar function. Within the picornaviridae the virus which has the closest resemblance to HAV is the EMCV of mice in which there is a 28% amino acid identity in protein 2C and 25% with 3C. The function of 2C is not known but it seems to be related to RNA replication and it also contains the determinant for sensitivity or resistance to the drug guanidine.[45] Like EMCV and foot-and-mouth disease virus (FMDV), which are also resistant to guanidine,[46] HAV has a glycine at position 198 of protein 2C;[27] by contrast, PV and human rhinovirus, which are both sensitive to guanidine, have an asparagine at this position[45] (see Figure 6). Cohen and his colleagues have also shown that the active sites identified in other picornaviruses of the protease protein, 3C (a cysteine and a histidine) are conserved in HAV and that the active site in protein 3D, the viral polymerase, consisting of two aspartate residues surrounded by 12 hydrophobic amino acids, is also conserved in HAV.

Because of sequence homology in some parts of the genome it has been possible to make

```
3C Active Site and 3C/3D

EMCV  KGWCGSALLA DLGGSKKILGIHSAGSMGIAAASIVSQEMIRAVVNAFEPQGALER LPDG
HAV   PGMCGGALVSSNQSIQNAILGIHVAGGNSILVAKLVTQEMFQNIDKKIESQRIMK  VEFT
FMDV  AGYCGGAVLA KDGADTFIVGTHSAGGNGVGYCSCVSRSMLLKMKAHIDPEPHHEGLIVDT

3D Active Site                  2C:Region of Guanidine Resistance Marker

EMCV  VKVLSYGDDLLVAT            EMCV  DGYENQFAAIMDDLGQNPDGSDFTTFCQMVS
HAV   LKILCYGDDVLIVF            HAV   DGYSGQLVCIIDDIGQNTTDEDWSDFCQLVS
RV2   LKIIAYGDDVIFSY            PV1   DGYKQQGVVIMDDLNQNPDGADMKLFCQMVS
```

FIGURE 6. Alignments of amino acids of selected peptides of HAV and other picornaviruses. Amino acids are designated by single-letter abbreviations; viral peptides are designated by standard nomenclature (see text); amino acids conserved with HAV are indicated by colons; cleavage sites are indicated by arrowheads; putative active sites and the guanidine resistance sites are indicated by open circles. (From Cohen, J. I., Ticehurst, J. T., Purcell, R. H., Buckler-White, A., and Baroudy, B. M., *J. Virol.*, 61, 50, 1987. With permission.)

cDNA probes from PV that will hybridize to the other enteroviruses.[47] Similar probes made from HAV will not hybridize to other picornaviruses even under very low-stringency conditions, again illustrating the highly divergent character of HAV.[64]

C. General Physical Structure

EM has demonstrated that HAV has the same morphological features as other picornaviruses; studies by Siegl[39] using Markham rotation suggest, however, that the virus has only 32 capsomeres rather than the usual 60. However, since all other aspects of HAV seem to closely resemble other picornaviruses it is assumed that the model of 60 capsomeres also holds for HAV. The resolution of this discrepancy awaits further analysis.

1. X-Ray Crystallography

The crystallographic structure of three important picornaviruses have recently been solved. Rossmann and colleagues have published the structure of rhinovirus 14[6], Hogle and colleagues reported the structure of PV type 1[5], and recently Luo has reported the structure of mengovirus.[7] In these elegant experiments crystals of purified virus were examined by X-ray diffraction and the three-dimensional structure of the viruses was determined at a resolution of 2.9 to 3.0 Å. These studies plus data on the amino acid sequences of the capsid proteins (which have been derived from nucleotide sequences) have revealed several important features of picornavirus structure. Most of the four capsid proteins of PV could be located by this technique, however, the first 20 NH_2-terminal amino acids of VP1, seven amino acid of VP2, and 12 amino acid of VP4 which are all in the interior of the particle, and the four COOH-terminal amino acids of VP3 which are believed to be on the exterior of the virion were not located. No densities could be assigned to the viral RNA or to the VPg. These molecules, as well as those portions of the capsid proteins that could not be precisely located, are presumed to be spacially disordered relative to the icosahedrally symmetric protein shell.

The dimensions of PV were determined by crystallography; the outer diameter of the capsid was approximately 28 nm and the thickness of the capsid about 3 nm. Two sets of surface projections, the largest of which at the 5-fold axis is 2.5 nm, gives a maximum particle diameter of 33 nm.

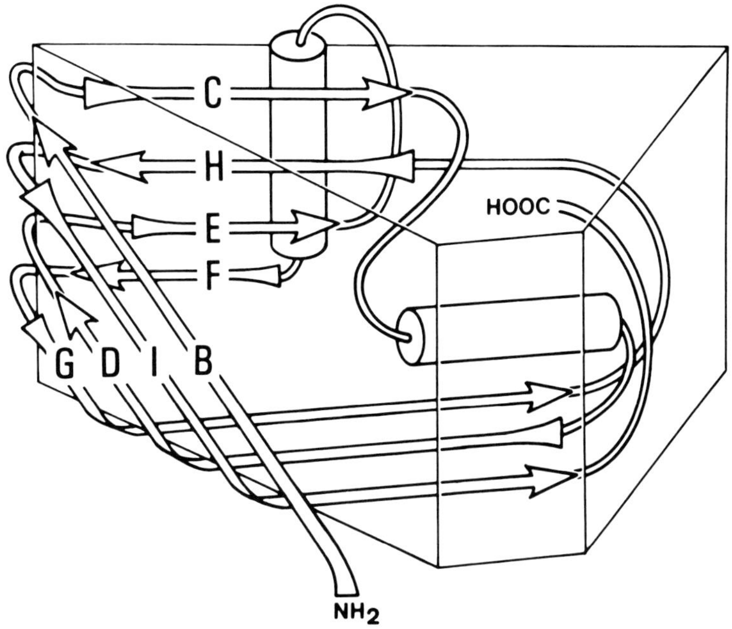

FIGURE 7. A schematic ribbon diagram of the three-dimensional structure of the three major picornavirus capsid proteins. These proteins consist of a core composed of an eight-stranded antiparallel beta barrel with two flanking helices. The overall structure is wedge-shaped. (From Hogle, J. M., Chow, M., and Filman, D. J., *Science*, 229, 1358, 1986. With permission.)

2. Organization of the Capsid

The three major capsid proteins of PV1, RV14, and mengovirus have the same general structure (Figure 7) and are very similar to the capsid proteins[48] of some of the small spherical plant viruses (e.g., tomato bushy stunt virus,[49] southern bean mosaic virus,[50] and satellite tobacco necrosis virus[51]). The three major capsid proteins of picornaviruses each consist of cores that are made up of an eight-stranded antiparallel beta barrel with two flanking helices (see Figure 7) — four strands form the front and bottom surfaces in a large twisted beta sheet, and four strands form the back surface of the barrel. The front and back surfaces are joined at one end by four loops that give the structure a wedge-like shape. The three wedge-shaped structures (VP1, VP2, and VP3) fit together to form the protomer subunit. Five protomers form a pentamer and 12 pentamers form the complete particle (Figure 8).

The closed end of the wedge of VP1 is pointed at the fivefold axis of symmetry while the closed ends of VP2 and VP3 alternate around the threefold axis. The amino termini of the three capsid peptides extend to the interior of the particle and VP4 appears to be the cleaved off amino terminal extension of VP2 and is located primarily inside the particle. VP4 is lost from PV particles if they are eluted off cells after attachment or after heating at 55 to 60°C.[52,53] How this internal protein is lost has not been explained but presumably some structural rearrangement must occur.

It is known that the cleavage of VP0 into VP2 and VP4 is the last maturation cleavage to occur and since this occurs inside the particle the reaction is probably autocatalytic. Rossmann noted that while RV14 has a serine in position 10 of VP2 very close to the carboxy-terminus of VP4, unlike other serine proteases there is no adjacent histidine to act as a proton donor. It is possible that the virion RNA could act as an alternative proton donor allowing cleavage of VP0 to occur at the time the RNA is inserted into the capsid.

This hypothesis has been expanded on by Arnold et al.[36] They have made alignments of

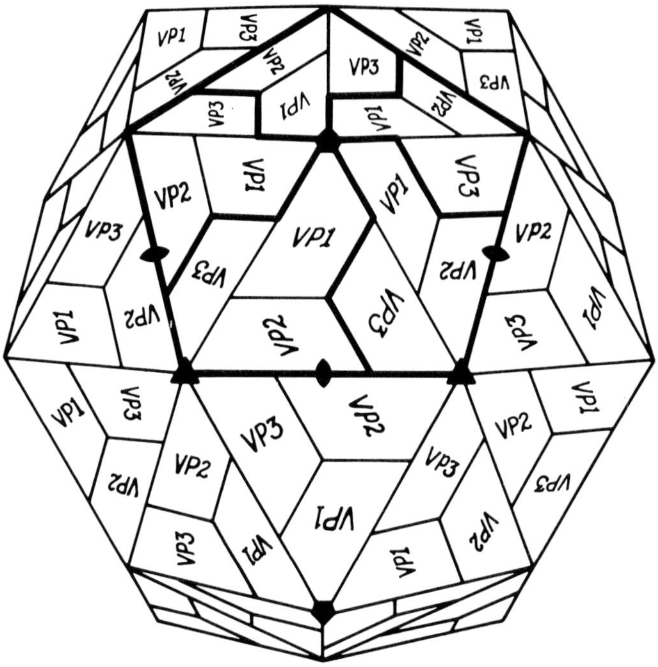

FIGURE 8. A diagrammatic representation of the surface structure of a picornavirus icosahedron. The relative positions of the three major capsid proteins forming the protomer subunit are shown. The narrow end of the wedge-shaped peptides point at the axes with VP1 surrounding the 5′-fold axis, and VP2 and VP3 alternating about the 3′-fold axes. Five protomers form the pentamers and 12 pentamers make up the viral capsid. VP4 is an internal protein cleaved from the amino terminus of VP2 and is not seen on the surface. (From Rossmann, M. G., Arnold, E., Erickson, J. W., Frankenberger, E. A., Griffith, J. P., Hecht, H.-J., Johnson, J. G., Kamer, G., Luo, M., Mosser, A. G., Rueckert, R. R., Sherry, B., and Vriend, G., *Nature (London),* 317, 145, 1985. With permission.)

the amino acids in the region of the VP4/VP2 junction of PV, HRV, EMCV, FMDV, and HAV and found a high degree of homology among all the picornaviruses near the amino terminus of VP2. All the picornaviruses studied except HAV and FMDV have the hypothetically protease-active serine in a similar location. HAV and FMDV so have a serine located several amino acids further down from the position in the other viruses. However, this position is thought to be spacially very close to the position occupied by the serine in the majority of picornaviruses.

In the picornaviruses which have been studied the interior of the capsid is formed by the bottom surfaces of the beta barrels of VP1, VP2, and VP3. The amino termini of VP1, VP2, and VP3 along with VP4 form a network that links the protomers together to form pentamers. The amino-termini of five subunits of VP3 form a twisted tube flanked on its lower surface by five two-stranded antiparallel beta sheets of the amino terminus of VP4. This twisted tube is thought by Hogle to be important for the stability of the pentamer.

The generally planar surfaces of the virion icosahedron are interrupted by sequence insertions that protrude from the particle. Rossmann et al.[6] noted that large protrusions in VP1, VP2, and VP3 result in the formation of a deep canyon on the viral surface which may represent the site of attachment to the cellular receptor in a lock (viral-binding site) and key (cellular receptor) fashion (see Figure 9). This canyon rim, which contains several antigenic sites, is formed by the separation of the VP1 subunits (in the north) that cluster

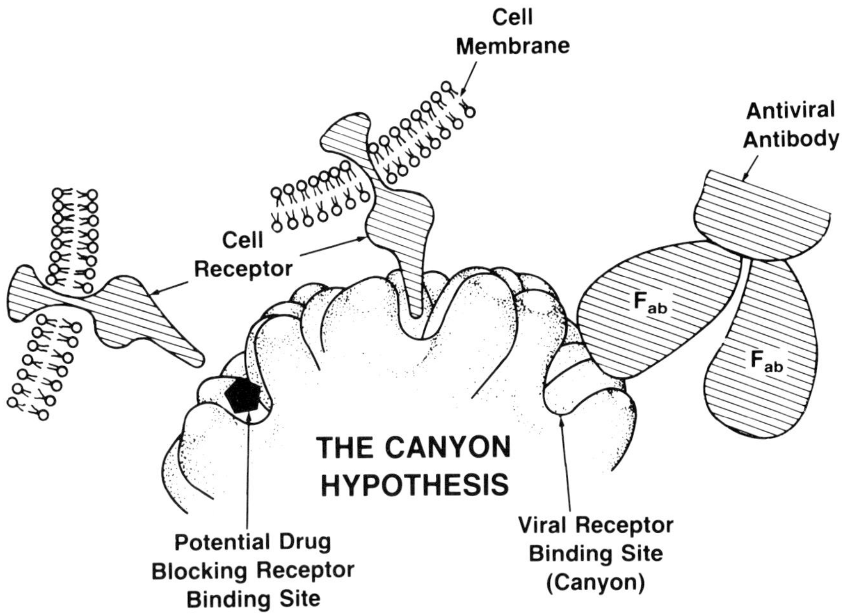

FIGURE 9. A schematic representation of a portion of the surface of a picornavirus demonstrating the "canyon" formed largely by VP1 and VP3. This canyon is thought to represent the binding site for the cellular receptor. The estimated dimensions of the rhinovirus canyon are 25 Å deep and 12 to 30 Å wide which is too small to permit the entrance of an antibody molecule. It may be possible to design chemicals that will bind in the canyon and block the attachment of the virus to cell receptors. (Modified from Luo, M., Vriend, G., Kamer, G., Minor, I., Arnold, E., Rossmann, M. G., Boege, V., Scraba, D. G., Duke, G. M., and Palmenberg, A. C., *Science*, 235, 182, 1987. With permission.)

about the pentamer axis from the surrounding VP2 and VP3 subunits (in the south). The interior walls of the canyon are made up of VP1 to the north and VP1 and VP3 to the south. The canyon is 25 Å deep and 12 to 30 Å wide, and therefore, is too small for an antibody molecule to fit inside.

Luo et al.[7] have shown that the receptor-binding site in mengovirus is substantially different from that observed in HRV14. The canyon of mengovirus is partially filled by two loops in VP1 and a modified "puff" in VP2 resulting in a pit 22 Å deep and 30 Å wide.

In a recent report Smith and colleagues[54] used X-ray diffraction to study two chemicals, WIN51711 and WIN52084 (Sterling Winthrop), that prevent rhinovirus uncoating but not cellular attachment or penetration, and are known to stabilize picornaviruses against alkaline and heat denaturation and the loss of VP4. These compounds consist of a 3-methylisoxazole group (I group) connected by a seven-member aliphotic chain to a 4-oxazolinylphenoxy group (OP group). When the compounds were reacted with the crystalline form of HRV14 it was found that the I group was inserted into a hydrophobic pocket in the interior of the VP1 beta barrel while the OP group covered the entrance to an ion channel in the floor of the canyon. The authors speculate that the drugs may inhibit viral disassembly by preventing the collapse of the VP1 hydrophobic pocket or by blocking the flow of ions into the virus interior. They also were able to demonstrate a molecular basis for differences in the activity of these drugs against different picornaviruses as small differences in molecular structures were shown to greatly change the binding properties of the drugs.

This type of detailed structural and molecular information may permit the design of drugs that would bind precisely to different parts of the virion. One of the most likely sites of attack would be the canyon that is thought to represent the cellular attachment site. A drug

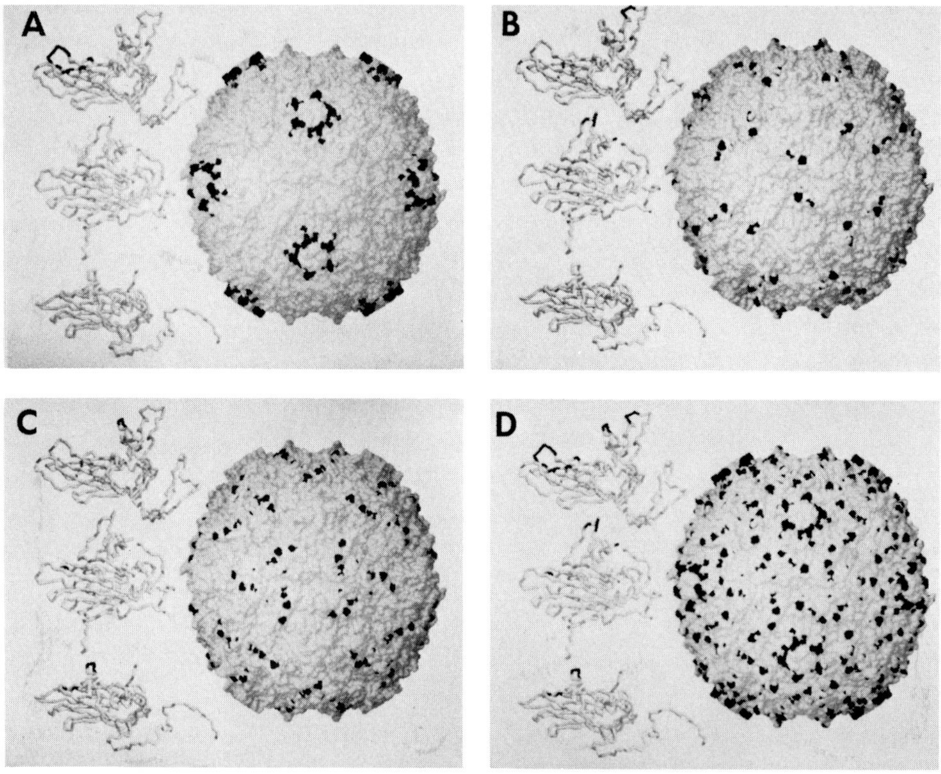

FIGURE 10. Three major antigenic clusters found on PV as determined by sequence analysis of neutralization-resistant mutants. The location of these mutations were positioned on the crystallographically determined three-dimensional representation of the PV peptides as well as on the surface of the virion. The antigenic sites are depicted in black. Panel A shows Cluster 1, Panel B Cluster 2, Panel C Cluster 3, and Panel D all three clusters combined. (From Hogle, J. M., Chow, M., and Filman, D. J., *Science,* 229, 1358, 1986. With permission.)

that would block this site may prevent viral attachment to cells and thereby prevent infection (see Figure 9).

3. Antigenic Sites

The externally protruding loops and puffs of PV and HRV that were revealed by these crystallographic studies have been shown to include many of the important antigenic sites on the virions.[5,6] A number of neutralization-escape mutants of PV have been sequenced.[55,56] The mutation sites of these viruses are located on the major surface projections of the virion and can be grouped into four major clusters (Figure 10). Cluster 1 is made up of loops of VP1 which form the peak around the fivefold axis. Cluster 2 is formed from the top loop of the barrel of VP2 and the internal insertion in VP3. Cluster 3 has contributions from the internal insertions of VP1 and VP2 and the carboxy-terminus of VP2. Cluster 4 contains residues from a loop in the carboxy-terminal region of VP1. All the neutralization-escape mutants that have been studied have their mutations in the likely antigenic sites and not at distant sites. It has been proposed that mutations distant from the antibody-binding site could alter its conformation and hence its ability to bind antibodies.

Very similar results had been obtained with HRV14. Four neutralizing sites, referred to as NIm-1A, NIm-1B, NIm-II and NIm-III, have been identified[5] (see Figure 11). These sites are similar to those identified in PV though some differences exist. The important point is that these sites, originally defined by neutralizing monoclonal antibodies and mapped to the amino acid sequence of the capsid proteins, have been located by crystallography on

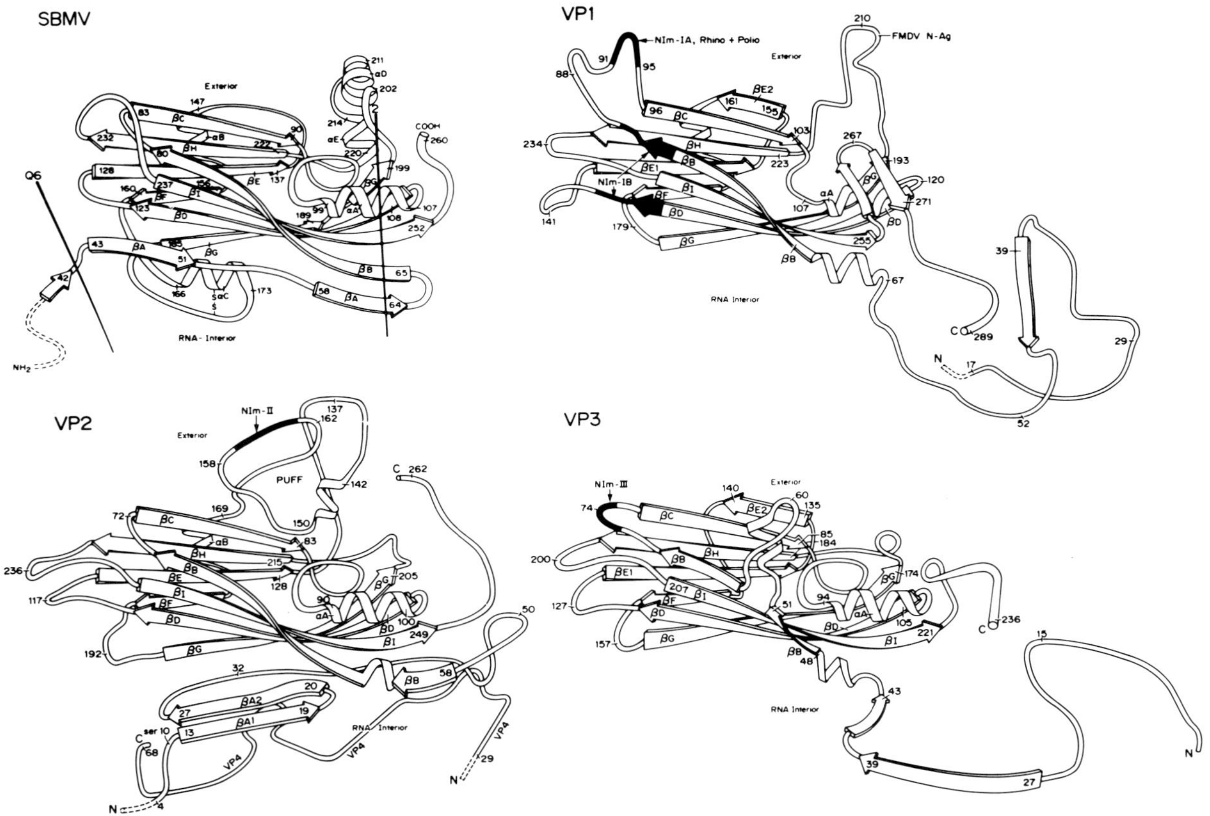

FIGURE 11. Ribbon diagrams of the three major capsid proteins of HRV-14 compared with the major capsid protein of southern bean mosaic virus (SBMV). The major sites to which neutralizing antibodies are directed are shown in black. It can be seen that, in general, the antigenic sites are on external loops or puffs. However, not all these surface projections seem to be important antigenic sites. (From Rossmann, M. G., Arnold, E., Erickson, J. W., Frankenberger, E. A., Griffith, J. P., Hecht, H.-J., Johnson, J. G., Kamer, G., Luo, M., Mosser, A. G., Rueckert, R. R., Sherry, B., and Vriend, G., *Nature (London)*, 317, 145, 1985. With permission.)

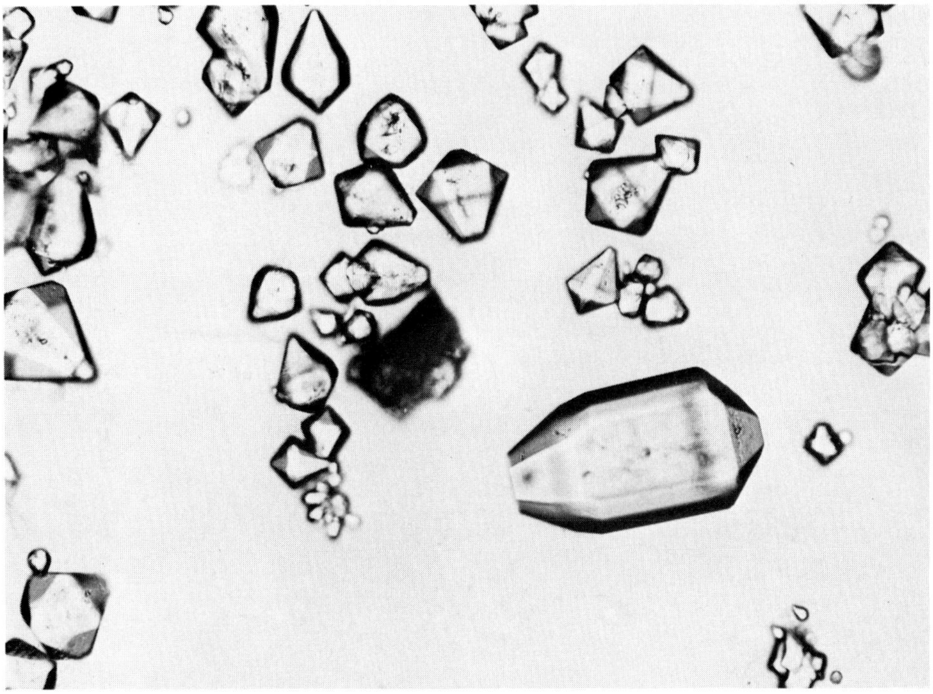

FIGURE 12. Crystals of HAV. Milligram quantities of cell culture-derived HAV was purified to homogeneity. Repeated pelleting of this purified virus resulted in the spontaneous formation of crystalline HAV. These crystals were birefringent in nature when examined with plane-polarizing light. The largest crystals measured approximately 0.15 mm in length. (C. M. Wheeler, personal communication; photomicrograph by Ron Howard.)

the exterior of the particle. Not all the external loops have been shown to be important antigens and not all the important antigens are located on the surface of the virion. Antibodies to synthetic peptides representing the NH_2-terminus of VP1 or the bottom loop of VP1 of PV, both of which have been shown to be completely within the interior of the virion, have been shown to have neutralizing activity.[57] This result has not been explained but has important implications for our understanding the native structure of the virus and the mechanism of neutralization.

While HAV has many of the features of other picornaviruses, it has several which are relatively unique. How closely the crystallographic structure of PV and HRV resemble HAV cannot be predicted until sufficient quantities of HAV are available to perform the crystallographic analysis. Wheeler and colleagues have recently produced milligram quantities of virus and have found crystals thought to be composed of HAV (see Figure 12).[81] Because HAV grows slowly and to a relatively low titer, producing and purifying the required amount of virus (more than 10 mg) for the crystallographic studies is a huge task. However, it is felt that such an effort would be rewarding in terms of the scientific information obtained.

4. *Location of Antigenic Sites of HAV*

The antigenic structure of HAV has not been studied to the same extent as the antigens of PV or several other picornaviruses. Monoclonal antibodies to HAV have been produced in several laboratories[58-60] and several of these have been shown to neutralize viral infectivity in cell culture. However, at this time it has not been possible to precisely map the epitopes recognized by these monoclones. In addition, Hughes and his colleagues[61] have shown that antibody raised both purified VP1 and VP3 can neutralize HAV while cross-linking experiments have shown that the neutralizing monoclonal antibody that they produced bind specifically to VP1. Finer localization of the epitope has not been made.

Stapleton and Lemon[62] have analyzed the binding characteristics of eight neutralizing monoclonal antibodies by competition assays and found that the antibodies seem to recognize epitopes within a closely related antigenic site. Two antibodies (K3-4C8 and K3-2F2) seemed to bind to very closely related or overlapping epitopes while a third antibody (B5-B3) did not compete with the first two, suggesting it recognizes a second epitope. However, a third group of monoclonal antibodies (6A5, 1B9, 2D2, 3E1, and K2-4F2) all competed at varying levels with the binding of the three other neutralizing monoclones to HAV. In addition, all of these monoclones could compete partially with the binding of polyclonal antibodies to HAV, and combinations of the B5-B3 monoclone with either K3-4C8, K2-4F2, or 6A5 could completely block the binding of polyclonal antibodies to HAV. These data suggest that there is one immunodominant antigenic cluster on HAV, in contrast to PV and HRV where three or four antigenic clusters have been identified.

Stapleton and Lemon[62] have also produced mutants of HAV that resist neutralization by these same monoclonal antibodies and show greatly reduced neutralization by polyclonal antibodies. These data again suggest a single immunodominant region on HAV. Lemon is presently sequencing the RNA from these mutants which should enable the precise mapping of this immunodominant, neutralization site of HAV.[82]

Synthetic oligopeptides have also been used to identify the important antigenic sites on HAV. Similar oligopeptides have been useful for studying epitopes of other picornaviruses. Chow[57] and colleagues studied antibodies to synthetic oligopeptides that spanned the entire VP1 protein. They found that while most of the anti-VP1 peptides bound to the virus, only antibodies to peptides in four distinct regions neutralized viral infectivity in vitro. These neutralizing antibodies were directed against amino acid regions 24 to 40, 86 to 103, 161 to 181, and 286 to 302, of the 302 amino acid VP1.

A similar approach has been used to study HAV neutralization sites. We have studied antibodies produced by Bittle and Hogle to synthetic peptides spanning the VP1 region of HAV. However, none of these antibodies has significant neutralizing activity (Feinstone et al. unpublished).

Emini and colleagues[63] showed that rabbits that were primed with PV VP1 synthetic peptides of amino acids 11 to 17, 70 to 75, and 73 to 103, and then inoculated with whole HAV (HAV does not replicate in rabbits) rapidly developed an IgG but not IgM neutralizing-antibody response to HAV. While the regions of PV VP1 did not have close amino-acid sequence homology with HAV VP1 they did share structural similarities according to computer analysis. It is not known if this priming effect has any value for predicting the important neutralizing epitopes of HAV.

If the neutralizing epitopes of HAV can be identified, it may be possible to use combinations of synthetic peptides as vaccines or diagnostic antigens. It would seem unwise to base a vaccine on only a single small peptide as the development of neutralization-escape mutants suggests that the virus can alter its major antigenic sites without affecting its ability to infect cells and replicate within them.[62]

III. APPLICATIONS OF THIS TECHNOLOGY

A. Use of Molecular Probes for Diagnosis of HAV Infection

The availability of cloned HAV cDNA has permitted the detection of HAV with molecular probes. Ticehurst et al.[64] have shown that HAV can be detected in clinical samples by molecular hybridization of cDNA with HAV RNA that has been extracted from the specimen and blotted onto nitrocellulose paper using a ''slot blot'' apparatus. This technique is approximately tenfold more sensitive than RIA or IEM, and is capable of detecting about 10^3 $TCID_{50}$ of HAV or 10^4 to 10^5 genome equivalents of HAV RNA.

While this technique was highly sensitive, occasional problems occur with specificity, especially when testing for HAV shedding in fecal specimens. It was presumed that the probe was contaminated with bacterial plasmid or genomic sequences that hybridized with similar sequences in some stool samples. However, even after extensive purification of the probe the problem was not totally eliminated, necessitating the careful use of properly selected controls.

When serial fecal samples from a marmoset infected with HAV were analyzed by hybridization and RIA, HAV was detected for several days longer by hybridization. As measured by RIA, stool shedding of HAV declined rapidly following the appearance of serum antibody on day 17 and became undetectable by day 25. However, when the same stools were examined by slot-blot hybridization, shedding of virus could be detected until at least day 40.

Tassopoulos et al.[65] used this technique to study HAV excretion in children and adults in Greece who had acute HAV infections. HAV was detected more frequently in stool samples by molecular hybridization (56.7%) than by RIA (24.3%), again demonstrating the higher sensitivity of molecular hybridization compared to antibody-based methods.

Jansen et al.[66] have combined the specificity of antibody capture and the sensitivity of cDNA hybridization to detect HAV in fecal specimens. This assay seems to eliminate much of the nonspecificity due to the presence of bacterial sequences in fecal samples by absorbing HAV from the sample onto a solid phase coated with monoclonal anti HAV. Ticehurst[64] found this technique to be two- to threefold less sensitive than techniques based on extraction of the total RNA from a fecal sample.

While it is not likely that molecular hybridization techniques will supplant the presently reliable serological methods for the diagnosis of acute hepatitis A (see Chapter 8), these techniques are valuable for certain laboratory and epidemiologic studies where high sensitivity detection of HAV is necessary. Although infectivity assays (primate inoculation or cell culture) may be the most sensitive detection methods available, they are not always practical because not all strains of HAV can be grown readily in cell culture and nonhuman primates are expensive and in limited supply.

B. Alternative Approaches to Vaccine Development

Several groups have recently produced experimental killed and live attenuated hepatitis A vaccines from virus grown in cell culture (see Chapter 11). While these classic approaches to vaccine development are likely to be successful, there are several potential problems. Besides possible immunogenicity problems, killed hepatitis A vaccines may be very expensive to produce due to the relatively low yield of HAV in cell culture. It may be possible to produce synthetic hepatitis A antigens in large quantity and cheaply through modern biotechnology.

Live vaccines have the major problems of safety and genetic stability. Recently Cohen et al.[67] have cloned and sequenced an attenuated mutant of the HM175 strain of HAV (see Chapter 11) produced by 32 passages of the virus in African green monkey kidney cells and compared it to the parental wild-type virus.[27] He found 24 nucleotide differences, spread throughout the genome, comprising five base deletions and two base substitutions in the 5′ noncoding region, one substitution in the 3′ noncoding region, and 16 base substitutions in the coding region. Twelve of these changes resulted in amino acid changes and seven of these resulted in amino acids of different charges that are more likely to alter the structure of protein. While nucleotide changes were observed in each of the capsid proteins except VP4, amino acid changes were only noted in VP1 and VP2. There were nucleotide changes in all the nonstructural protein-coding regions except 3C and these changes resulted in amino acid changes in all the nonstructural proteins except 3A.

Thus, a relatively few nucleotide (25) and amino acid (12) changes are responsible for the cell culture adaptation and attenuation characteristics of this virus. However, it will require further analysis to determine exactly which of these mutations are responsible for the phenotype of this virus.

Racaniello[68] has shown that the full-length cloned cDNA of PV can be transfected into permissive cell culture and complete infectious virus particles produced. Similar results have also been reported for other PV clones[69] and rhinoviruses.[70] Cohen has produced full-length cDNA clones of both the wild-type HM175 and the tissue-culture attenuated strain by ligating subgenomic clones together. He has recently shown that the full-length cDNA derived from the tissue-cluture adapted virus is infectious.[83] These constructs will be useful for analyzing individual or groups of mutations. In addition, the availability of infectious cDNA should permit the engineering of site-specific mutations and recombinations that may have desired characteristics such as attenuation, high in vitro replication, or even tissue specificity.

C. Better Understanding of the Replicative Strategy

One of the central problems of the virology of HAV concerns its replication, especially how it differs from other picornaviruses and how these differences relate to the pathogenesis of the disease. While many of the replication events have not been studied (see Chapter 6), the general process is assumed to resemble other picornaviruses.

There are at least two distinctive features of HAV replication in vitro; first, the virus grows much more slowly than many other picornaviruses; and second, it generally does not cause lysis of its host cell. These two features may also occur in vivo and may be important factors in the pathogenesis of hepatitis A.

Slow viral replication and lack of cell lysis may both be related to the same viral characteristic(s) or may have unrelated causes. In growth-cycle experiments Anderson et al.[71] have demonstrated that even a strain of HAV which had been adapted to rapid growth in cell culture has a 12-hr lag phase (see Figure 9, Chapter 6) followed by a 12-hr logarithmic phase, which is much slower than PV in which both phases are complete in less than 8 hr. These workers found little difference in the rate of penetration and uncoating of PV and HAV but noted a delayed appearance of viral RNA in HAV infected cells compared to PV infected cells, leading them to conclude that the block in HAV replication occurs at the level of RNA replication.

Most picornaviruses shut off host-cell protein synthesis to allow for efficient translation of viral-gene products and subsequently viral-RNA replication. Exactly how picornaviruses disrupt the normal cellular protein synthesis is not completely known and may involve different mechanisms for different picornaviruses.[72] A cellular protein, p220, that forms part of the cap-binding protein complex is cleaved in cells infected by PV and rhinovirus.[73,74] The 5′ cap structure, $m^7G(5')ppp(5')N$, is a nearly ubiquitous feature of eukaryotic mRNA and the cap-binding protein complex seems to be important for efficient translation of mRNA has the covalently linked protein, VPg. It seems logical that if the cap-binding protein complex is disrupted in infected cells, the viral mRNA may have an advantage over host mRNA in ribosomal binding. It has been demonstrated however, that the 3C protease of PV does not cleave the p220.[76] Recently it has been shown that protein 2A is also a protease[77] and that a PV with a mutation in this region failed to either shut down host protein synthesis or cleave p220.[77,78] It remains possible that the viral infection activates some cellular factor that in turn cleaves P220.

Unlike PV and HRV (and probably the other enteroviruses) EMCV does not cleave p220 but does shut off host-protein synthesis.[79] The mechanism that EMCV utilizes to shut down protein synthesis is not known but proteolytic cleavage of initiation factors does not seem to be involved. It is possible that the viral RNA simply out competes host RNA for ribosomal binding, but other mechanisms such as phosphorylation of inition factors have also been hypothesized.

It is also not known what role the shut off of host-protein synthesis plays in the production by promoting initiation complex formation.[75] Neither the viral RNA (which can also act as a messenger molecule) nor viral mRNA of picornaviruses is capped at its 5′ end but instead of the viral cytopathic effect and cell lysis. The PV mutant that fails to shut off the host and did not cleave p220 has a small plaque phenotype. Certainly a cell which is not making its own proteins efficiently cannot survive long. However, many other factors such as the effect of the virus on the cytoskeleton have been proposed as mechanisms for development of CPE.

HAV seems to replicate in cell culture and probably in vivo as well without affecting the host-protein synthesis.[18] The virus also does not lyse the host cell nor cause any easily detectable CPE. It would be hard to escape the conclusion that these two facts are related. If the host cell is not sufficiently inhibited then viral replication must compete with the cellular protein synthesis. If viral proteins are made in small quantity the viral RNA replication, which depends on the systhesis of the viral RNA polymerase, is also likely to be retarded. The central question then becomes: why does HAV infection not shut down host-protein systhesis as efficiently as other picornaviruses? It is likely that this question can be answered through the use of molecular techniques. Although it is difficult to analyze the effect of HAV infection on p220 (simply because HAV replicates so poorly), it may be possible to obtain synthetic HAV 2A protease and compare it with the 2A protease of PV. It may be possible to answer these questions by making recombinant viruses[80] from portions of HAV and PV or EMCV utilizing infectious cDNA, and thereby analyze which genes are important for these functions. Such experiments are potentially hazardous and should only be undertaken after the most careful consideration and consultation with the proper regulatory authorities.

More acceptable experiments may involve the use of eukaryotic expression systems in which HAV infection could be complemented. For instance the 2A protease gene of PV could be expressed in a cell permissive for HAV replication. One could test if the HAV replication is enhanced or CPE produced by the presence of the PV protease.

Molecular biology, X-ray crystallography, and other techniques of modern virology combined with improved biologic and serologic systems will continue to reveal the nature of this interesting, unusual and important picornavirus.

REFERENCES

1. **Cooper, P. D., Agol, V. I., Bachrach, H. L., Brown, F., Ghendon, Y., Gibbs, A. S., Gillespie, J. H., Lonberg-Holm, K., Mandel, B., Melnick, J. L., Mohanty, S. B., Povey, R. C., Rueckert, R. R., Schaffer, F. L., and Tyrrell, D. A. J.,** Picornaviridae: second report, *Intervirology,* 10, 165, 1978.
2. **Gust, I. D., Coulepis, A. G., Feinstone, S. M., Locarnini, S. A., Moritsugu, Y., Najera, R., and Siegl, G.,** Taxanomic classification of hepatitis A virus, *Intervirology,* 20, 1, 1983.
3. **Rueckert, R. R.,** Picornaviruses and their replication, in *Virology,* Fields, B. N., Ed., Raven Press, New York, 1985, 705.
4. **Racaniello, V. R. and Baltimore, D.,** Molecular cloning of poliovirus cDNA and determination of the complete nucleotide sequence of the viral genome, *Proc. Natl. Acad. Sci. U.S.A.,* 78, 4887, 1981.
5. **Hogle, J. M., Chow, M., and Filman, D. J.,** Three-dimensional structure of poliovirus at 2.0 Å resolution, *Science,* 229, 1358, 1986.
6. **Rossmann, M. G., Arnold, E., Erickson, J. W., Frankenberger, E. A., Griffith, J. P., Hecht, H.-J., Johnson, J. E., Kamer, G., Luo, M., Mosser, A. G., Rueckert, R. R., Sherry, B, and Vriend, G.,** Structure of a human common cold virus and functional relationship to other picornaviruses, *Nature (London),* 317, 145, 1985.
7. **Luo, M., Vriend, G., Kamer, G., Minor, I., Arnold, E., Rossmann, M. G., Boege, U., Scraba, D. G., Duke, G. M., and Palmenberg, A. C.,** The structure of mengo virus at atomic resolution, *Science,* 235, 182, 1987.
8. **Wimmer, E.,** Genome-linked proteins of viruses, *Cell,* 28, 199, 1982.

9. **Brown, F., Newman, J. F. E., Stott, J., Porter, A., Frisby, D., Newton, C., Carey, N., and Fellner, P.**, Poly C in animal viral RNAs, *Nature (London)*, 251, 342, 1974.
10. **Palmenberg, A. C.**, Genome organization, translation and processing in picornaviruses, in *The Molecular Biology of Positive Strand RNA Viruses*, Rowlands, D. J., Mahy, B. W. J., Mayo, M., Eds., Academic Press, New York, 1985.
11. **Nicklin, M. J. H., Toyoda, H., Murray, M. G., and Wimmer, E.**, Proteolytic processing in the replication of polio and related viruses, *Biotechnology*, 4, 33, 1986.
12. **Coulepis, A. G., Locarnini, S. A., and Gust, I. D.**, Iodination of hepatitis A virus reveals a fourth structural polypeptide, *J. Virol.*, 35, 572, 1980.
13. **Tratschin, J. D., Siegl, G., and Frösner, G. G.**, Characterization and classification of virus particles associated with hepatitis A. III. Structural proteins, *J. Virol.*, 38, 151, 1981.
14. **Provost, P. J., Wolanski, B. S., Miller, W. J., Ittensohn, O. L., McAleer, W. J., and Hilleman, M. R.**, Physical, chemical and morphologic dimensions of human hepatitis A virus strain CR326, *Proc. Soc. Exp. Biol. Med.*, 148, 532, 1975.
15. **Siegl, G. and Frösner, G. G.**, Characterization and classification of virus particles associated with hepatitis A. II. Type and configuration of nucleic acid, *J. Virol.*, 26, 48, 1978.
16. **Coulepis, A. G., Tannock, G. A., Locarnini, S. A., and Gust, I. D.**, Evidence that the genome of hepatitis A virus consists of single-stranded RNA, *J. Virol.*, 37, 473, 1981.
17. **Feinstone, S. M., Moritsugu, Y., Shih, J. W.-K., Gerin, J. L., and Purcell, R. H.**, Characterization of hepatitis A virus, in *Viral Hepatitis*, Vyas, G., Cohen, S. N., and Schmid, R., Eds., Franklin Institute Press, Philadelphia, 1978, 41.
18. **Locarnini, S. A., Coulepis, A. G., Westaway, E. G., and Gust, I. D.**, Restricted replication of human hepatitis A virus in cell culture: intracellular biochemical studies, *J. Virol.*, 37, 216, 1981.
19. **Ticehurst, J. R., Racaniello, V. R., Baroudy, B. M., Baltimore, D., Purcell, R. H., and Feinstone, S. M.**, Molecular cloning and characterization of hepatitis A virus cDNA, *Proc. Natl. Acad. Sci. U.S.A.*, 80, 5885, 1983.
20. **Von der Helm, K., Winnacker, E. L., Deinhardt, F., Frösner, G., Gauss-Müller, V., Bayerl, B., Scheid, R., and Siegl, G.**, Cloning of hepatitis A virus genome, *J. Virol. Methods*, 3, 37, 1981.
21. **Najarian, R., Caput, D., Gee, W., Potter, S. J., Renard, A., Merryweather, J., Van Nest, G., and Dina, D.**, Primary structure and gene organization of human hepatitis A virus, *Proc. Natl. Acad. Sci. U.S.A.*, 82, 2627, 1985.
22. **Venuti, A., Di Russo, C., del Grosso, N., Patti, A.-M., Ruggeri, F., De Stasio, P. R., Martiniello, M. G., Pagnotti, P., Degener, A. M., Midulla, M., Panà, A., and Perez-Bercoff, R.**, Isolation and molecular cloning of a fast-growing strain of human hepatitis A virus from its double-stranded replicative form, *J. Virol.*, 56, 579, 1985.
23. **Ross, B. C., Anderson, B. N., Coulepis, A. G., Chenoweth, M. P., and Gust, I. D.**, Molecular cloning of cDNA from hepatitis A virus strain HM-175 after multiple passages *in vivo* and *in vitro*, *J. Gen. Virol.*, 67, 1741, 1986.
24. **Ovchinnikov, I. A., Sverdlov, E. D., Tsarev, S. A., Arsenian, S. G., Rokhina, T. O., Chizhikov, V. E., Petrov, N. A., Prikhod'ko, G. G., Blinov, V. M., Vasilenko, S. K., Sandakhchiev, L. S., Kusov, I. I., Grabko, V. I., Fleer, G. P., Balayan, M. S., and Drozdov, S. G.**, Sequence of 3372 nucleotide units of RNA of the hepatitis A virus, coding the capsids VP4-VP1 and some non-structural proteins, *Dokl. Akad. Nauk SSSR*, 285, 1014, 1985.
25. **Linemeyer, D. L., Menke, J. G., Martin-Gallardo, A., Hughes, J. V., Young, A., and Mitra, S. W.**, Molecular cloning and partial sequencing of hepatitis A viral cDNA, *J. Virol.*, 54, 247, 1985.
26. **Baroudy, B. M., Ticehurst, J. R., Miele, T. A., Maizel, J. V., Jr., Purcell, R. H., and Feinstone, S. M.**, Sequence analysis of hepatitis A virus cDNA coding for capsid proteins and RNA polymerase, *Proc. Natl. Acad. Sci. U.S.A.*, 82, 2143, 1985.
27. **Cohen, J. I., Ticehurst, J. T., Purcell, R. H., Buckler-White, A., and Baroudy, B. M.**, Complete nucleotide sequence of wild-type hepatitis A virus: comparison with different strains of hepatitis A virus and other picornaviruses, *J. Virol.*, 61, 50, 1987.
28. **Larsen, G. R., Semler, B. L., and Wimmer, E.**, Stable hairpin structure within the 5′-terminal 85 nucleotides of poliovirus RNA, *J. Virol.*, 37, 328, 1981.
29. **Currey, K. M., Peterlin, B. M., and Maizel, J. V., Jr.**, Secondary structure of poliovirus RNA: correlation of computer-predicted with electron microscopically observed structure, *Virology*, 148, 33, 1986.
30. **Evans, D. M. A., Dunn, G., Minor, P. D., Schild, G. C., Cann, A. J., Stanway, G., Almond, J. W., and Maizel, J. V., Jr.**, Increased neurovirulence associated with a single nucleotide change in a non-coding region of the Sabin type 3 poliovaccine genome, *Nature (London)*, 314, 548, 1985.
31. **Kozak, M.**, Compilation and analysis of sequences upstream from the translational start site in eukaryotic mRNAs, *Nucleic Acids Res.*, 12, 857, 1984.
32. **Rueckert, R. R. and Wimmer, E.**, Systematic nomenclature of picornavirus proteins, *J. Virol.*, 50, 957, 1984.

33. **Diamond, D. C., Wimmer, E., Helm, K. V. D., and Deinhardt, F.,** The genomic map of hepatitis A virus: an alternate analysis, *Microbiol. Pathogenesis,* 1, 217, 1986.
34. **Ticehurst, J. R.,** Hepatitis A virus: clones, cultures, and vaccines, *Semin. Liver Dis.,* 6, 46, 1986.
35. **Toyoda, H., Nicklin, M. J. H., Murray, M. G., Anderson, C. W., Dunn, J. J., Studier, F. W., and Wimmer, E.,** A second virus-encoded proteinase involved in proteolytic processing of poliovirus polyprotein, *Cell,* 45, 761, 1986.
36. **Arnold, E., Luo, M., Vriend, G., Rossmann, M. G., Palmenberg, A. C., Parks, G. D., Nicklin, M., and Wimmer, E.,** Implication of the picornavirus capsid structure for polyprotein processing, *Proc. Natl. Acad. Sci. U.S.A.,* in press, 1987.
37. **Wheeler, C. M., Robertson, B. H., Van Nest, G., Dina, D., Bradley, D. W., and Fields, H. A.,** Structure of the hepatitis A virion: peptide mapping of the capsid region, *J. Virol.,* 58, 307, 1986.
38. **Weitz, M., Baroudy, B. M., Maloy, W. L., Ticehurst, J. R., and Purcell, R. H.,** Detection of a genome-linked protein (VPg) of hepatitis A virus and its comparison with other picornaviral VPgs, *J. Virol.,* 60, 124, 1986.
39. **Siegl, G.,** Structure and biology of hepatitis A virus, in *Viral Hepatitis 1982 International Symposium,* Szmuness, W., Alter, H. J., and Maynard, J. E., Eds., Franklin Institute Press, 1982, 13.
40. **Weitz, M. and Siegl, G.,** Variation among hepatitis A virus strains. I. Genomic variation detected by T_1 oligonucleotide mapping, *Virus Res.,* 4, 53, 1985.
41. **Gust, I. E., Lehmann, N. I., Crowe, S., McCrorie, M., Locarnini, S. A., and Lucas, C. R.,** The origin of the HM-175 strain of hepatitis A virus, *J. Infect. Dis.,* 151, 365, 1985.
42. **Lemon, S. M., LeDuc, J. W., Binn, L. N., Escajadillo, A., and Ishak, K. G.,** Transmission of hepatitis A among recently captured Panamanian owl monkeys, *J. Med. Virol.,* 10, 25, 1982.
43. **Lemon, S. M., Chao, S. F., Jansen, R. W., Binn, L. N., and LeDuc, J. W.,** Genomic heterogeneity among human and non-human strains of hepatitis A virus, *J. Virol.,* in press, 1987.
44. **Burke, D. S. and Heisey, G. B.,** Wild Malaysian cynomolgous monkeys are exposed to hepatitis A virus, *Am. J. Trop. Med. Hyg.,* 33, 940, 1984.
45. **Pincus, S. E., Diamond, D. C., Emini, E., and Wimmer, E.,** Guanidine-selected mutants of poliovirus: mapping of point mutations to polypeptide 2C, *J. Virol.,* 57, 638, 1986.
46. **Siegl, G. and Eggers, H. J.,** Failure of guanidine and 2-(α-hydroxybenzyl) benzimidazole to inhibit replication of hepatitis A virus *in vitro, J. Gen. Virol.,* 61, 111, 1982.
47. **Hyypia, T, Stålhandske, P., Vainionpää, and Pettersson, U.,** Detection of enteroviruses by spot hybridization, *J. Clin. Microbiol.,* 19, 436, 1984.
48. **Argos, P., Kamer, G., Nicklin, M. J. H., and Wimmer, E.,** Similarity in gene organization and homology between proteins of animal picornaviruses and a plant comovirus suggest common ancestry of these virus families, *Nucleic Acids Res.,* 12, 7251, 1984.
49. **Harrison, S. C., Olson, A. J., Schutt, C. E., and Winkler, F. K.,** Tomato bushy stunt virus at 2.9 Å resolution, *Nature (London),* 276, 368, 1978.
50. **Abad-Zapatero, C., Abdel-Meguid, S. S., Johnson, J. E., Leslie, A. G. W., Rayment, I., Rossmann, M. G., Suck, D., and Tsukihara, T.,** Structure of southern bean mosaic virus at 2.8 Å resolution, *Nature (London),* 286, 33, 1980.
51. **Liljas, L., Unge, T., Jones, T. A., Fridborg, K., Lövgren, S., Skoglund, V., and Strandberg, B.,** Structure of satellite tobacco necrosis virus at 3.0 Å resolution, *J. Mol. Biol.,* 159, 93, 1982.
52. **DeSena, J. and Mandele, B.,** Studies in the *in vitro* uncoating of poliovirus. II. Characterization of the membrane modified particle, *Virology,* 78, 554, 1977.
53. **Maizel, J. V., Phillips, B. A., and Sammers, D. F.,** Composition of artificially produced and naturally occurring empty capsids of poliovirus type I, *Virology,* 32, 692, 1967.
54. **Smith, T. J., Kremer, M. J., Luo, M., Vriend, G., Arnold, E., Kamer, G., Rossmann, M. G., McKinlay, M. A., Diana, G. D., and Otto, M. J.,** The site of attachment in human rhinovirus 14 for antiviral agents that inhibit uncoating, *Science,* 233, 1286, 1986.
55. **Diamond, D. C., Jameson, B. A., Bonin, J., Kohara, M., Abe, S., Itoh, H., Komatsu, T., Anita, M., Kuge, S., Nomoto, A., Osterhaus, A. D. M. E., Crainic, R., and Wimmer, E.,** Antigenic variation and resistance to neutralization in poliovirus type 1, *Science,* 229, 1090, 1985.
56. **Minor, P. D., Ferguson, M., Evans, D. M. A., Almond, J. W., and Icenogle, J. P.,** Antigenic structure of polioviruses of serotypes 1, 2 and 3, *J. Gen. Virol.,* 67, 1283, 1986.
57. **Chow, M., Yabrov, R., Bittle, J., Hogle, J., and Baltimore, D.,** Synthetic peptides from four separate regions of the poliovirus type 1 capsid protein VP1 induce neutralizing antibodies, *Proc. Natl. Acad. Sci. U.S.A.,* 82, 910, 1985.
58. **MacGregor, A., Kornitschuk, M., Hurrell, J. G. R., Lehmann, N. I., Coulepis, A. G., Locarnini, S. A., and Gust, I. D.,** Monoclonal antibodies against hepatitis A virus, *J. Clin. Microbiol.,* 18, 1237, 1983.
59. **Dawson, G. J., Decker, R. H., Norton, D. K., Bryce, W. H., Whittington, R. O., Tribby, I. I., and Mushahwar, I. K.,** Monoclonal antibodies to hepatitis A virus, *J. Med. Virol.,* 14, 1, 1984.

60. **Hughes, J. V., Stanton, L. W., Tomassini, J. E., Long, W. J., and Scolnick, E. M.,** Neutralizing monoclonal antibodies to hepatitis A virus: partial localization of a neutralizing antigenic site, *J. Virol.*, 52, 465, 1984.
61. **Hughes, J. V. and Stanton, L. W.,** Isolation and immunizations with hepatitis A viral structural proteins: induction of antiprotein, antiviral, and neutralizing responses, *J. Virol.*, 55, 395, 1985.
62. **Stapleton, J. T. and Lemon, S. M.,** Neutralization escape mutants define a dominant immunogenic neutralization site on hepatitis A virus, *J. Virol.*, in press, 1987.
63. **Emini, S. A., Hughes, J. V., Perlow, D. S., and Boger, J.,** Induction of hepatitis A virus-neutralizing antibody by a virus-specific synthetic peptide, *J. Virol.*, 55, 836, 1985.
64. **Ticehurst, J., Feinstone, S. M., Chestnut, T., Tassopoulos, N. C., Popper, H., and Purcell, R. H.,** Detection of hepatitis A virus RNA by molecular hybridization, *J. Clin. Microbiol.*, 1987, in press.
65. **Tassopoulos, N. C., Papaevangelou, G. J., Ticehurst, J. R., and Purcell, R. H.,** Fecal excretion of Greek strains of hepatitis A virus in patients with hepatitis A and in experimentally infected chimpanzees, *J. Infect. Dis.*, 154, 56, 1986.
66. **Jansen, R. W., Newbold, J. E., and Lemon, S. M.,** Combined immunoaffinity cDNA-RNA hybridization assay for detection of hepatitis A virus in clinical specimens, *J. Clin. Microbiol.*, 22, 984, 1985.
67. **Cohen, J. K., Rosemblum, B., Ticehurst, J. R., Daemer, R., Feinstone, S., and Purcell, R. H.,** Complete nucleotide sequence of an attenuated hepatitis A virus: comparison with wild-type virus, *Proc. Natl. Acad. Sci. U.S.A.*, 1987, in press.
68. **Racaniello, V. R. and Baltimore, D.,** Cloned poliovirus complementary DNA is infectious in mammalian cells, *Science*, 214, 916, 1981.
69. **Semler, B. L., Dorner, A. J., and Wimmer, E.,** Production of infectious poliovirus from cloned cDNA is dramatically increased by SV40 transcription and replication signals, *Nucleic Acids Res.*, 12, 5123, 1984.
70. **Mizutani, S. and Colonno, R. J.,** In vitro synthesis of an infectious RNA from cDNA clones of human rhinovirus type 14, *J. Virol.*, 56, 628, 1985.
71. **Anderson, D. A., Locarnini, S. A., Ross, B. C., Coulepis, A. G., Anderson, B. N., and Gust, I. D.,** Single-cycle growth kinetics of hepatitis A virus in BSC-1 cells, in *Positive Stranded RNA Viruses*, Rueckart, R., Ed., Alan R. Liss, New York, 1987, in press.
72. **Ehrenfeld, E.,** Poliovirus-induced inhibition of host-cell protein synthesis, *Cell*, 28, 435, 1982.
73. **Etchison, D., Hansen, J., Ehrenfeld, E., Edery, I., Sonenberg, N., Milburn, S., and Hershey, J. W. B.,** Demonstration *in vitro* that eucaryotic initiation factor 3 is active but that cap-binding protein complex is inactive in poliovirus-infected HeLa cells, *J. Virol.*, 51, 832, 1984.
74. **Etchison, D. and Fout, S.,** Human rhinovirus 14 infection of HeLa cells results in the proteolytic cleavage of the p220 cap-binding complex subunit and inactivates globin mRNA translation in vitro, *J. Virol.*, 54, 634, 1985.
75. **Shatkin, A.,** mRNA Cap binding proteins: essential factors for initiating translation, *Cell*, 40, 223, 1985.
76. **Lee, K. A. W., Edery, I., Hanecak, R., Wimmer, E., and Sonenberg, N.,** Poliovirus protease 3C (P3-7c) does not cleave P220 of the Eucaryotic mRNA cap-binding protein complex, *J. Virol.*, 55, 489, 1985.
77. **Bernstein, H. D., Sonenberg, N., and Baltimore, D.,** Poliovirus mutant that does not selectively inhibit host cell protein synthesis, *Molec. Cell. Biol.*, 5, 2913, 1985.
78. **Bernstein, H. D., Sarnow, P., and Baltimore, D.,** Genetic complementation among poliovirus mutants derived from an infectious cDNA clone, *J. Virol.*, 60, 1040, 1986.
79. **Mosenkis, J., Daniels-McQueen, S., Janovec, S., Duncan, R., Hershey, J. W. B., Grifo, J. A., Merrick, W. C., and Thach, R. E.,** Shutoff of host translation by encephalomyocarditis virus infection does not involve cleavage of the eucaryotic initiation factor 4F polypeptide that accompanies poliovirus infection, *J. Virol.*, 54, 643, 1985.
80. **Semler, B. L., Johnson, V. H., and Tracy, S.,** A chimeric plasmid from cDNA clones of poliovirus and coxsackievirus produces a recombinant virus that is temperature-sensitive, *Proc. Natl. Acad. Sci. U.S.A.*, 83, 1777, 1986.
81. **Wheeler, C.,** personal communication.
82. **Lemon, S. M.,** personal communication.
83. **Cohen, J.,** personal communication.

School[7] and have been widely reported in studies of outbreaks of hepatitis A in institutions and in family groups.[10-13]

The single most important factor affecting the clinical expression of disease appears to be age. It has been estimated that from 90 to 95% of infections in children under the age of 5 years are silent and that the proportion of infections which are associated with symptoms and signs increases with age.[14] In the Greenland epidemic of 1970 to 1974, the frequency with which hepatitis was detected in susceptible children varied from 1% in the first year of life to 24% by the age of 15 years,[15] data which is in accord with the low rates of clinically apparent infection in children attending day-care centers in the U.S.[16] By contrast, relatively high rates of clinical disease are seen among adults who are exposed to the virus. In the well-documented outbreak which occurred in a college football team in Worchester, Mass. in the early 1970s, 90 previously healthy team members and coaches developed symptoms of hepatitis or transient elevations of alanine aminotransferase (ALT) after drinking contaminated tap water.[17] Thirty-two team members became jaundiced and an additional 58 developed transient nausea, vomiting, or diarrhea accompanied by transient ALT elevations or had ALT elevations alone. As this outbreak took place prior to the discovery of HAV, the data was interpreted to indicate a high attack rate (93%) with a ratio of clinical/subclinical infections of 1:2. Retrospective analysis of sera collected during this outbreak has shown, however, that only those people who were jaundiced developed serological evidence of infection with HAV.[18] None of the so-called subclinical infections were confirmed serologically, indicating that infection in adults is usually associated with disease. These data have been confirmed by two other studies of naturally[19] and experimentally[20] infected adults in which about 90% of those who had serologically confirmed infections developed symptoms or signs of disease.

Case fatality rates are usually derived from groups of patients who have been hospitalized, however, as most infections with HAV are not sufficiently severe to result in the patient being admitted to hospital, these figures grossly overestimate the risk of a fatal outcome. Even so, such figures as exist indicate that liver failure and death are rare, occurring in less than 1 in 500 hospitalized patients. In the largest hospital-based series that has been reported, 2174 patients with serologically confirmed hepatitis A (who were admitted to establish the diagnosis or for social reasons) only three died — a case fatality rate of 0.14%.[21]

B. Clinical Features

Hepatitis A is an acute or subacute infection with generalized systemic manifestations which is accompanied at times by jaundice. The course of the disease is variable and appears to be affected by a number of factors such as age, gender, and the physical condition of the patient at the onset of disease. In childhood the disease is usually anicteric and frequently asymptomatic — infection being detected only by biochemical evidence of liver dysfunction or by serological tests. Increased severity is marked by a variety of vague complaints such as fatigue, malaise, anorexia, and nausea. These symptoms are usually short lived and followed by a complete recovery.

1. Prodromal Illness

Patients who develop jaundice are usually able to describe a short prodromal illness lasting from several days to a week or more which is characterized by increasing fatigue, malaise, loss of appetite, nausea, and vomiting. These symptoms are rarely severe enough to cause the patient to stay home from work or to seek medical attention. The patient may complain of abdominal discomfort or tenderness over the liver. In one study in which most of the subjects were children, the mean duration of the prodrome was 5 days; however in 14% of patients, symptoms were detected for 10 or more days prior to the onset of jaundice.[22] Similarly, in a group of British soldiers with hepatitis, the prodromal period was usually less than a week, but exceeded this duration in 12% of patients.[23]

Table 1
SYMPTOMS REPORTED BY PATIENTS WITH HEPATITIS A

Symptom	Percent	Symptom	Percent
Appearance of dark urine	94	Drowsiness	49
Lassitude and loss of appetite	90	Irritability	43
Nausea	87	Itching	42
Weakness	77	Constipation	29
Fever	75	Diarrhea	25
Vomiting	71	Arthralgia	21
Headache	70	Sore throat	20
Abdominal discomfort	65	Running nose	14
Paler feces	52	Cough	7
Myalgia	52		

2. *Major Symptoms and Signs*

Flu-like symptoms are common in the early stages of the illness; fever (37.5 to 38.5°C) occurs in about three quarters of patients and may be accompanied by chills; headache is common but usually mild. Loss of appetite is a very common symptom, with the patient not only losing his appetite but often finding the sight or smell of food, especially fatty foods, nauseating. Vomiting may occur but is usually neither severe nor protracted. The cause of the nausea and vomiting is not known and has been attributed to inflammation of the mucosa of the stomach and duodenum which can be detected by gastroscopy and X-ray studies in both naturally acquired and experimentally infected infections.[24,25] In addition, patients with hepatitis often lose interest in cigarettes and find cigarette smoke offensive. Occasionally, hepatitis A may present in an atypical manner either with diarrhea[26] or fever, cough, or coryza.[27] Such manifestations are more common among young children.

In most patients the first objective sign of illness is the appearance of dark urine (due to bilirubinuria) followed within a few days by paler colored feces and a yellowish discoloration of the sclera, skin, and mucous membranes. Most patients seek medical attention at this stage.

Paler than usual feces are observed in about half the patients during the first week or so of illness. The color usually returns in the second or third week and is a good marker of clinical improvement. Many patients develop itching, which can occasionally be severe enough to require antipruritics, but is usually mild and self limiting. Arthralgia may occur although it is considerably less common than in patients with hepatitis B.

The major symptoms reported by 415 patients involved in outbreaks of hepatitis A, investigated by the Centers for Disease Control, are shown in Table 1.

On physical examination, the patient's liver may be enlarged and sufficiently tender to make palpitation difficult; in adults the liver is usually enlarged to 12 to 14 cm in the vertical axis, has a firm consistency, and a sharp or rounded edge. The spleen is palpable in 5 to 15% of patients and spider naevi may appear on the trunk. These disappear during convalescence and do not indicate chronic liver disease. Otherwise, the physical findings in patients with hepatitis A are generally unremarkable.

Cardiac involvement is rare, although patients with acute hepatitis may have bradycardia and ECGs may show prolongation of the P-R interval and some mild T-wave depression. These changes resolve rapidly during convalescence.[28]

C. General Laboratory Tests

1. *Liver Function Tests*

The clinical diagnosis of acute viral hepatitis can be confirmed by liver function tests which usually include testing the patient's urine for bilirubin and serum for total and direct bilirubin, ALT, aspartate aminotransferase (AST), and alkaline phosphatase levels.[29] The laboratory findings in HAV infection are usually indistinguishable from other forms of viral hepatitis. The most popular and useful tests for detecting parenchymal damage are serum aminotransferase levels which can be performed in most general laboratories. ALT and AST levels are sensitive indicators of liver damage, rising rapidly during the incubation period of the disease and often reaching peak levels at or within a week of the time the patient seeks medical attention.[30] Levels range from 50 to 2000 IU/ℓ, occasionally reaching 10 to 20,000 IU/ℓ, and are usually greater than 500 IU/ℓ — in general, the height of the enzyme elevation is not related to outcome.

In the hepatocytes, ALT is found exclusively in the cytosol of the cell whereas AST is found mainly in the mitochondria. Inflammation of the liver and altered permeability of the plasma membrane leads to a loss of enzymes from the cytosol so that in acute uncomplicated hepatitis, ALT levels are usually higher than levels of AST. High levels of AST are found when severe tissue necrosis is occurring. As the acute illness subsides, transaminase levels usually rapidly return to normal — although elevated levels may continue for several weeks or even months, this is usually followed by a complete recovery.

Alkaline phosphatase levels are useful to monitor the degree of cholestasis and to assist the clinician differentiate between acute viral hepatitis and other causes of jaundice. In hepatitis A, they are usually mildly elevated (less than twice normal). Elevated levels of this enzyme are found in some other viral infections involving the liver such as infectious mononucleosis.

2. Serum Immunoglobulin Levels

Quantitative and qualitative changes in serum immunoglobulins were first noted in some patients with hepatitis in the 1960s.[31,32] The first attempt to relate immunoglobulin levels to the type of hepatitis came from the work of Giles and her colleagues who examined serial sera collected from children who had been experimentally infected with the MS-1 and MS-2 strains of hepatitis.[33] In children who received MS-1, elevated levels of serum IgM were detected within a few days of the onset of biochemical hepatitis; these levels gradually returned to normal over the next 4 or 5 weeks. By contrast, in children who received MS-2, IgM levels remained unchanged throughout the course of the disease. These findings were rapidly confirmed in patients with naturally acquired infections.[34,35] In a prospective study of patients with acute viral hepatitis admitted to Fairfield Hospital in Melbourne, elevated levels of IgM were detected in more than three quarters of patients with hepatitis A.[36] The mean IgM level in the first specimen obtained from each patient after admission was 443 mg/dℓ compared with a mean of 148 (50 to 253) mg/dℓ among controls. The mean levels of IgM among patients with hepatitis A were significantly higher than among patients with hepatitis B (270 mg/dℓ) or HNANB (197 mg/dℓ).

The ratio of IgG/IgM in sera collected during the acute phase of the illness may be useful in distinguishing infection with HAV from the sporadic form of HNANB. Patients with the former disease tend to have high levels of IgM and normal levels of IgG whereas in sporadic HNANB the reverse is the case. In patients with hepatitis an IgG/IgM ratio of greater than six usually indicates that the patient has HNANB (see Figure 1).

While it is known that the IgM response is polyclonal and that only a small fraction of the antibody is directed against HAV, the nature and function of the remaining antibodies are unknown. It has been speculated that these antibodies are stimulated by enteric bacteria which would not be antigenic if the liver was functioning normally.

3. Other

A number of serum proteins and clotting factors are synthesized in the liver. In patients

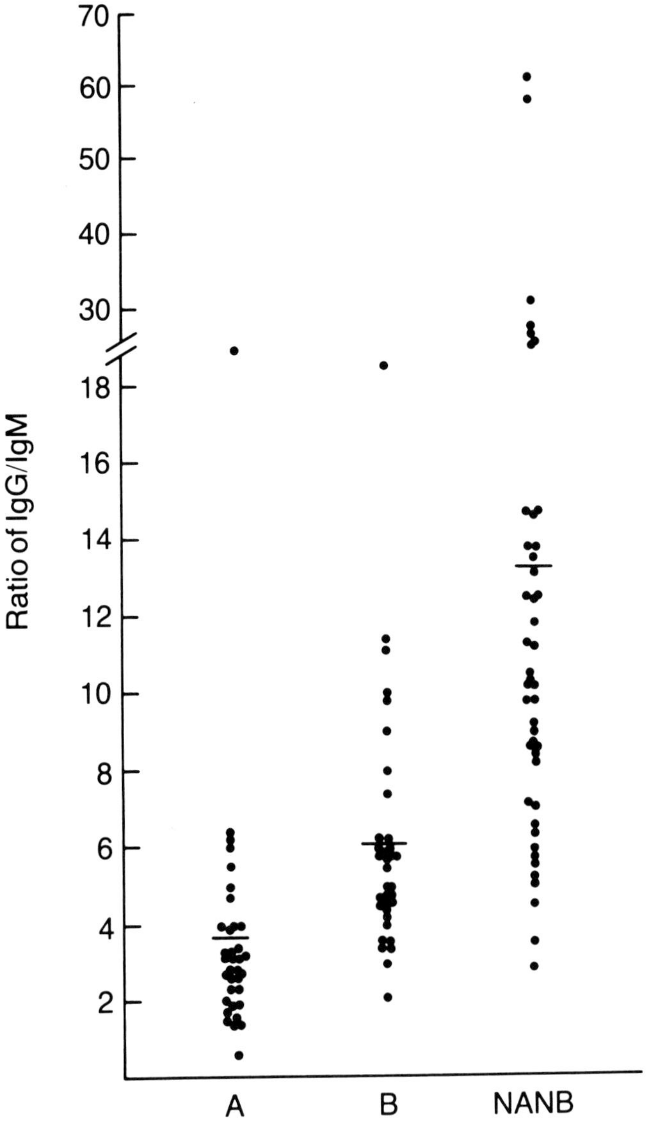

FIGURE 1. Ratio of IgG/IgM levels in patients with acute viral hepatitis admitted to Fairfield Hospital, Melbourne. Note most patients with HNANB were intravenous drug users and were probably suffering from the parenterally transmitted form of the disease. (From Zhuang, H., Kaldor, J., Locarnini, S. A., and Gust, I. D., *Gastroenterology,* 82, 549, 1982. With permission.)

with acute hepatitis A, electrophoretic separation of serum proteins usually reveal no major changes; fibrinogen levels may be depressed and minor coagulation abnormalities can often be detected; however these changes are rarely accompanied by bleeding.[37] These factors are not usually measured unless the patient becomes extremely ill.

Examination of the blood usually reveals a mild lymphocytosis and the presence of occasional atypical mononuclear cells (see Figure 2). The hemoglobin and hematocrit are usually normal[38] although hemolytic anemia (especially in patients with glucose-6-phosphate deficiency) and aplastic anemia are recognized but uncommon complications.[39-41]

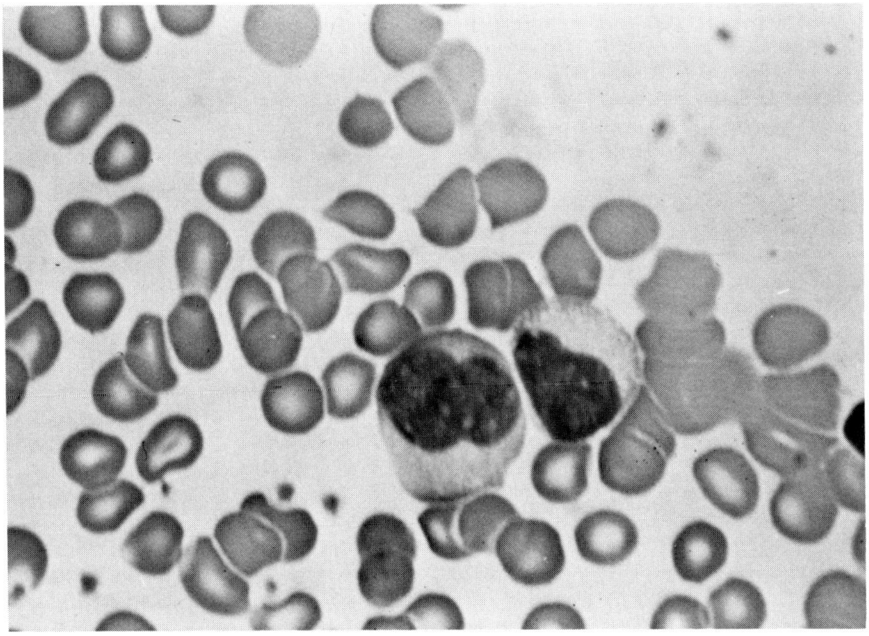

FIGURE 2. A peripheral blood smear from a patient with acute hepatitis A showing the presence of atypical mononuclear cells. (Photo kindly provided by Dr. Greg Buist.)

In 1978, Villarejos and his colleagues reported that elevated levels of antibodies to single-stranded DNA could be regularly detected in the sera of patients with acute viral hepatitis and provided a reliable marker of parenchymal liver damage.[42] Subsequent studies in other laboratories, using more sensitive assays and coded panels of sera, failed to confirm these observations.[43]

In patients who develop fulminant hepatitis, albumin concentrations and measurements of clotting activity are useful indicators of the severity of the illness, prognosis, and response to therapy.

D. Specific Diagnostic Tests

A range of specific serological tests have been developed for confirming infection with HAV. These include detection of virus particles,[44] viral antigens,[45] or specific nucleic acid[46] in the patient's feces; or detecting a characteristic antibody response, either a rising titer of antibodies in sera collected early and late in the illness;[44,47] or detecting the presence of specific IgM in acute-phase sera.[48]

1. Detection of HAV Particles in Feces

HAV particles are shed into the feces during the incubation period of the disease and for several days or weeks after the onset of symptoms.[49] These particles may occasionally be detected by direct EM, however, because of their low concentration and the presence of other particulate material; they are most commonly sought by IEM.

In this technique, 5 to 10% extracts of feces in buffered saline are clarified by low-speed centrifugation and the supernatant fluid reacted with human or animal sera known to contain high titers of anti-HAV. After incubation, the resulting immune complexes are deposited by high-speed centrifugation, the pellet resuspended, placed on a coated E-M grid, negatively stained and examined by transmission EM. Virus particles may be visualized as single particles surrounded by a halo of antibody, immune complexes, or as crystalline arrays. Depending upon the time at which the specimen is collected, one or several hundred particles may be seen per EM grid square. As the distribution of particles may be uneven, it is usual to search a fixed number of grid squares (a minimum of 10 to 20 squares per 400-square grid) before declaring a sample to be negative.

IEM, as initially practiced, was extremely expensive on reagents as reaction volumes of several milliliters of extract and antibody were used. However, in recent years the advent of a range of new centrifuges has made it possible to perform this technique with small volumes of reagents and even to deposit the complexes directly onto the grid. Particles can usually be detected when they are present in a concentration of 10^5 to $10^6/m\ell$ in the starting extract.

While extremely valuable as an experimental tool, IEM is time consuming, requires an experienced microscopist, and access to an EM. In recent years this technique has been largely superseded.

2. Detection of HAAg in Feces

HAAg can be detected in the feces by solid-phase radioimmunoassay (SPRIA)[45,50] and enzyme-linked immunoabsorbent assay.[51] These techniques have essentially the same sensitivity and both are more sensitive then IEM. When serial fecal specimens from patients with acute hepatitis A were examined by SPRIA, HAAg was detected in 45% of specimens collected within the 1 week of the onset of illness (as measured from the onset of dark urine). In the second week of illness, the detection rate fell to 11%. HAAg was rarely detected in the third or fourth week after the onset of symptoms.[49] These findings have been confirmed by other workers. Thus, while the detection of HAAg in a patient's feces confirms the diagnosis of hepatitis A, its absence does not preclude the diagnosis.

3. Detection of HAV RNA in the Feces

With the successful isolation of HAV in cell culture and the production of large quantities of virus, it became possible to apply modern molecular biological techniques to this virus. Several groups have succeeded in cloning the viral cDNA and have produced labeled probes which have proved useful for detecting viral nucleic acid in a variety of specimens.[46,52] Molecular hybridization has been used to detect viral nucleic acid in serial fecal specimens from patients with hepatitis A. Extracts of feces were applied to nitrocellulose paper and reacted with ^{32}P-labeled sequences of cloned HAV, cDNA. This technique has been found to be more sensitive than RIA for detection of HAAg and to detect the shedding of virus in the feces for a longer period of time.

In a recent study,[52] of 74 children and adults with sporadic hepatitis A admitted to the Athens Infectious Diseases Hospital, serial fecal specimens were obtained from each patient on admission and at 3 to 5 days afterward. Suspensions (20% w/v) of each sample were prepared in phosphate buffered saline (PBS) and after clarification by centrifugation at 3000 $\times$ g for 30 min tested for HAAg by SPRIA. Suspensions (5% in 10 m*M* Tris-HCl, 0.1 *M* NaCl, 0.1% NP 40) of each sample were also treated to extract HAV RNA and hybridized with radiolabeled cloned HAV cDNA. HAAg was detected in 18 of 74 (24.3%) patients by RIA and in 42 of 74 (56.7%) by hybridization. HAV RNA was detected more frequently in the second or third fecal specimen than HAAg; moreover, serial titration of 18 samples which were positive by both tests showed that HAV RNA was usually detectable at 4- to 10-fold higher dilution than HAAg.

This technique is useful for detecting the presence of HAV and for investigating patterns of viral shedding, but as with tests for viral particles or viral antigens, only positive results are of clinical significance.

4. Detection of Rising Titers of Anti-HAV

A variety of techniques are available for detecting antibodies to HAV including IEM,[44] complement fixation (CF),[53] immune-adherence hemagglutination (IAHA),[54] SPRIA,[45] enzyme-linked immunoabsorbent assay (EIA),[51] and immunofluorescence (IF).[55] The latter techniques are quite extensively used for detection of infection with HAV and are commercially available from a number of suppliers.

The IAHA test is not widely used in clinical laboratories because the antibodies are usually not detected until several weeks after the onset of symptoms. Unlike the EIA or RIA tests which detect all types of immunoglobulin and become positive early in the illness. IAHA does not detect HAV-specific IgM.

It is only possible to demonstrate a rising titer of antibodies in subjects who are studied prospectively or in patients seen early in their infection. Antibodies to HAV (detectable by RIA or EIA) develop early in the illness and can often be detected while antigen is still being shed into the feces.[56] The titer of antibody rises rapidly, often reaching peak levels within 1 to 2 weeks of the onset of symptoms. For this reason it may be difficult to demonstrate a rising titer especially if the patient does not present until the second week of illness. The corollary to this is that if a patient presents with acute viral hepatitis and anti-HAV cannot be detected in the first serum specimen collected, it is unlikely that he is suffering from hepatitis A. In exceptional circumstances, such as during experimental infections or when hepatitis A develops in patients who are having blood specimens collected regularly for other purposes, elevated levels of ALT may be detected for short periods prior to the development of specific antibody.

Antibodies to HAV persist for many years and probably for life. Their presence indicates previous infection with HAV and immunity to the disease. In our experience it is possible to detect rising titers of antibody in about two thirds of patients with acute hepatitis A; in the remaining third, high levels of antibody are already present at presentation. For this reason and because of the value of class-specific assays in confirming a diagnosis of rubella we began to seek evidence of a specific IgM response and to develop assays capable of detecting HAV-specific IgM.

5. Detection of HAV-Specific IgM

The first indication that specific IgM was produced during acute infection with HAV came from the work of Locarnini and his colleagues in Melbourne[57] who reacted fecal extracts from patients with hepatitis A with their own acute-phase sera and searched for virus-antibody complexes by IEM. They noted that the HAV particles in the resultant complexes were separated by large antibody molecules, some of which formed staple-like bridges, suggesting that the predominant population of antibodies was IgM.

Subsequently this group was able to demonstrate the specificity of this reaction by immunoferritin staining and to detect two populations of antibodies with sedimentation coefficients of 19 S and 7 S, respectively. 19 S Antibodies predominated early in the illness and were rapidly replaced by 7 S antibodies (see Figure 3).[48]

Following these observations, this group and others developed a variety of approaches to detection of HAV-specific IgM by RIA and EIA.[58] The most widely used assays involve attachment of μ-chain specific antihuman IgM to a solid phase which is then incubated with a patient's serum to allow the separation of any IgM molecules present. Following the addition of purified HAAg, the presence of anti-HAV IgM is detected by the addition of labeled anti-HAV IgG (see Figure 4). This test has proved to be the single most useful assay for confirming a diagnosis of hepatitis A, and is invariably positive at the time of presentation and can be performed on a single serum specimen.

In general, rheumatoid factors do not react in these "capture" type IgM assays but false positives due to RF can be virtually eliminated by using Fab IgG fragments as the detector

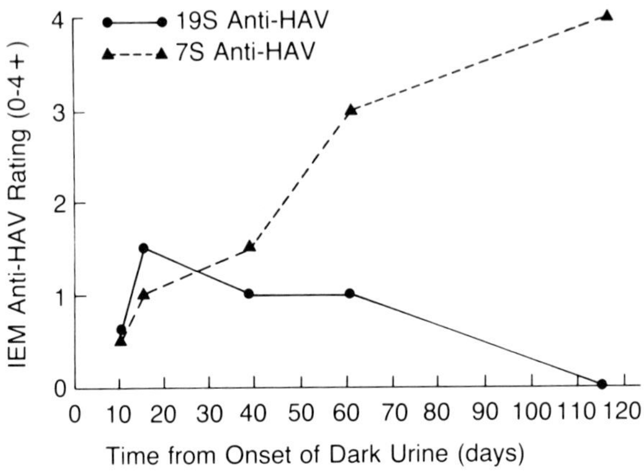

FIGURE 3. Development of anti-HAV IgM and IgG detected by IEM following acute infection with hepatitis A. (From Locarnini, S. A., Ferris, A. A., Lehmann, N. F., and Gust, I. D., *Intervirology*, 8, 309, 1977. With permission.)

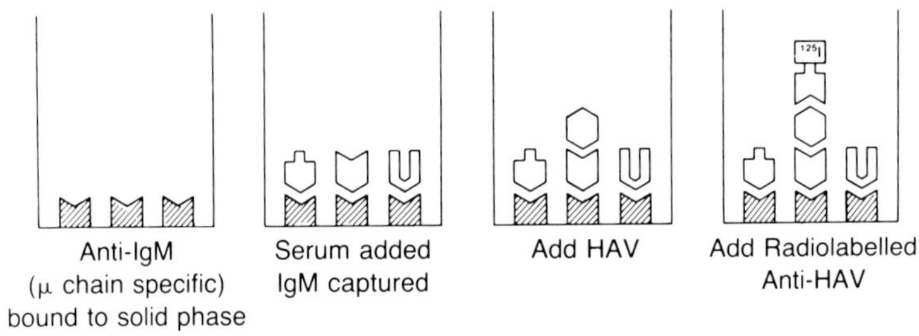

FIGURE 4. Diagram illustrating the principles of IgM-capture immunoassays for detection of hepatitis A specific IgM.

antibody. Furthermore, sensitivity of these tests is usually adjusted by dilution of the patient's serum (up to 1:4000). This further reduces false positives due to RF.

Recently a simple immunofluorescent assay for anti-HAV IgM, using HAV-infected cells as the substrate has been described.[55] This assay is easy to perform and is suitable for laboratories which are familiar with this technique and less conversant with EIA or RIA.

Although a number of strains of HAV have been adapted to cell culture, the isolation rate from specimens collected from hospitalized patients is relatively low, so that in contrast to many other infections with enteroviruses, isolation of the virus in cell culture is not yet a practical method of confirming the diagnosis.

E. Histological Findings

Liver biopsy is rarely indicated to establish a diagnosis because other tests are available and this procedure is associated with a small but finite risk to the patient. When biopsies have been performed the morphological changes which have been observed are similar to those seen in other forms of acute hepatitis.[59,60] (see Figure 5).

Hepatocytes are generally swollen, have indistinct plasma membranes, and enlarged nuclei, an appearance commonly referred to as ballooning degeneration. The cytoplasma of the cells is relatively featureless. Some cells become shrivelled and eosinophilic and there is a diffuse accumulation of mononuclear cells throughout the lobules. Breakdown products of damaged cells enter the sinusoids where they are ingested by Kupffer cells.

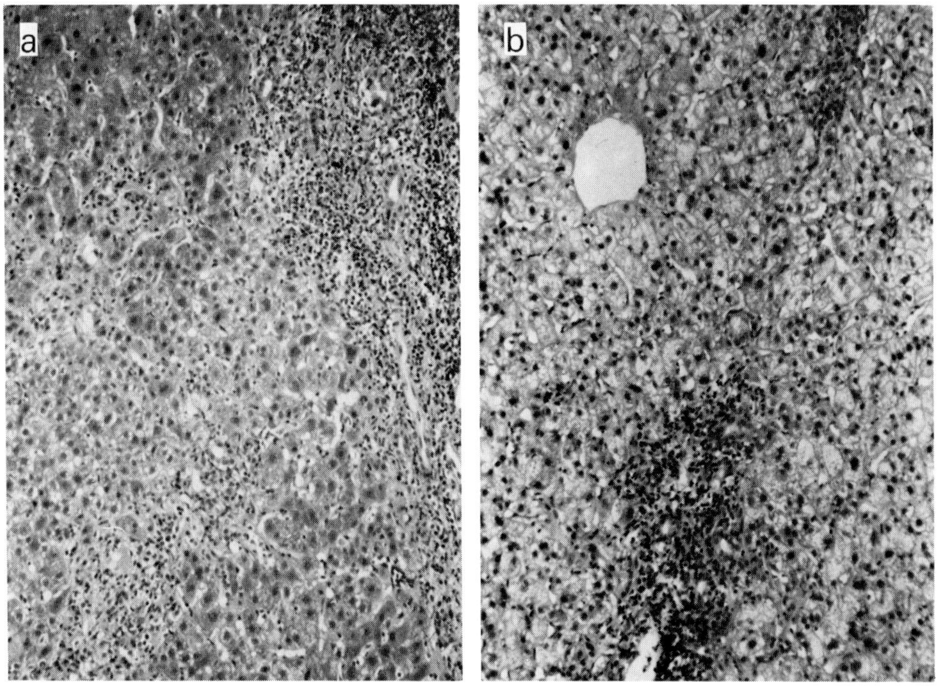

FIGURE 5. Low-power view of biopsies obtained from adults with acute hepatitis A showing typical changes (see text). (Pictures kindly provided by Dr. Hans Popper, Mt. Sinai School of Medicine, New York.)

The portal tracts are enlarged by edema and inflammatory cells. In general, regeneration occurs rapidly and the liver returns to normal within 8 to 12 weeks. Hepatitis A antigen and HAV particles can be detected in the cytoplasm of infected cells by immunofluorescence, immunoperoxidase staining, or thin-section EM (see Chapter 6).

F. Outcome

1. Usual Course

Most studies on the duration of illness are biased by the fact that they are based on hospitalized cases, which usually represent most forms of the disease. In addition, until relatively recently the duration of hospitalization was often prolonged because of an ill-founded belief that early activity increased the chance of sequelae or a relapse of symptoms.

The duration of illness varies, but by the third week most patients feel better, have lost their hepatomegaly, and have serum transferase levels which are normal or nearly normal. In many patients the appearance of jaundice is associated with a rapid resolution of symptoms. Such patients often find it curious that they continue to work while feeling ill but are confined to bed when they feel much better! Prolonged periods of enzyme elevation are occasionally encountered, but this is unusual.[5,61,62]

Some patients, particularly students preparing for exams and athletes, become depressed at the prospect of a prolonged period of illness; others may develop a more severe depression requiring treatment. Although depression is quite common it is usually mild and self limiting.

2. Cholestatic Form

A cholestatic form of hepatitis A characterized by fever, marked pruritis, prolonged jaundice, and weight loss is occasionally encountered. In a recent report by Gordon et al.[63] six cases were described in detail; peak serum bilirubin levels ranged from 12 to 29 mg/

dalton, with jaundice lasting from 12 to 18 weeks. Aminotransferase levels fell early in the course of the illness and were below 500 IU/ℓ for most of the icteric phase. Liver biopsies were performed on several patients and revealed marked centrilobular cholestasis and portal inflammation. Although in each instance the disease resolved spontaneously, it has been suggested that a short course of steroids may shorten the period of cholestasis and enable the patient to resume normal activities more quickly.[64,65]

3. Relapses

Although acute icteric hepatitis A is usually a self-limiting disease, relapses have occasionally been reported.[66-68] It is difficult to place much weight on the studies which were carried out prior to the availability of aminotransferase tests and specific assays for HAV infection and in which relapse rates of between 1 to 18% were described. In several of these settings more than one type of hepatitis virus was circulating and it is possible that the "relapses" were actually second attacks due to different agents. With the advent of specific laboratory tests for infection with HAV, HBV, CMV, and EBV it has been possible to demonstrate that most "relapses" of hepatitis are due to infection with another virus.[69] There remains however, a small group of patients with serologically confirmed hepatitis A in whom relapses are documented and who have no evidence of infection with another hepatotoxic agent.

The first evidence of this was provided by Weir in 1981[70] who described two cases and by Bamber[71] who observed relapses in 3 of 88 patients with hepatitis A admitted to hospital in London. These observations were confirmed by Gruer et al.[72] who described five patients with hepatitis A who were readmitted to hospital with a second epidodes of hepatitis 7 to 10 weeks later; in each instance the illness resolved rapidly. An additional group of patients in which hepatitis A appeared to relapse were described by Chiriaco et al.[73] from Padua, Italy. Between January 1983 and July 1984 this pattern was observed in 23 children admitted to Brindisi Hospital with acute viral hepatitis. In each case the diagnosis of hepatitis A was established on the basis of a characteristic clinical picture supported by biochemical tests and the presence of hepatitis A-specific IgM in acute-phase sera. All were negative for HBsAg, anti-HBV, EBV, and CMV. Each patient's symptoms and jaundice subsided within 5 to 30 days (mean 13 days) as their ALT levels fell to normal or near normal levels; however, each developed a second rise in transaminase levels reaching 210 to 2365 IU/ℓ (mean 710) from 15 to 90 days (mean 31 days) later. Sera collected during these episodes showed no serological evidence of infection with HBV, CMV, or EBV. In all but four patients the "relapses" were anicteric and all patients made uneventful recoveries and had normal LFTs when reviewed 6 months later. No attempt was made to detect shedding of HAV in the feces during the acute phase of the illness or during relapses. In 1985 Jacobsen and his colleagues in Boston[62] described three patients in Boston with serologically confirmed hepatitis A who developed elevated transaminases 1 to 2 months after their initial illness. None of these cases was associated with a rise in anti-HAV IgM titer and no attempt was made to detect HAV in the patients' feces. It is possible that some of the patients had dual infections with HAV and HNANB which were either acquired simultaneously or during the period of hospitalization.

Recently Sjogren and her colleagues[74] have provided evidence that relapses are associated with a reactivation of infection and shedding of HAV in the feces. Of 256 patients with serologically confirmed hepatitis A in Rosario, Argentina, 17 (6.6%) developed a relapse of symptoms and a significant increase in ALT levels 30 to 90 days after the onset of their primary illness. Detailed follow up, including serial fecal samples, was possible in seven patients aged between 8 and 27 years. Seven fecal specimens obtained from four patients during their relapses apparently contained 27-nm virus-like particles which could be aggregated by guinea pig and human sera containing anti-HAV, but not by preinoculation or

Table 2
OUTCOME OF ACUTE VIRAL HEPATITIS IN PATIENTS ADMITTED TO FAIRFIELD HOSPITAL, MELBOURNE BETWEEN JANUARY 1971 AND JUNE 1983

Age (years)	Patients	Deaths
0—9	524	1
10—19	506	1
20—29	745	0
30—39	290	0
40—49	67	0
≧50	42	1
Total	2174	3

preinfection sera which did not contain specific antibody. Fecal specimens collected later in convalescence were uniformly negative. The specificity of the reaction was demonstrated by blocking experiments in which the fecal extracts were incubated with preinfection and convalescent sera from an experimentally infected chimpanzee prior to test. The presence of HAV in these specimens was confirmed by nucleic acid hybridization using a ^{32}P-labeled complementary DNA probe.

In the absence of specimens collected during the patients' first illness and early convalescence, it is not possible to determine whether these patients had unusually long periods of virus shedding or whether these findings represent a reactivation of infection.

Other complications of hepatitis A such as acute renal failure and polyneuritis, myelitis, or encephalopathy have also been described[75,76] but appear to be extremely rare.

G. Fulminant Hepatitis

Fulminant hepatitis is a rare complication of hepatitis A infection and only a small number of well documented cases exist in the literature. The case fatality rate of patients whose illness is sufficiently severe to merit hospitalization is less than 0.2%. In the largest series reported to date, that at Fairfield Hospital, Melbourne, the case fatility rate in 2174 consecutive, virologically confirmed cases was 0.14%[21] (see Table 2). Surprisingly, two of the deaths occurred in patients who were less than 20 years old, an age at which deaths due to hepatitis are extremely rare. Both patients had previously been in good health and neither had any predisposing factors for severe illness. The mean time from the onset of dark urine to the first signs of encephalopathy was 7 days and from encephalopathy to death, 6 days.

A number of common-source, water-borne outbreaks of hepatitis have been documented in developing countries which have been associated with alarmingly high case-fatality rates, especially in pregnant women. These outbreaks have been demonstrated to be due to an agent or agents other than HAV (see Chapter 12).

Fulminant hepatitis A is characterized by increasing severity of jaundice, deterioration in liver function, drowsiness, fetor hepaticus, and eventually encephalopathy and coma.[77] Mortality appears to be correlated with age, with survival being rare over the age of 50 years.[78] Danger signs include excitability, irritability, insomnia, sleepiness, confusion, and severe vomiting. One of the most ominous signs is a rapid decrease in size of the liver.

Laboratory evidence of deteriorating liver function includes a prolongation of the prothrombin time, an increase in serum bilirubin levels, and sometimes a fall in very high aminotransferase levels. Fever develops in most patients and leukocytosis and neutrophilia are common. Gastrointestinal bleeding may occur as a result of a fall in serum fibrinogen levels, impaired synthesis of clotting factors, or diffuse intravascular coagulation. Renal failure may occur, usually as a terminal event.

At autopsy the liver is usually pale and flabby (see Figure 6). When examined by light microscopy there is a virtually complete destruction of the hepatic parenchyma with only a reticulin framework and portal tracts remaining. Occasionally small groups of surviving hepatocytes can be seen close to portal tracts — these usually show evidence of regenerative changes. The spaces between sinusoids are filled with necrotic material and scavenger cells but perhaps surprisingly there is little histological evidence of a vigorous inflammatory response.

There is no evidence of fibrosis and patients with hepatitis A who survive hepatic coma do not appear to develop cirrhosis.

IV. MANAGEMENT

As there is no specific treatment for hepatitis A, management is basically supportive and is aimed at making the patient comfortable and maintaining nutritional and electrolyte balance.

Bed rest is recommended during the acute phase of the illness, though its value has been debated since the time of Hippocrates and remains scientifically unproven. The notion that early, even vigorous physical activity has an adverse effect on the course of the disease appears to be a myth.[79,80] Nor is there any scientific evidence for the adverse effects of ingestion of modest quantities of alcohol during the convalescent phase of the illness.

The notion that early activity could have an adverse effect appears to have arisen from the experience with American troops during World War II.[81] When troops with hepatitis A were permitted early ambulation, relapses of symptoms appeared to be more common. As a result, the average period of bed rest recommended by the army rose from 30 days in 1943 to 1944, to 60 days in 1947 to 1949, to almost 90 days in 1949 to 1981.[82]

Subsequent studies, however, suggest that the data obtained with debilitated troops cannot be extrapolated to the general community.[79,83] Not only does the duration of bed rest not seem to influence the outcome of the disease but there is no evidence that strenuous physical activity during the early convalescent period is harmful.

These days most physicians suggest a period of bed rest without the need for prolonged inactivity; hospitalization is sometimes the only way to provide adequate rest, especially for parents of young families. In general it is best to be guided by how the patient feels and allow him to gradually increase his activities as he begins to feel better.

The need or otherwise for dietary restrictions has also been debated, however, these days most physicians allow patients to eat whatever they like provided that the diet contains adequate amounts of calories, protein, and fluid. In the first few days this may be difficult to achieve because of nausea and vomiting, although the patient is usually able to take fluids.

Patients with hepatitis A usually choose to avoid fatty foods because they make them nauseated. There is no evidence, however, that a high-fat diet reduces the rate of recovery.[79] It has been traditional to recommend that patients recovering from hepatitis abstain from alcohol for several months, presumably on the grounds that alcohol is hepatotoxic and may interfere with the process of regeneration. While it is clear that the regular intake of alcohol is associated with liver damage, there is no clinical data to support the concept that early consumption of modest quantities of alcohol affects recovery from the disease.[83]

A variety of heroic measures have been used for the treatment of hepatic failure. These include exchange transfusions,[86] extracorporeal transfusion,[86] and plasmapheresis.[88] None of these are of proven efficacy. Treatment with massive doses of steroids or immune serum globulin has not been effective.[89]

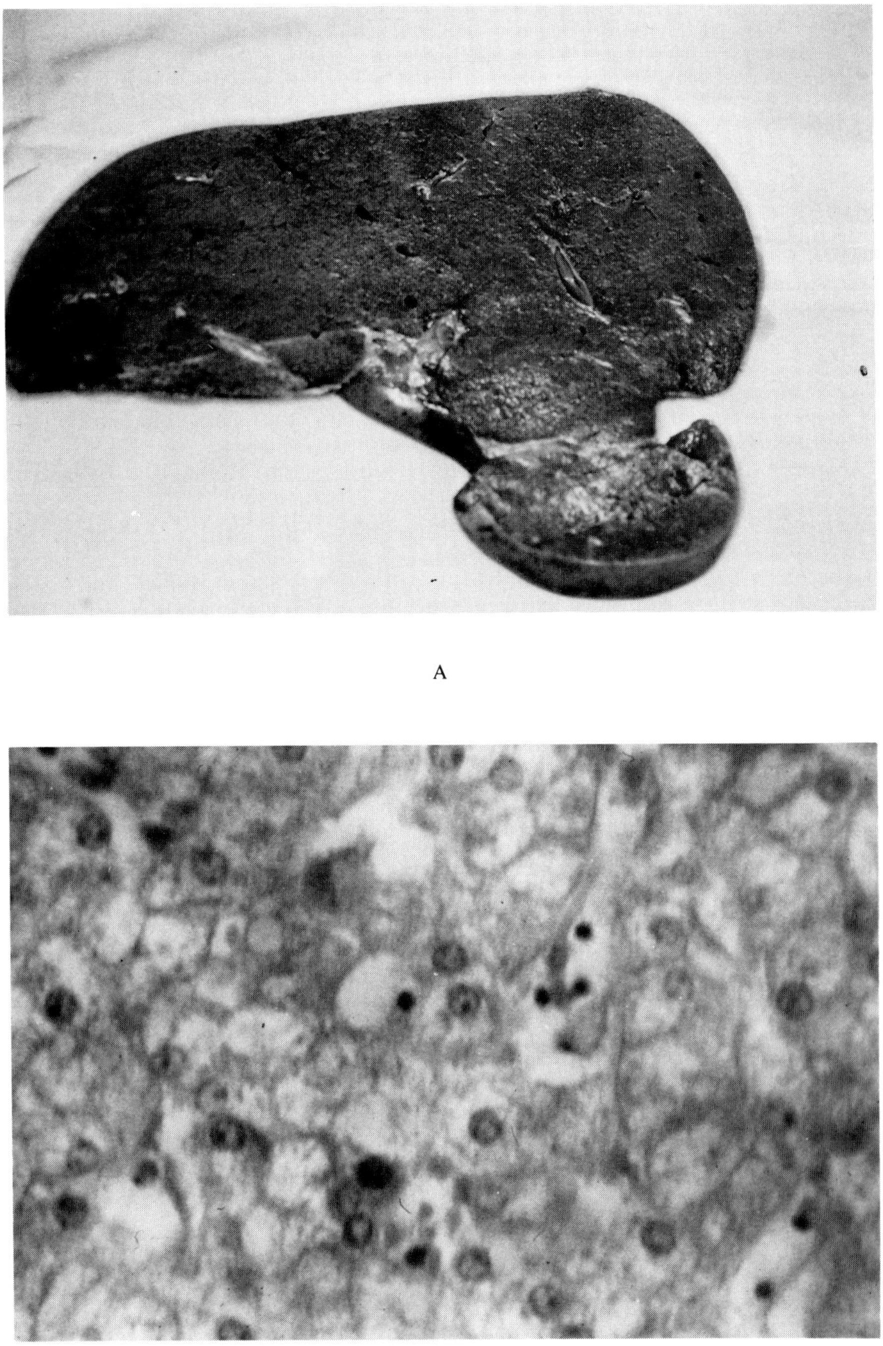

A

B

FIGURE 6. (A) Characteristic post-mortem appearance of the liver; and (B) typical histological changes in a patient with fatal hepatitis A. (Photo kindly provided by Dr. G. Buist.)

REFERENCES

1. **Lindstedt, F.,** Beitrag zur Kenntnis Icterus Catarrhalis mit besonderer Rucksicht auf die Incubationszeit dessen epidemischen formen, *Nord. Med. Ark.*, 51, 583, 1919.
2. **Picles, W. N.,** Epidemic catarrhal jaundice: an outbreak in Yorkshire, *Brit. Med. J.*, 1, 944, 1930.
3. **Fraser, R.,** A study of epidemic catarrhal jaundice, *Can. J. Public Health*, 22, 396, 1981.
4. **Dienstag, J. L., Gust, I. D., Lucas, C. R., Wong, D. C., and Purcell, R. H.,** Mussel-associated viral hepatitis, type A: serological confirmation, *Lancet*, 1, 561, 1976.
5. **Routenberg, J. A., Dienstag, J. L., Harrison, W. O., Kilpatrick, M. E., Hooper, R. R., Chisari, F. V., Purcell, R. H., and Forness, M. L.,** Food borne epidemic of hepatitis A: clinical and laboratory features of acute and protracted illness, *Am. J. Med. Sci.*, 278, 123, 1979.
6. **Havens, W. P., Jr.,** Hepatitis, *Medicine*, 27, 279, 1948.
7. **Krugman, S., Ward, R., Giles, J. P., Bodansky, O., and Jacobs, A. M.,** Infectious hepatitis: detection of virus during the incubation period and in clinically inapparent infections, *N. Engl. J. Med.*, 261, 729, 1959.
8. **Krugman, S. and Giles, J. P.,** Viral hepatitis: new light on an old disease, *JAMA*, 212, 1019, 1970.
9. **MacCallum, F. O.,** Early studies of viral hepatitis, *Br. Med. Bull.*, 28, 105, 1972.
10. **DeRitis, F., Mallucci, L., Coltorti, M., Giusti, G., and Caldera, M.,** Anicteric virus hepatitis in a closed environment as shown by serum transaminase activity, *Bull. WHO*, 20, 589, 1959.
11. **Goldberg, D. M. and Campbell, D. R.,** Biochemical investigation of outbreak of infectious hepatitis in a closed community, *Br. Med. J.*, 2, 1435, 1962.
12. **Krugman, S., Ward, R., and Giles, J. P.,** Infectious hepatitis: studies on the effect of gamma globulin and on the incidence of inapparent infection, *JAMA*, 174, 823, 1960.
13. **Schneider, A. J. and Mosley, J. W.,** Studies of variations of glutamicoxaloacetic transaminase in serum in infectious hepatitis, *Pediatrics*, 24, 367, 1959.
14. **Hollinger, F. B., Melnick, J. L., and Robinson, W. S.,** *Clinical Features in Viral Hepatitis*, Raven Press, New York, 1985, 72.
15. **Skinhoj, P., Gluud, C., and Ramsoe, K.,** Traveller's hepatitis: origin and characteristics of cases in Copenhagen 1976—1978, *Scand. J. Infect. Dis.*, 13, 1, 1981.
16. **Hadler, S. and Erben, J.,** *Viral Hepatitis*, Szmuness, W., Alter, H. J., and Maynard, J. E., Eds., Franklin Institute Press, Philadelphia, 1982, 740.
17. **Morse, L. J., Bryan, J. A., Hurley, J. P., Murphy, J. F., O'Brien, T. F., and Wacker, W. E. C.,** The Holy Cross College football team hepatitis outbreak, *JAMA*, 219, 706, 1972.
18. **Friedman, L. S., O'Brien, T. F., Morse, L. J., Chang, L. W., Wacker, W. E. C., Ryan, D. M., and Dienstag, J. L.,** Revisiting the Holy Cross football team hepatitis outbreak (1969) by serological analysis, *JAMA*, 254, 774, 1985.
19. **Benenson, M. W., Takafuji, E. T., Bancroft, W. H., Lemon, S. M., Callahan, M. C., and Leach, D. A.,** A military community outbreak of hepatitis type A related to transmission in a child care facility, *Am. J. Epidemiol.*, 112, 471, 1980.
20. **Decker, R. H., Overby, L. R., Ling, C. M., Frosner, G., Deinhardt, F., Boggs, J.,** Serological studies of transmission of hepatitis A in human, *J. Infect. Dis.*, 139, 74, 1979.
21. **McNeill, M., Hoy, J. F., Richards, M. J., Lehmann, N. J., Dimitrikakis, and M., Gust, I. D.,** Etiology of fatal hepatitis in the Melbourne, *Med. J. Aust.*, 2, 637, 1984.
22. **Molner, J. G. and Meyer, K. F.,** Jaundice in Detroit, *Am. J. Public Health*, 30, 509, 1940.
23. **Zuckerman, A. J.,** The clinical and laboratory features of acute hepatitis in the Royal Air Force, *Mon. Bull. Health Lab. Serv.*, 24, 340, 1965.
24. **Havens, W. P., Jr., Kushlan, S. D., and Green, M. R.,** Experimentally induced infectious hepatitis: roentgenographic and gastroscopic observations, *Arch. Intern. Med.*, 79, 457, 1947.
25. **Knight, W. A. and Cogswell, R. C.,** Preliminary observations of the gastric mucosa in patients with infectious hepatitis, *JAMA*, 128, 803, 1945.
26. **Ellsbury, E., Owosekum, O., Mackey, K., Ten Eyck, D., and Mahoney, L. E.,** Outbreak of viral hepatitis in the staff of a pediatric ward — California, *Morb. Mort. Wkly. Rep.*, 26, 77, 1977.
27. **Storch, G., McFarland, L. M., Kelso, K., Heilman, C. J., and Caraway, C. T.,** Viral hepatitis associated with day-care centers, *JAMA*, 242, 1514, 1979.
28. **Hoaglund, C. L. and Shank, R. E.,** Infectious hepatitis: a review of 200 cases, *JAMA*, 130, 615, 1946.
29. **Clermont, R. J. and Chalmers, T. C.,** The transaminase tests in liver disease, *Medicine, (Baltimore)*, 46, 197, 1967.
30. **Boggs, J. D., Melnick, J. L., Conrad, M. E., and Felsher, B. F.,** Viral hepatitis: clinical and tissue culture studies, *JAMA*, 214, 1041, 1970.
31. **Lee, F. I.,** Immunoglobulins in viral hepatitis and active alcoholic liver disease, *Lancet*, 2, 1043, 1965.

32. **Wolheim, F. A.,** Immunoglobulins in the course of viral hepatitis and in cholestatic and obstructive jaundice, *Acta Med. Scand.,* 183, 473, 1968.
33. **Giles, J. P. and Krugman, S.,** Viral hepatitis: immunoglobulin response during the course of the disease, *JAMA,* 208, 497, 1968.
34. **Iwarson, S. and Holmgren, J.,** Patterns of immunoglobulins in acute viral hepatitis: relation to hepatitis-associated antigen, *J. Infect. Dis.,* 125, 178, 1972.
35. **Peters, C. J. and Johnson, K. M.,** Serum immunoglobulins in acute viral hepatitis: relation to hepatitis-associated antigen, *Clin. Exp. Immunol.,* 11, 381, 1972.
36. **Zhuang, H., Kaldor, J., Locarnini, S. A., and Gust, I. D.,** Serum immunoglobulin levels in acute A, B and non-A, non-B hepatitis, *Gastroenterology,* 82, 549, 1982.
37. **Humphrey, M. J., Hansen, R. C., Bradley, D. W., Nagle, R. B., and Stillman, A. E.,** Severe clotting abnormalities and transient hepatic fibrosis in acute anicteric hepatitis A, *J. Pediatr.,* 95, 987, 1979.
38. **Conrad, M. E., Schwartz, F. D., and Young, A. A.,** Infectious hepatitis — a generalized disease: a study of rival, gastrointestinal and hematologic abnormalities, *Am. J. Med.,* 37, 789, 1964.
39. **Hans Barger, E. A., Jr., and Hyun, B. H.,** Acute hemolytic aneuria in viral hepatitis: report of two cases, *Med. Mon.,* 90, 134, 1963.
40. **Hagler, L., Pastore, R. A., and Bergin, J. J.,** Aplastic anemia following viral hepatitis: report of two fatal cases and literature review, *Medicine (Baltimore),* 54, 139, 1975.
41. **Parker, M. L.,** Aplastic anaemia and infectious hepatitis, *Lancet,* 2, 216, 1971.
42. **Villarejos, V. M., Arquembourg, P. C., Visona, K. A., and Gutierrez, A.,** Antibodies to single-stranded DNA: an aid to diagnosis of viral hepatitis, *J. Med. Virol.,* 2, 359, 1978.
43. **Gust, I. D., Feinstone, S. M., Purcell, R. H., and Alter, H. J.,** Detection of antibodies to single-stranded DNA in naturally acquired and experimentally induced viral hepatitis, *J. Med. Virol.,* 6, 1, 1980.
44. **Feinstone, S. M., Kapikian, A. Z., and Purcell, R. H.,** Hepatitis A: detection by immune electron microscopy of a virus-like antigen associated with acute illness, *Science,* 182, 1026, 1973.
45. **Hollinger, F. B., Bradley, D. W., Maynard, J. E., Dressman, G. R., and Melnick, J. L.,** Detection of hepatitis A viral antigen by radioimmunoassay, *J. Immunol.,* 115, 1464, 1975.
46. **Jansen, R. W., Newbold, J. E., and Lemon, S. M.,** Combined immunoaffinity CDNA-RNA hybridization assay for detection of hepatitis A virus in clinical specimens, *J. Clin. Microsc.,* 221, 984, 1985.
47. **Locarnini, S. A., Gust, I. D., Ferris, A. A., Stott, A. C., and Wong, M. L.,** A prospective study of acute viral hepatitis with particular reference to hepatitis A, *Bull. WHO,* 54, 199, 1976.
48. **Locarnini, S. A., Ferris, A. A., Lehmann, N. I., and Gust, I. D.,** The antibody response following hepatitis A infection, *Intervirology,* 8, 309, 1977.
49. **Coulepis, A. G., Locarnini, S. A., Lehmann, N. I., and Gust, I. D.,** Detection of hepatitis A virus in the feces of patients with naturally acquired infections, *J. Infect. Dis.,* 141, 151, 1980.
50. **Purcell, R. H., Wong, D. C., Moritsugu, Y., Dienstag, J. L., Routenberg, J. A., and Boggs, J. D.,** A microtitre solid phase radioimmunoassay for hepatitis A antigen and antibody, *J. Immunol.,* 116, 349, 1976.
51. **Locarnini, S. A., Garland, S. M., Lehmann, N. I., Pringle, R. C., and Gust, I. D.,** Solid-phase enzyme-linked immunoabsorbent assay for detection of hepatitis A virus, *J. Clin. Microsc.,* 8, 277, 1978.
52. **Tassopoulos, N. C., Papavangelou, G. J., Ticehurst, J. C., and Purcell, R. H.,** Fecal excretion of Greek strains of hepatitis A virus in patients with hepatitis A and in experimentally infected chimpanzees, *J. Infect. Dis.,* 154, 231, 1986.
53. **Provost, P. J., Ittensohn, O. L., Villarejos, V. M., and Hilleman, M. R.,** A specific complement-fixation test for human hepatitis A employing CR326 antigen: diagnosis and epidemiology, *Proc. Soc. Exp. Biol. Med.,* 148, 962, 1975.
54. **Miller, W. J., Provost, P. J., McAleer, W. J., Ittensohn, O. L., Villarejos, V. M., and Hilleman, M. R.,** Specific immune adherence assay for human hepatitis A antibody: application to diagnostic and epidemiologic investigations, *Proc. Soc. Exp. Biol. Med.,* 149,254, 1975.
55. **Anderson, D. A., Coulepis, A. G., Chenoweth, M. P., and Gust, I. D.,** Indirect immunofluorescence assay for the detection of hepatitis A virus-specific serum immunoglobulins, *J. Clin. Microsc.,* 24, 163, 1986.
56. **Gust, I. D.,** Recent developments in hepatitis A, *Pathology,* 10, 299, 1978.
57. **Locarnini, S. A., Ferris, A. A., Stott, A. C., and Gust, I. D.,** The relationship between a 27 nm virus-like particle and hepatitis A as demonstrated by immune electron microscopy, *Intervirology,* 4, 110, 1974.
58. **Locarnini, S. A., Coulepis, A. G., Stratton, A. M., Kaldor, J., and Gust, I. D.,** Solid-phase enzyme immunoassay for detection of hepatitis A specific immunoglobulin M, *J. Clin. Microsc.,* 9, 495, 1979.
59. **Bianchi, L., De Groote, J., Desmet, V. J., Gedigk, P., Gorb, G., Popper, H., Poulsen, H., Scheuer, P. J., Schmid, M., Thaler, H., and Welper, W.,** Morphological criteria in viral hepatitis, *Lancet,* 1, 333, 1971.
60. **Ishak, K. G.,** Viral hepatitis: the morphologic spectrum, in *The Liver,* Gall, E. A. and Mostofi, F. K., Eds., Williams & Wilkins, Baltimore, 1972, 218.

61. **Krugman, S., Ward, R., and Giles, J. P.,** The natural history of infectious hepatitis, *Am. J. Med.*, 32, 717, 1962.
62. **Jacobsen, I. M., Dienstag, J. L., and Nath, B. J.,** Relapsing viral hepatitis type A, *J. Med. Virol.*, 16, 163, 1985.
63. **Gordon, S. C., Reddy, K. R., Schiff, L., and Schiff, E. R.,** Prolonged intrahepatic cholestatis secondary to acute hepatitis A, *Ann. Intern. Med.*, 101, 635, 1984.
64. **Schiff, L.,** The use of steriods in liver disease, *Medicine, (Baltimore)*, 45, 565, 1966.
65. **Williams, R. and Billing, B. H.,** Action of steriod therapy in jaundice, *Lancet*, 2, 392, 1961.
66. **Havens, W. P., Jr.,** Infectious hepatitis in the Middle East: a clinical review of 200 cases seen in a military hospital, *JAMA*, 126, 17, 1944.
67. **Hoaglund, C. L. and Shank, R. E.,** Infectious hepatitis: a review of 200 cases, *JAMA*, 130, 615, 1946.
68. **Krugman, S., Giles, J. P., and Hammond, J.,** Infectious hepatitis: evidence for two distinctive clinical, epidemiological and immunological types of infection, *JAMA*, 200, 365, 1967.
69. **Koff, R. S. and Galambos, J.,** Viral hepatitis, in *Diseases of the Liver*, Schiff, J., Schiff, E. R., Eds., Lippincott, Philadelphia, 1982, 547.
70. **Weir, W. R. C., Mellor, J. A., Smith, H., and Tyrell, D. A. J.,** Significance of hepatic enzyme levels at discharge in acute viral hepatitis, *J.J.D.*, 3, 309, 1981.
71. **Bamber, M., Thomas, H. C., Bannister, B., and Sherlock, S.,** Acute type A, B and non-A, non-B hepatitis in a hospital population in London: clinical and epidemiological features, *Gut*, 24, 561, 1983.
72. **Gruer, L. D., McKendrick, M. W., Beeching, N. J., and Geddes, A. M.,** Relapsing hepatitis associated with hepatitis A virus, *Lancet*, 2, 163, 1982.
73. **Chiriaco, P., Guadalupi, C., Armigliato, M., Bortolotti, F., and Realki, G.,** Polyphasic course of hepatitis type A in children, *J. Infect. Dis.*, 153, 378, 1986.
74. **Sjogren, M. H., Tanno, H., Fay, O., Sileoni, S., Cohen, B. D., Burke, D. S., and Feighny, R. J.,** in press, 1987.
75. **Johnston, C. L. W., Schwartz, M., and Wansborough-Jones, M. H.,** Acute inflammatory polyradiculoneuropathy following type A viral hepatitis, *Postgrad. Med. J.*, 57, 647, 1981.
76. **Bromberg, K., Newhall, D. N., and Peter, G.,** Hepatitis A and meningocephalitis, *JAMA*, 247, 815, 1982.
77. **Saunders, S. J., Seggie, J., Kirsch, R. E., and Terblanche, J.,** Acute liver failure, in *Liver and Biliary Disease*, Wright, R., Alberti, K. G. M. M., Karran, S., and Millward -Sandler, G. H., Eds., W.B. Saunders, Philadelphia, 1979, 569.
78. **Trey, C.,** The fulminant hepatic failure surveillance study: brief review of the effects of presumed etiology and age of survival, *Can. Med. Assoc. J.*, 106, 525, 1972.
79. **Chalmers, T. C., Eckhardt, R. D., Reynolds, W. E., Cigarroa, J. G., Jr., Deane, N., Reifenstein, R. W., Smith, C. W., and Davidson, C. S.,** The treatment of acute infectious hepatitis: controlled studies of the effect of diet, rest and physical reconditioning on the acute course of the disease and on the evidence of relapses and residual abnormalities, *J. Clin. Invest.*, 34, 1163, 1955.
80. **Repsher, L. H. and Freebern, R. K.,** Effects of early and vigorous recovery exercise on recovery from infectious hepatitis, *N. Engl. J. Med.*, 281, 1393, 1969.
81. **Capps, R. B. and Barker, M. H.,** The management of infectious hepatitis, *Ann. Intern. Med.*, 26, 405, 1945.
82. **Havens, W. P., Jr.,** Infectious hepatitis, *Medicine, (Baltimore)*, 27, 279, 1948.
83. **Gardner, H. T., Rovelstad, R. A., Moore, D. J., Streitfield, F. A., and Knowlton, M.,** Hepatitis among American occupation troops in Germany, *Ann. Intern. Med.*, 30, 1009, 1949.
84. **Trey, C., Burns, D. G., and Saunders, S. J.,** Treatment of hepatic coma by exchange blood transfusion, *N. Engl. J. Med.*, 274, 473, 1966.
85. **Burnell, J. M., Dawborn, J. K., Epstein, R. B., Gutman, R. A., Leinbach, G. E., Thomas, E. D., and Volwiler, W.,** Acute hepatic coma treated by cross circulation or exchange transfusion, *N. Engl. J. Med.*, 276, 935, 1967.
86. **Bosman, S. C. W., Terblanche, J., Saunders, S. T., Harrison, G. G., and Barnard, C. N.,** Cross circulation between man and baboon, *Lancet*, 2, 583, 1968.
87. **Parbhoo, S. P., Kennedy, J., James, I. M., Chaistrey, L. J., Ajdukiewicz, A., Brock, P. J., Xanalatos, C., Sayer, P., and Sherlock, S.,** Extracorporeal pig-liver perfusion in treatment of hepatic coma due to fulminant hepatitis, *Lancet*, 1, 659, 1971.
88. **Lepore, M. J. and Martel, A. J.,** Plasmaphoresis with plasma exchange in hepatic coma: methods and results in five patients with fulminant hepatic necrosis, *Ann. Intern. Med.*, 72, 165, 1970.
89. **Evans, A. E., Sprinz, H., and Nelson, R. S.,** Adrenal hormone therapy in viral hepatitis. III. The effect of ACTH and cortisone in severe and fulminant cases, *Ann. Intern. Med.*, 38, 1148, 1953.

Chapter 9

EPIDEMIOLOGY

I. GENERAL

A. Morbidity Data

Prior to the development of specific diagnostic tests for detection of current or past infection with HAV, our understanding of the epidemiology of hepatitis A was based largely on data obtained from the few countries in which the disease was notifiable and in which detailed records were kept. This data were supplemented by the observations of physicians and epidemiologists, mostly during epidemics and a limited number of studies in experimentally infected volunteers.

Although hepatitis has been a notifiable disease in some countries for 40 or 50 years, many epidemiologists have little enthusiasm for the data which have been collected. Compliance with government regulations varies greatly from country to country and is notoriously poor. In many areas the absence of reliable laboratory support may make it impossible for a clinician to distinguish hepatitis from other forms of jaundice, and hepatitis A from other types of viral hepatitis. Even in countries with relatively good reporting systems, major differences exist in the notification rates from city to city, district to district, and even from doctor to doctor. The ability of a country to obtain reliable data on the incidence of infectious diseases appears to be related not only to its socioeconomic level, but the emphasis and support provided to public health services and whether doctors and primary health-care workers are provided with encouragement or incentives to notify cases of disease. For example, of 167 countries currently associated with the World Health Organization only 100 regularly provide data on the prevalence of viral hepatitis and much of this information is extremely unreliable. Excellent data on the incidence of viral hepatitis are available from Eastern Europe,[1] where there is considerable interest in the disease and hospitalization is either usual or compulsory, and from Scandinavia. Norway, Sweden, Finland, and Denmark were among the first countries to make infectious hepatitis a notifiable disease, with Denmark having data which extend back to 1928.

In many developed countries reporting is less reliable and grossly underestimates the frequency of the disease. For example, in Israel in 1963 to 1964, the incidence of hepatitis notified to health authorities was 82 per 100,000.[2] Information derived from health insurance reports completed by the patients' doctors indicated that true evidence of the disease was nearly 250 per 100,000, i.e., at least three times more common than had been thought. Similar data have been obtained in many other developed countries; a study carried out in Boston in the late 1960s showed that only 12% of cases were reported to the City Health Department, and that while intensive education campaigns would result in a doubling of notifications, doctors soon slipped back into their previous habits.[3] At a conservative estimate, the true incidence of clinical hepatitis A in most developed countries is at least five times greater than is reported while the prevalence of infection is many times higher.

In many developing countries data on the incidence of viral hepatitis are either not collected or not reliable due to deficiencies in the number and training of the staff required to gather and process the information, and lack of access to appropriate laboratory tests. In some countries there is a tendency to classify any jaundiced patient as having hepatitis. One of us was recently invited to investigate a cluster of cases of "fatal hepatitis" which were reported to have occurred in Nauru, a remote Pacific island. It transpired that none of the patients had had hepatitis, but as jaundice had been observed in each prior to death (which was usually due to septicemia), the major illness recorded on the death certificate was hepatitis!

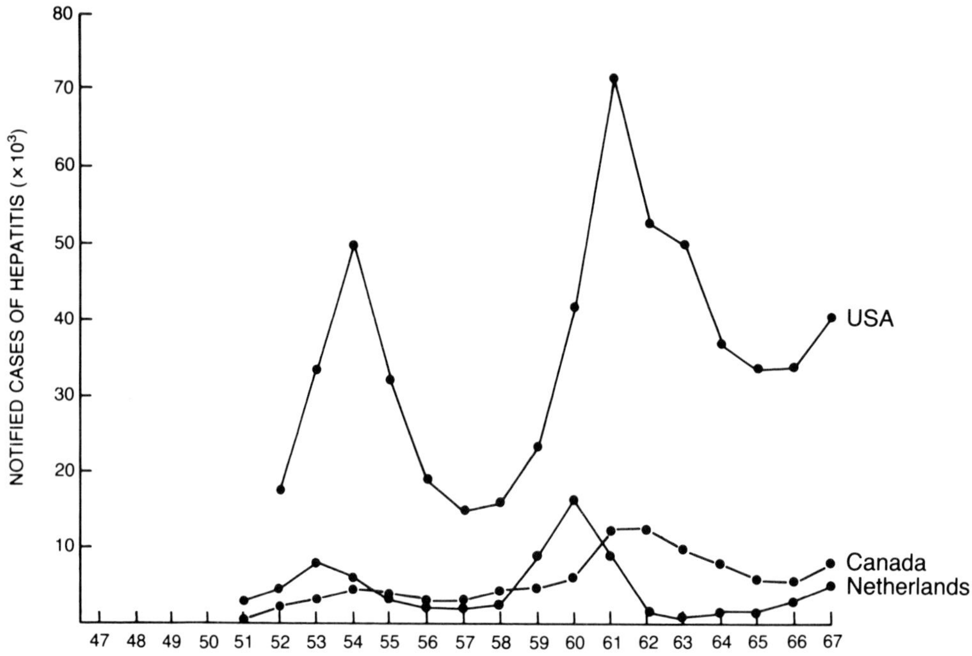

FIGURE 1. Cyclical pattern in notifications of acute viral hepatitis in the U.S., Canada, and the Netherlands from 1951 to 1967. (From notification data provided to W.H.O.)

Because of differences in the quality of the information gathered in different countries, it is virtually impossible to compare the data obtained without further detailed analysis. For example, from 1960 to 1964 the incidence of hepatitis in Czechoslovakia, which has a very complete reporting system, was reported as 263 per 100,000 — three times higher than the incidence reported in Israel over the same period. Data from health insurance statistics show that the true incidence of disease in Israel was about 250 per 100,000, which is about identical to the incidence in Czechoslovakia. Had this information not been available the conclusions which would have been drawn from the comparative figures would have been wrong and could have led to mistakes in public health policy.

Despite these limitations, data obtained from notifications are of considerable value as a guide to the prevalence of hepatitis in a particular country and give reliable information on seasonal patterns, age and sex distribution, and secular changes. The proportion of illnesses reported to health authorities from year to year tends to remain relatively constant so that it is possible to recognize major trends in that population. An exception to this rule which is occasionally seen, occurs when a major overhaul of the notification system produces a large increase in notifications which may temporarily obscure a real decline in the incidence of the disease. The rapid rise in notified cases of hepatitis A which occurred in Cuba from 1961 to 1965 is an example of this phenomenon.[4]

The major features of the epidemiology of hepatitis A which are revealed by an analysis of morbidity data are (1) a cyclic pattern with peaks of incidence every 5 to 10 years; (2) an increase in the number of cases at certain times of the year; (3) a peak incidence among young children; (4) a high secondary attack rate among family members, and (5) a low case-fatality rate.[5] In addition, the disease can be prevented or reduced in severity by injection of human normal immunoglobulin (HNIG). It should be noted that not all these features are seen in every country.

1. Epidemic Waves

In many developed countries with temperate climates, hepatitis A appears to occur in

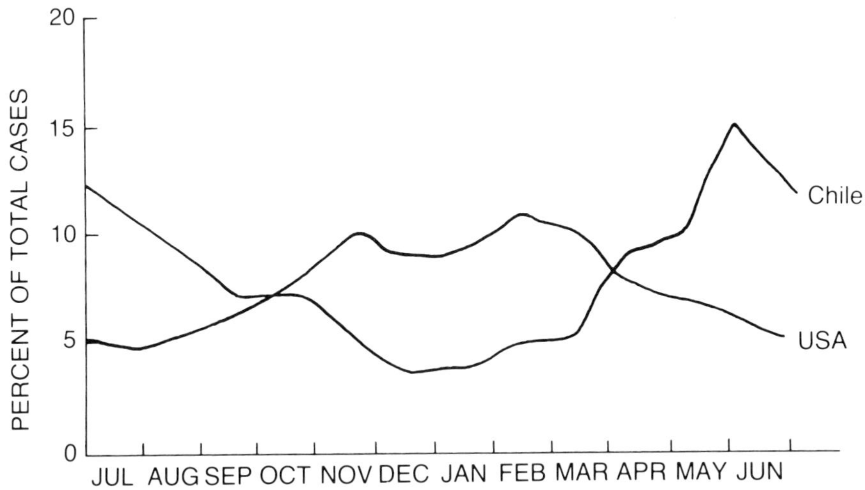

FIGURE 2. Seasonal pattern of hepatitis notifications in two countries in the northern hemisphere (U.S.) and southern hemisphere (Chile) from 1965 to 1977. (From notification data provided to W.H.O.)

epidemic waves at intervals of about 7 years (see Figure 1), suggesting that this interval is required for a sufficiently large population of susceptibles to develop to sustain an outbreak. In the U.S. and Canada, major waves of the disease occurred in 1954 and 1961; in Australia in 1956 and 1961; in Denmark in the immediate postwar period and the mid 1950s; and in the Netherlands in 1954 and 1960.[4] In recent years as the incidence of disease in these countries has declined, this periodicity has been lost presumably because improved living conditions have made it difficult for the virus to maintain itself in the population or to spread widely when introduced from outside. By contrast, in some countries such as Cuba and Chile in which standards of hygiene and sanitation are improving rapidly, alternating periods of higher and lower incidence are now a major feature of the epidemiology of the disease.

2. *Seasonal Patterns*

Hepatitis A is reported to have a seasonal pattern, especially in temperate climates where it is usually more common in the late autumn and early winter.[4,6] These patterns can be distorted by unusual events such as large outbreaks of the disease or the impact of cases among tourists who have journeyed to warmer countries during the winter.

The seasonal cycle is exemplified by data from Canada and the U.S. where the peak incidence of the disease occurs between February and March, and Argentina and Chile which, because they are in the southern hemisphere, have their peaks about 6 months later (see Figure 2). These patterns are not seen in all temperate climates; in Australia and New Zealand for example, the seasonal pattern is less clear cut with the disease being, if anything, more common in the summer months.[4]

In tropical and semitropical countries the patterns differ with either no seasonal peaks (El Salvador, Indonesia, and Guinea) or peaks in the autumn (Cuba, Puerto Rico, Egypt, and Vietnam). There appears to be no simple common factor underlying these differences, the reasons for which are unknown.

3. *Age Distribution*

The age distribution of cases of hepatitis A would be expected to follow the pattern of other enteric viral infections such as poliomyelitis and to be affected by levels of hygiene and sanitation. Traditionally, in developed countries the highest incidence of hepatitis A has been seen among children of primary school age, with the second highest incidence in older children and young adults.[7] As standards of hygiene rise one would expect that the mean age of infection and disease would increase — a phenomenon which is now being recognized in many countries.

For example, in Australia hepatitis A became a major public health problem in the first and second decades following World War II. The disease was seen mainly in children and young adults. For example, in 1969 to 1970, 49% of patients with hepatitis A admitted to Fairfield Hospital, Melbourne were less than 15 years of age, 33% were aged between 15 and 29 years and only 18% were 30 years or older. Over the next few years not only did the incidence of hepatitis A decline but the proportion of cases among children fell to 22% while the proportion of infections occurring in the 15- to 29-year age group rose to 61%.[8] (See Figure 5b.)

4. Gender Distribution

There appears to be no data that gender is important in hepatitis A infections except under unusual circumstances. Higher rates of infection are seen in some groups of men, such as sewage workers and homosexuals, whose occupation or sexual preferences bring them into contact with human feces.[9,10] The high case-fatality rates reported among pregnant women during some water-borne epidemics of hepatitis[11-13] now seem likely to be due to infection with an agent or agents other than HAV (see Chapter 12).

5. Other

Other characteristics of hepatitis A revealed from notification data are (1) an apparently higher prevalence of disease in rural areas in some countries; (2) a high secondary attack rate in household members; (3) the occasional occurrence of outbreaks following consumption of fecally contaminated food or water; and (4) the value of HNIG in protecting travelers to endemic areas and reducing the incidence of disease among close contacts of cases. Several of these aspects are addressed in greater detail elsewhere in the book.

B. Major Patterns of Infection

The pattern or infection varies from country to country and seems to be largely dependent upon standards of hygiene, sanitation, and the mobility of the population. Studies of the age-specific prevalence of anti-HAV have revealed two major and one minor patterns of infection (see Figure 3).

1. Pattern A

In most developing countries with large populations, infection with HAV is hyperendemic. Under these circumstances most children become infected early in life, so that by the age of 10 years or so, virtually the entire population has been exposed to the virus and are immune (Pattern A). As infections in early childhood tend to be silent or extremely mild, hepatitis is not a clinical problem and the existence of a reservoir of infection is only detected by serological surveys of the occurrence of the disease in susceptible visitors.

Hepatitis A has long been recognized as a problem among missionaries and Peace Corp workers who work for long periods in developing countries even though the disease itself may be rare among the local population. In a study by Cline et al. of 558 American missionaries who had spent at least 12 months overseas, the annual attack rate was almost 2%, with the highest rate, 4%, being observed in those under the age of 30 years.[69] Similar studies among missionaries serving in Africa revealed annual attack rates of between 2 and 5.5% depending on location, an incidence which is more than 400 times higher than that observed in comparable groups stationed in the U.S.[70] Although no serology is available from these studies, the high secondary attack rates within families (from 9 to 26%) and the

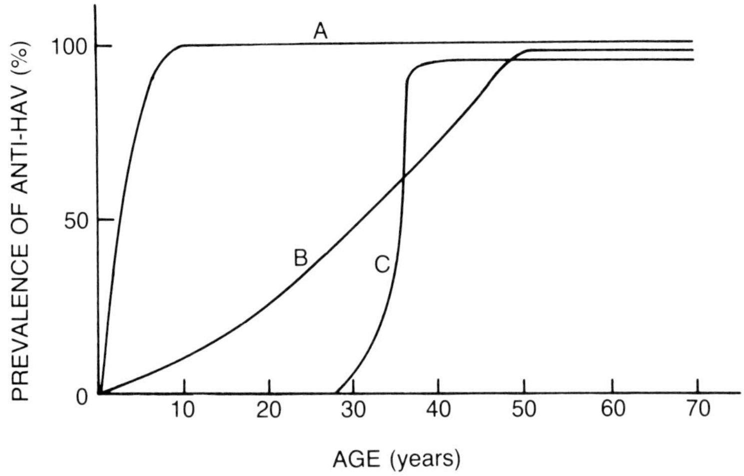

FIGURE 3. The three major patterns of the age-specific prevalence of anti-HAV. (From Gust, I. D., Vyas, G. N., Dienstag, J. L., and Hofnagle, J. N., Eds., Grune & Stratton, Orlando, 1984, 417. With permission.)

interval between primary and secondary cases (2 to 6 weeks) suggest that the disease was hepatitis A. In both studies the incidence of hepatitis remained high for several years, the cumulative risk of acquiring the disease during a 10-year stay in an endemic area being estimated at about 45%. Because of these findings most missionaries and other workers required to spend long periods in developing countries usually receive injections of gamma globulin at 4- to 6-month intervals. The absence of readily available running water for cleansing hands after defecation seems to be important in facilitating transmission of infection.[23]

2. Pattern B

In most developed countries the age-specific prevalence of anti-HAV has a sigmoidal shape (see Pattern B, Figures 3 and 4) suggesting that as living standards improve, there is a gradual decline in the spread of infection in the community. In some countries such as Sweden, Norway, Switzerland, and Japan, antibodies to HAV are now virtually absent among people in the first two or three decades of life suggesting that the disease is no longer endemic. Statistical analysis of these data indicates that in Scandinavia the incidence of hepatitis A has been declining since 1930. Direct evidence of the declining prevalence of infection has been obtained by studies of the age-specific prevalence of anti-HAV in panels of sera collected recently and similar panels collected many years ago (see Figure 5A).[71,72]

In other developed countries such as Holland, France, Germany, and Australia, the decline in infection has been less dramatic which may reflect differences in the rate at which living conditions have improved in those countries, or the effect of the influx of migrants or guest workers from countries in which hepatitis A infection is extremely common. The decline in infection with HAV in these countries is reflected not only in a decline in the prevalence of antibodies among young people, but a decline in the number of notifications of hepatitis A and a switch in the peak incidence of the disease from early childhood to early adult life[73] (see Figure 5B). The absence of naturally acquired immunity means that tourists from these countries are at risk of developing hepatitis A during travel to endemic countries. In some European countries a high proportion of the cases of hepatitis A which are currently seen are acquired during travel overseas.

In developing countries, as standards of hygiene and sanitation improve and the prevalence of other enteric infections declines, clinical hepatitis A often emerges as a major problem.

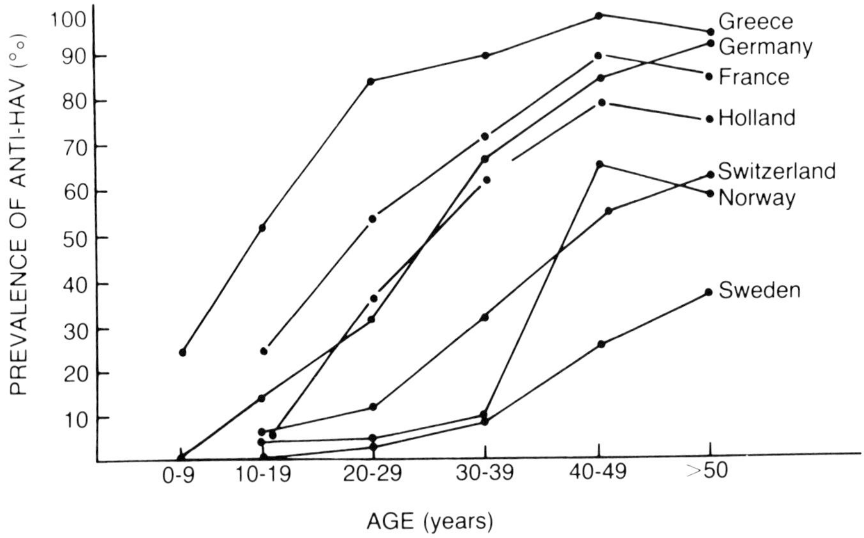

FIGURE 4. The age-specific prevalence of anti-HAV in serum in European countries. (From Frosner, G. G., Roggendorf, M., Frosner, H., Gerth, H. J., Borst, V., Blochinger, G., and Schmid, W., *Viral Hepatitis,* Szmuness, W., Alter, H. J., and Maynard, J. E., Eds., Franklin Institute Press, Philadelphia, 1982, 157. With permission.)

This situation is often a major concern to public health officials who worry that their programs are failing. The explanation for this apparent paradox is simple — as the incidence of infection with HAV declines, an increasing proportion of the population reach adult life without having encountered the virus and developing immunity. Infections which occur at this stage tend to be clinically apparent, so although the incidence of infection may have declined dramatically, the incidence of disease actually increases. Once hepatitis A becomes a problem it may take 10 to 20 yr of improved living conditions before the incidence of the disease begins to decline. This pattern was demonstrated in countries such as Australia and New Zealand in the 1950s and 1960s and in many parts of Western Europe immediately after World War II, and is now being seen in Greece, China, Cuba, Chile, and several other countries in Europe, Southeast Asia, and Latin America.

3. *Pattern C*

The third pattern of age-specific prevalence of anti-HAV (see Pattern C, Figure 3) is occasionally seen in closed communities in which an epidemic of hepatitis A has occurred and most of the susceptible population has been infected. Lacking any means of maintaining itself, the disease disappears until the virus is subsequently reintroduced into the community. When this occurs the resulting outbreak is confined to children born since the previous outbreak and susceptible adults who did not reside in the community at the time of the last outbreak.

Although it is convenient to think of these patterns as mutually exclusive they are oversimplifications of what is often a complex epidemiological situation. In many large countries all three patterns may exist among different groups of the population, e.g., hepatitis A may be unknown among the wealthiest sections of the community, may be a declining problem among manual workers, but may be endemic among groups of poor emigrants who live in overcrowded conditions with poor hygiene or in certain large institutions.

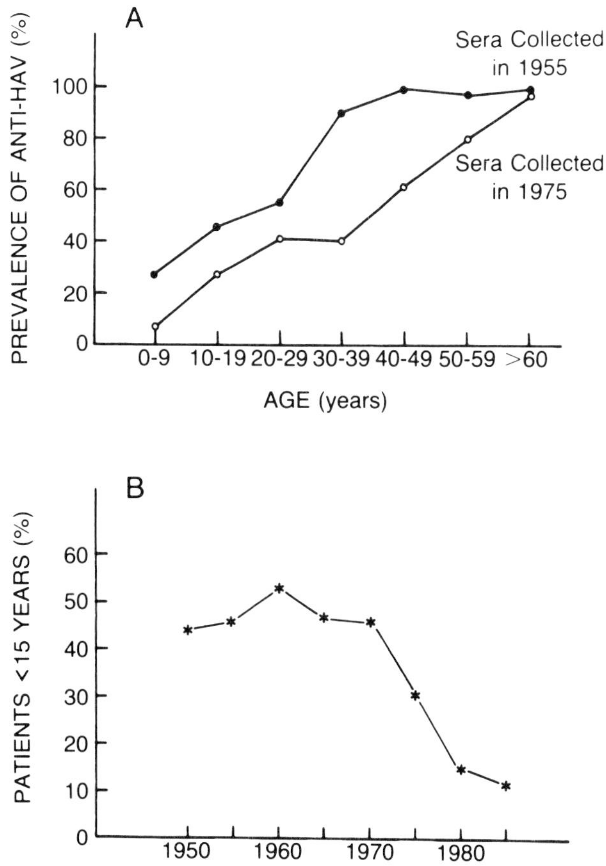

FIGURE 5. (A) Changing patterns in the epidemiology of hepatitis A in Australia. The age-specific prevalence of anti-HAV in two populations of urban dwellers in Melbourne, sampled in 1954 to 1955 and 1975. (From Gust, I. D., Lehmann, N. I., and Lucas, C. R., *J. Infect. Dis.*, 138, 425, 1978. With permission.) (B) The declining proparation of cases of hepatitis in children less than 15 years of age admitted to Fairfield Hospital, Melbourne.

II. MODE OF TRANSMISSION

A. Duration of Infectivity

Data on the length of infectivity of patients with hepatitis A are based largely on studies in human volunteers supplemented by epidemiological observations and, to a limited extent, by studies in experimentally infected animals and in cell culture.

1. Shedding of Virus in the Feces

a. Volunteer Studies

The best information about the shedding of infectious virus in the feces comes from the studies at Willowbrook State School (see Chapter 1).[14] Three children who developed hepatitis 39, 41, and 45 days after oral administration of a serum pool containing the MS-1 strain of HAV, had fecal specimens collected on days 11 and 25. The specimens collected on each of these dates were pooled and administered to further volunteers (see Figure 6). The pool of feces collected on day 11 failed to transmit infection to any of 8 children while the pool collected on day 25 transmitted infection to 4 of 13 subjects. Thus, HAV could

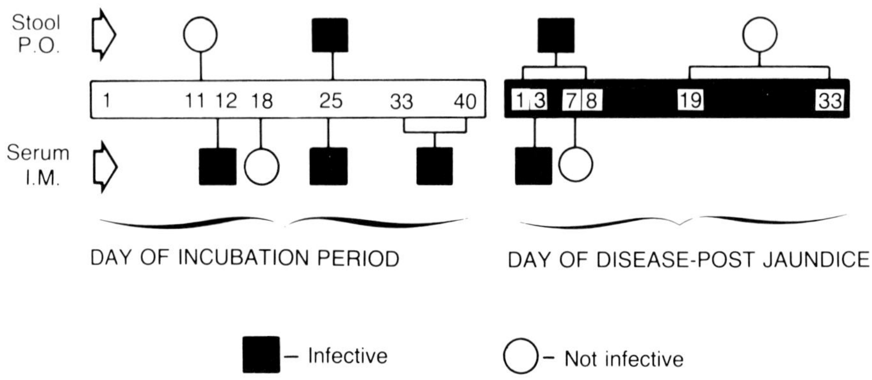

FIGURE 6. Schematic illustration of the period of infectivity of serum and feces during the course of experimentally induced viral hepatitis type A. (From Krugman, S., Ward, R., and Giles, J. P., *Am. J. Med.*, 32, 717, 1962. With permission.)

not be demonstrated in feces collected 28 to 35 days prior to the onset of jaundice, but was present in samples collected 14 to 21 days before jaundice. The infectivity of feces collected immediately prior to or in the acute phase of illness was first demonstrated by Havens and Neefe during World War II[15-17] and has been amply confirmed since. Feces collected on the first and third day after the onset of jaundice and a pool of specimens collected between the first and eighth day after the onset of jaundice have been shown to be infectious for volunteers.[14,18] On the other hand, specimens collected during the convalescent phase of the disease have consistently failed to transmit disease. Studies have been reported in which fecal samples from 12 different patients collected from 19 to 43 days after the onset of jaundice were administered to volunteers. In no instance did a single volunteer develop hepatitis.[16,19,20]

b. Prolonged Shedding

The question of whether chronic or prolonged fecal shedding of HAV may occur has been controversial. Neefe et al.[19] collected fecal specimens from three men who developed persistent symptoms accompanied by abnormal liver function tests and mild abnormalities on liver biopsy after attacks of hepatitis A. A filtrate of 23 specimens collected from these subjects 92 to 342 days after the onset of illness failed to induce hepatitis in a group of volunteers. On the other hand, Stokes and her colleagues claimed to detect chronic shedding of HAV in two children with laboratory evidence of chronic hepatitis[20] and postulated that this was the cause of continued infections among children newly admitted to an orphanage.[21] In these studies, suspensions of feces collected 5 and 15 months after hepatic abnormalities were first detected in the children were administered orally to two groups of four volunteers. In each instance one of four subjects developed jaundice after incubation periods of 22 and 26 days. Although there is no reason to doubt these results and there is good epidemiological evidence that hepatitis A was circulating in the orphanage,[21,22] these data do not confirm the existence of a carrier state, as transmission was demonstrated on only a single occasion. It is possible that the two children were initially infected with one of the agents of HNANB which was responsible for their first attack of hepatitis and subsequent abnormal liver function tests, and then suffered mild infections with HAV. Alternatively, the illnesses which developed in the two children who were fed these fecal extracts may have been unrelated to the inocula but naturally acquired from other children in the institution. As no specimens remain from these studies it is not possible to resolve these questions.

Several other lines of evidence suggest that if chronic (or prolonged) shedding of HAV occurs, it must be a rare phenomenon. Several studies have demonstrated that in isolated

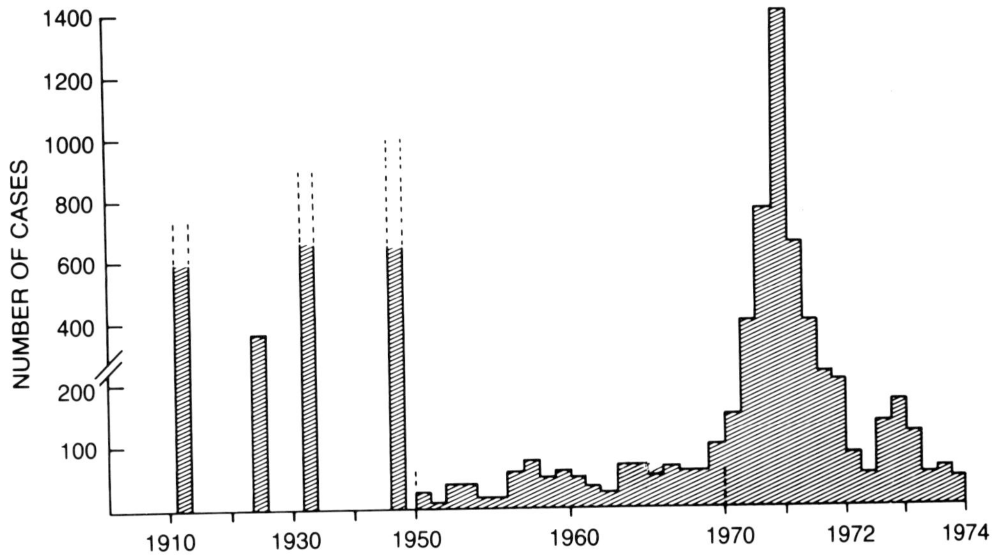

FIGURE 7. The occurrence of viral hepatitis in Greenland between 1910 and 1974. (From Skinhoj, P., Mikkelson, F., and Hillinger, G., *Am. J. Epidemiol.*, 105, 140, 1977. With permission.)

populations hepatitis A may occur in epidemics many years apart and that the disease disappears in the intervening years, suggesting that the virus is only capable of surviving in a large population in which large numbers of people are susceptible to infection. Maynard and his colleagues demonstrated a gap of 16 years between outbreaks in Fort Yukon, Alaska[23] while Skinhoj and McNair reported major epidemics of hepatitis A in Greenland and the Faeroe Islands in 1913, 1926, 1934, 1947 to 1948, and 1971 to 1972[24,25] (see Figure 7). Serological studies carried out following the latest epidemic support the idea that HAV can disappear completely from isolated populations. When sera which had been collected from Greenlanders in 1970 were tested for anti-HAV, the prevalence of antibodies in people born prior to 1947 was 93% compared with a prevalence of 3% in those born after 1948. Similarly in the 1971 to 1972 outbreak, the age-specific attack rates were between 46 and 70% in Greenlanders aged less than 24 years, but only 5% in those between 24 and 34 and 1% in those over 35 (see Figure 8). More than 80% of the cases in adults occurred in people who had not lived in epidemic areas during the previous outbreaks or who had immigrated to Greenland since 1948. Furthermore, Skinhoj and his colleagues were able to show that the handful of cases of hepatitis which continued to occur between epidemics, and which some authors have interpreted as evidence for endemicity,[26] were due to infection with other agents, principally HBV. Similar data has recently been obtained for the isolated Pacific islands of Ponape and Kosrae.[27]

The situation is not entirely resolved, as longitudinal studies in some Pacific island populations show evidence of continuing infection. Mosley and his colleagues have demonstrated seroconversions to anti-HAV among Tongan children, despite the fact that the prevalence of antibodies in this population is over 90%, a level which should be sufficient to provide herd immunity and prevent the spread of introduced infections.[28] Similarly, cross-sectional studies of the populations of several other Pacific islands performed in the authors' laboratories[29,30] indicate that the majority of the population of even the most remote islands become infected by the age of 15 years (see Figure 9). These data do not necessarily suggest the existence of a carrier state as they could be explained by considering the myriad Pacific islands as a single epidemiologic unit in which HAV was constantly reintroduced from outside or by prolonged survival of HAV in the environment.

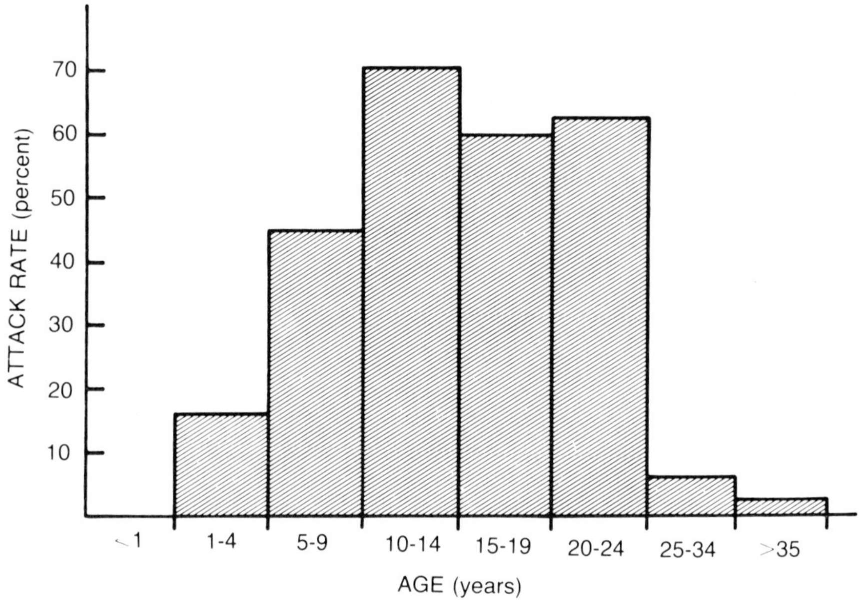

FIGURE 8. The age-specific attack rates among Eskimos in Greenland during the 1970 to 1974 epidemic. (From Skinhoj, P., Mikkelson, F., and Hillinger, B., *Am. J. Epidemiol.*, 105, 140, 1977. With permission.)

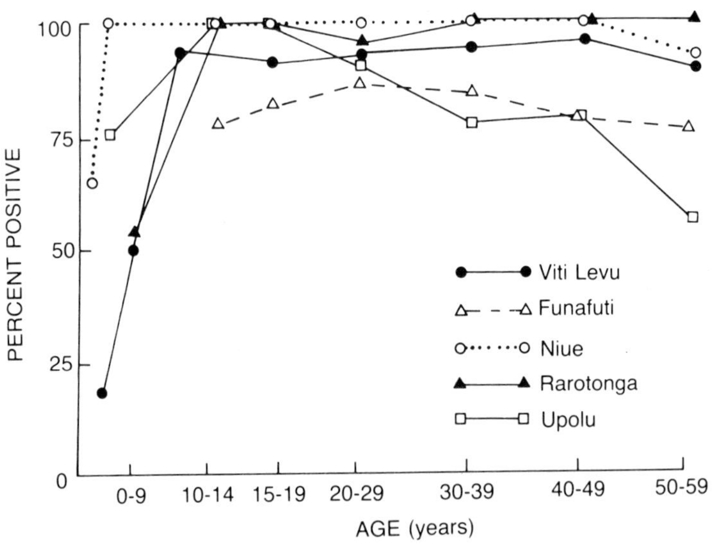

FIGURE 9. The age-specific prevalence of anti-HAV in five Pacific island populations: Fiji, Tuvalu, Niue, the Cook Islands, and Western Samoa. (From Gust, I. D., Lehmann, N. I., and Dimitrikakis, M., *Am. J. Epidemiol.*, 110, 237, 1979. With permission.)

Hepatitis A has been recognized as a particular problem in institutions for the mentally handicapped where overcrowding and poor standards of hygiene allow the rapid transmission of most enteric pathogens. In most institutions observed over prolonged periods, the pattern

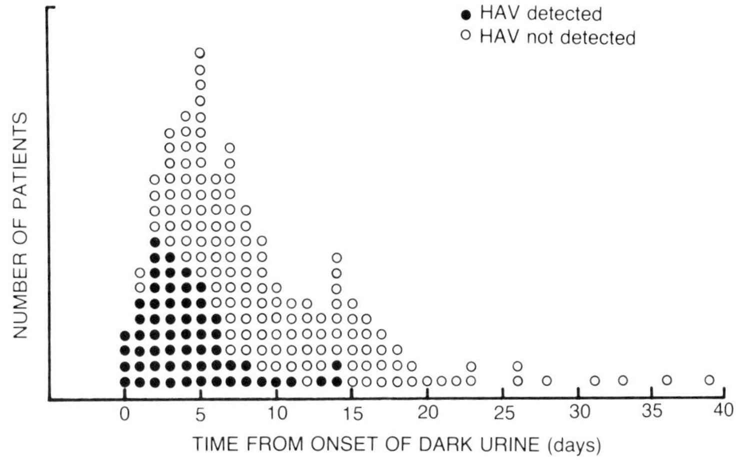

FIGURE 10. Detection of HAV in the first fecal sample collected from 200 consecutive patients with serologically confirmed hepatitis A. (From Coulepis, A. G., Locarnini, S. A., Lehmann, N. I., and Gust, I. D., *J. Infect. Dis.*, 141, 151, 1980. With permission.)

of disease has been consistent with occasional introductions of the virus by newly admitted patients, or occasionally by staff caring for them.[31-35] On the other hand, hepatitis was clearly endemic in two large institutions for the mentally retarded — the Willowbrook State School and the Suffolk Developmental Center,[36,37] both of which were characterized by extremely high rates of admission. If a chronic carrier state was common one would expect endemic hepatitis A to be the norm rather than the exception.

c. Detection of HAAg or Nucleic Acid

The most detailed data on the extent and duration of shedding of HAV in the feces have been obtained from studies using EM, RIA, or radioactive probes. When Dienstag et al.[38] examined fecal samples from two experimentally infected volunteers by IEM, virus-like particles were detected at least 5 days prior to the onset of biochemical hepatitis but had disappeared by the time that transaminase levels reached their peak. In a subsequent study by the same group, serial fecal samples from subjects involved in a point-source epidemic were examined by IEM.[39] While 36% of 42 specimens collected prior to peak ALT levels contained virus particles, none of 23 samples from 18 patients collected on the day at which enzyme levels peaked or the day after, were positive. Similar data were obtained by Rakela and Mosley who examined serial samples from six children involved in an outbreak of hepatitis A.[40]

The sensitivity of the methods used in these studies was probably not optimal, as a number of groups were soon able to demonstrate more prolonged periods of virus shedding. Locarnini and his colleagues in Melbourne detected HAV by IEM in 12 of 18 patients tested within 1 week of the onset of dark urine and 5 of 15 patients tested in the 2nd week.[41] Similar data were provided by Hall in the U.S.[42] and amplified by Coulepis and his colleagues.[43] When fecal specimens from 200 consecutive patients with naturally acquired sporadic hepatitis A were examined by IEM and RIA, virus particles were detected in 59. Of patients admitted within 1 week of the onset of dark urine, 45% were found to be shedding virus in their feces compared with 11% of those admitted in the second week of illness. HAV particles were not detected more than 14 days after the onset of dark urine (see Figure 10). The concentration of viral particles in each specimen (estimated by counting the number of particles in four EM grid squares) was shown to decline rapidly over the first 7 days of illness (see Figure 11). Failure to detect virus particles by IEM or viral antigen by RIA cannot

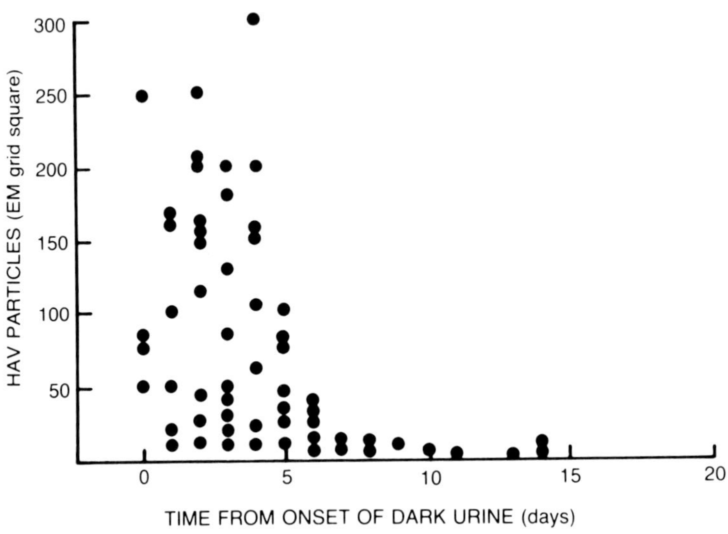

FIGURE 11. Correlation of the number of virus particles detected in the feces of patients with hepatitis A with the time of collection of the sample. (From Coulepis, A. G., Locarnini, S. A., Lehmann, N. I., and Gust, I. D., *J. Infect. Dis.*, 141, 151, 1980. With permission.)

be taken as evidence of the absence of infectious virus as both techniques have limited sensitivity. IEM is usually only positive when the concentration of virus particles in the starting material is $\geqq 10^6/m\ell$, and RIA has been found to be $\log^{5\text{-}6}$ fold less sensitive than animal inoculation for detection of infectious particles. It seems likely, therefore, that most patients continue to shed virus for up to 2 weeks after the onset of dark urine.

d. Implications for Hospital Staff

These findings have stimulated a lively debate on whether patients with hepatitis A should be isolated and whether special precautions are required to protect staff from infection.[44,45] While it is clear that most hospitalized patients continue to shed virus into their feces, their infectivity for staff and visitors appears to be extremely low. Among several thousand patients admitted to the Liver Unit of the University of Southern California Hospital over a 10-year period, approximately one in four or one in five had acute hepatitis A. Despite lack of segregation no evidence of spread of infection to other patients was observed.[28] Similar data were obtained by Papavangelou et al.[46] in a prospective study of 248 adults with viral hepatitis admitted to hospital in Athens. Despite the absence of strict infection control procedures, no evidence of nosocomial infection was obtained.

Nevertheless, since it is known that patients with hepatitis A continue to shed virus into their stools for up to 2 weeks after the onset of jaundice, and that a high proportion of the medical and nursing staff in developed countries lack immunity to HAV, special care should be exercised in the disposal of feces, and staff or relatives handling patients with hepatitis A should wash their hands after each visit. The major risk of transmission of hepatitis in a hospital occurs when patients admitted for other reasons happen to be in the late incubation period of the disease.[47] The risk is increased with children or patients who are incontinent. Two such outbreaks have been documented recently. In the first, 11 staff members who had contact with an 18-month-old child developed hepatitis A within 28 to 41 days of her admission. The child, who was incontinent, presented with *S. flexneri* dysentery and developed symptoms and signs of hepatitis in the second week in hospital.[48] In the second, four physicians attending a 21-month-old girl admitted to hospital with diarrhea and an

Table 1
REPORTS OF SEROLOGICALLY CONFIRMED TRANSFUSION-ASSOCIATED HEPATITIS A

Blood component	Interval between collection and onset of illness in the donor (days)	Interval between transfusion and onset of illness in the recipient (days)	Ref.
Whole blood	28	—	57
	18	22	58
	15	24	59
Packed RBC	7	31	60
Fresh-frozen plasma	7	30	61

amoebic liver abscess developed hepatitis A, 27 to 29 days later. The child had unexplained increases in her serum transaminase levels 14 days after admission. Her illness was confirmed as hepatitis A by the detection of hepatitis A specific IgM.[49]

2. Viremia

a. Studies in Volunteers and Recipients of Blood Products

HAV has been detected in the blood during the incubation period of the disease although the duration of viremia has not been clearly defined. Transmission of infection has been achieved with specimens collected as early as 25 days prior to the onset of jaundice[50] and with pools of serum collected 14 to 21 days and 3 to 7 days prior to jaundice[51] (see Figure 6). Differing results have been obtained with sera collected around the time of onset of jaundice suggesting that the duration of viremia varies from patient to patient.[14,16,52] In contrast to hepatitis B and HNANB there are few accounts of the transmission of HAV by blood transfusion. Francis et al. in 1946 described the occurrence of short-incubation-period hepatitis in the recipient of a unit of blood collected from a donor 11 days before the onset of jaundice,[53] and a similar case was described by Harden and his colleagues in 1955.[54] While the epidemiology is consistent with transmission of HAV, the absence of specific laboratory tests makes it impossible to exclude HNANB as a cause of either of these cases. An epidemic of hepatitis in a transplantation center apparently associated with contaminated preparations of platelets and initially thought to be hepatitis A, was shown by retrospective testing of stored sera, to have been HNANB.[55,56] Serologically confirmed hepatitis A has occasionally been documented following transfusion of whole blood, packed red blood cells, and fresh-frozen plasma collected from patients in the incubation period of the disease (see Table 1).[57-61] Interestingly, in two of the five documented cases, the donor had a history of recent overseas travel.

While it is clear that posttransfusion hepatitis A can occur, evidence that this is extremely rare comes from studies of the prevalence of anti-HAV in individuals who have received multiple transfusions. Studies in New York[62] and Athen[63] have failed to demonstrate an increased prevalence of anti-HAV in these groups or to show a correlation between frequency of transfusion and the prevalence of antibody. Similarly, no increased prevalence of antibody has been detected in the patients or staff of 15 dialysis units studied in the U.S.[64] Blood-borne infections therefore seem to play little role in transmission of HAV in nature, and blood donors are not routinely tested for the presence of antibodies.

b. The LAK Cell Epidemic

A fascinating outbreak of hepatitis, probably due to the presence of HAV in samples of "normal human serum," has recently been reported in a group of patients with cancer who were treated with interleukin-2 (IL-2) and lymphokine-activated killer cells (LAK cells) in a multicenter study sponsored by the N.I.H.[132]

Table 2
INTERIM ANALYSIS OF THE OUTBREAK OF HEPATITIS A IN PATIENTS RECEIVING LAK CELLS

A. Attack Rate in Different Centers

Center	Patients	Attack rates in susceptibles (%)
1	14	42
2	18	18
3	17	60
4	14	50
5	14	29
6	13	38

B. Attack Rate in Recipients of Different Bulks

Bulk used	Patients	Attack rates (%)
Week 22	13	85
Week 31	13	62
Both	6	50
Neither	24	0

The problem was first noted in August 1986 when one patient involved in the study and the spouse of another patient being treated at the same center developed acute hepatitis A. Serologic testing of all patients was undertaken together with a detailed epidemiological investigation by officers of the Centers for Disease Control (CDC). At the time of writing, 30 infections with hepatitis A have been confirmed serologically on the basis of seroconversion or development of specific IgM antibodies post treatment. At least two cases have been recognized in each center with attack rates ranging from 18 to 60% of those susceptible to infection (see Table 2A). In addition, three household contacts of patients under treatment developed hepatitis A. In each case the illness was acquired after their relative had completed his or her course of treatment.

As no local factors could account for the distribution of cases, suspicion focused on the medium used to maintain the LAK cells and the IL-2 to which all patients were exposed. IL-2 can be excluded as the cause of infection as it was produced by recombinent DNA technology in media free from human serum. By exclusion, the most probable source of contamination is the pooled "normal" human serum, used to supplement the medium in which the LAK cells were maintained. These sera were obtained from paid donors from a single supplier. All sera were tested for HBSAg and anti-human immunodeficiency virus (HIV), which was filtered, heated at 50°C for 30 min, then pooled to produce bulk lots. Aliquots were then added to other components to produce media lots in which the final concentration of serum was 2%.

Seven batches of media were suspected to have transmitted infection; five of these were derived from a week-22 bulk lot, and the other two from a week-31 lot (see Table 2B). The highest attack rate (81%) occurred in patients exposed to the week-22 lot. Although the two lots were prepared nearly 2 months apart and did not share the same sera, 20 donors contributed to both pools. While the epidemiological investigation has not been completed, the most probable explanation is that one of these donors was viremic at the time of donation.

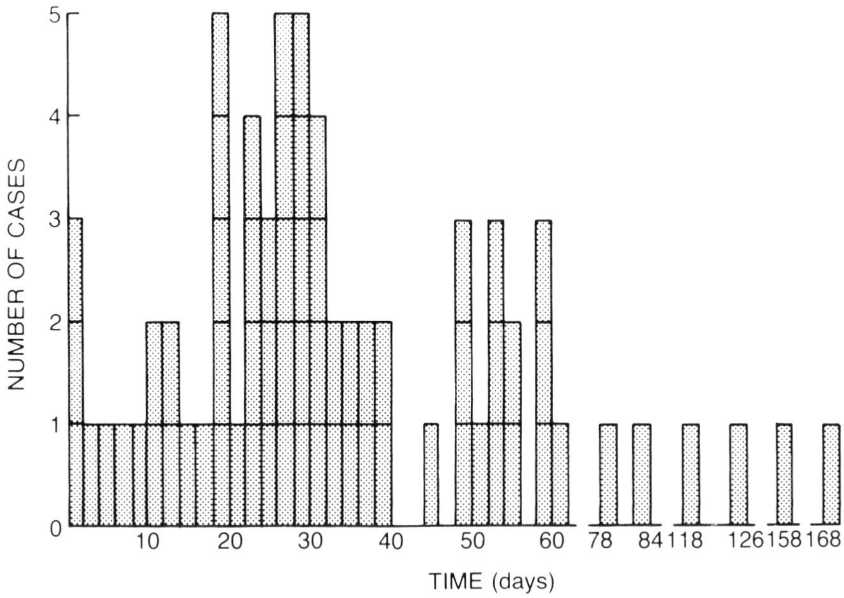

FIGURE 12. Intervals between the onset of symptoms in the index and subsequent cases in 37 families studied in the U.S. in 1951. (From Knight, V., Drake, M. E., Belden, E. A., Franklin, B. J., Romer, M., and Copple, L. O., *Am. J. Hyg.*, 59, 1, 1954. With permission.)

Since the original unit of plasma was diluted by other units in the pool (e.g., the week-31 bulk contained 171 units of plasma) and then diluted 1:50 in the culture medium, the high attack rate suggests that there may have been some amplification of the virus, perhaps by replication in the LAK cells. This possibility is now under investigation.

c. Other

While it is possible that HAV may occasionally be detected in sailva, urine, and nasopharyngeal secretions, there is no evidence that any of these is of major epidemiological significance.

B. Routes of Infection

1. Person to Person

The most important means of transmission of HAV is undoubtably from person to person via the fecal-oral route. Transmission is generally limited to close contacts, especially those within families. Young children are frequently involved in the spread of infection in households, as infections in this group are often silent and standards of hygiene are lower in children than among adults. There is good virological and epidemiological evidence that patients are at their most infectious during the incubation period of the disease.

Transmission of infection from children incubating the disease to susceptible contacts was postulated by Ward and his colleagues[36] to account for several cases of hepatitis A observed in uninoculated children at the Willowbrook State School. Direct evidence for this phenomenon was provided by Batten et al. in 1963.[65] Several children who acquired their infection institutionally were sent home for brief vacations and developed hepatitis shortly after their return. A number of household members whose only contact with the infected children had been 1 or 2 weeks prior to the onset of jaundice, subsequently developed acute hepatitis A. Similar data have been obtained from studies of individuals implicated in food-borne epidemics (reviewed by Mosley[28]). Family studies indicate that most household cases occur one incubation period or more after the index case,[66,67] indicating that the patient is most infectious just before the onset of symptoms (see Figure 12).

Table 3
PREVALENCE OF ANTIBODIES TO HAV AMONG HOMOSEXUAL AND HETEROSEXUAL MEN IN SEATTLE, WASHINGTON

	Number studied	Number with anti-HAV (%)
Homosexual		
Frequent oral-anal contacts	41	23 (50%)
Infrequent oral-anal contacts	55	13 (24%)
Total	96	36 (37%)
Heterosexual	57	7 (12%)

Hepatitis A is rarely spread by casual contact and a study of the distribution of cases in school and neighborhoods suggest that play contacts are more important than contacts in the classroom.[68] Because transmission requires close contact and the incubation period of the disease is several weeks, outbreaks of hepatitis A spread slowly through a community, taking several months to reach a peak and several more to recede. An excellent example is the outbreak which occurred in 1982 to 1983 in the tiny Pacific island of Kosrae.[27] This remote island in the Carolines has a population of about 5500 people of whom half are children less than 14 years of age. After a long period of absence hepatitis was recognized in June, 1982. The number of cases increased gradually, peaking in November/December and continuing until May of 1983. All villages were affected, with almost all the cases occurring in children under the age of 15 years.

a. Special Groups

i. Travelers

Several studies have documented the occurrence of hepatitis A among European travelers who spend their holidays in the southern Mediterranean countries or North Africa.[74-76] The risk of Danish travelers to North Africa acquiring hepatitis is estimated to be 32.5 per 100,000 airline passengers, and disease acquired in this way is estimated to be responsible for at least 20% of all cases of hepatitis seen in that country.[77] Similarly, Frosner et al. have estimated that the risk of Germans traveling or living in tropical countries, acquiring hepatitis A is nearly 40 times higher than among comparable groups living at home.[78]

Within developed countries the prevalence of infection with HAV is closely related to socioeconomic factors and is higher among people who live in poor, overcrowded conditions than among the middle classes (see Table 2).[79,80] In Greece the prevalence of infection has been shown to be related to family size and the degree of overcrowding as well as standards of hygiene and sanitation.[81]

ii. Male Homosexuals

An increased prevalence of infection has also been reported among some groups of homosexual men. In a study of homosexual and heterosexual men in Seattle, Corey and Holmes found that the prevalence of anti-HAV and annual incidence of hepatitis A was significantly higher among homosexuals and that the rate of infection seemed to be related to the frequency of oral-anal contact[82] (see Table 3).

An increased prevalence of anti-HAV among homosexual men has also been reported in Denmark[83] and England,[84] but not among homosexual men studied in New York[85] in 1974 or 1975, or Halifax, Nova Scotia.[86] These differences probably reflect the prevalence of hepatitis A in the respective communities at the times they were studied, since once HAV enters a community, homosexual practices enable it to spread rapidly. Since the original study was performed, hepatitis A has entered the New York homosexual population and currently poses a serious problem.

Because of the large number of sexual partners, one would predict that homosexual communities might be subject to occasional outbreaks of hepatitis A, especially in developed countries in which a significant number of homosexual men are still susceptible to infection. Several such outbreaks have been documented in recent years, especially in Scandinavia.[87,88] Although there is no evidence of spread of infection beyond the homosexual population, the fact that almost 20% of those infected in the Stockholm epidemic worked as food handlers indicates the potential for further spread.

iii. The Institutionalized

Several studies have indicated that infection with HAV is more common in large institutions for the mentally retarded.[34,89] It appears that the virus is intermittently introduced from outside and is able to spread rapidly because of difficulty in maintaining high standards of personal hygiene. There is currently no evidence that mentally retarded individuals with immune defects are more likely to become infected with HAV or more likely to shed the virus for prolonged periods.

A high prevalence of anti-HAV has also been noted in some prison populations[90] and some groups of i.v. drug users.[91,92] In the latter group it appears that infection is spread by overcrowding and poor hygiene rather than by the use of shared syringes.

iv. Daycare Centers

In the U.S., a significant feature of the epidemiology of hepatitis A is the occurrence of sporadic cases and outbreaks associated with day care centers for children. With the increasing number of single-parent families and families in which both parents work, large numbers of children are cared for in centers which provide a variety of services and vary widely in size, quality of care, and hygienic standards. In recent years hepatitis has become an increasing problem in these centers.[93] Over a 3-year period outbreaks were reported in 19 states; and in two cities, Phoenix and New Orleans, cases associated with day care centers were estimated to account for 30% and 13%, respectively, of all episodes of hepatitis A reported to health authorities.[94] The major features of these outbreaks are a low incidence of clinically apparent disease in children attending the center, a high rate of infection among adults, especially household contacts of the children. When epidemiological studies are carried out the majority of cases appear to follow contact with infected children less than 2 years of age (see Figure 13), presumably because many children of this age are not toilet trained.

Studies of the prevalence of infection in child care centers have shown that hepatitis is more likely to be a problem in centers with a large population (that is, those with the highest enrollment which are open for the longest hours per day), centers in which a high proportion of children are aged 2 years or under, and centers operated for profit.[94]

v. Primate Handlers

Outbreaks of hepatitis among handlers of recently imported juvenile chimpanzees were first reported by Hillis in 1963[95] (see Chapter 4). Since that time more than 35 outbreaks have been reported involving some 200 people involved in the capture, holding, transport, and care of nonhuman primates.[96]

The epidemiology of these outbreaks suggests that the animals usually acquire these infections after capture and then transmit it to their attendants. Typical of these outbreaks is the one described by Davenport et al. in which five of eight people who handled a chimpanzee became ill[97] (see Figure 14).

On May 7, 1964 a young (8 month) male chimpanzee was imported by a New York

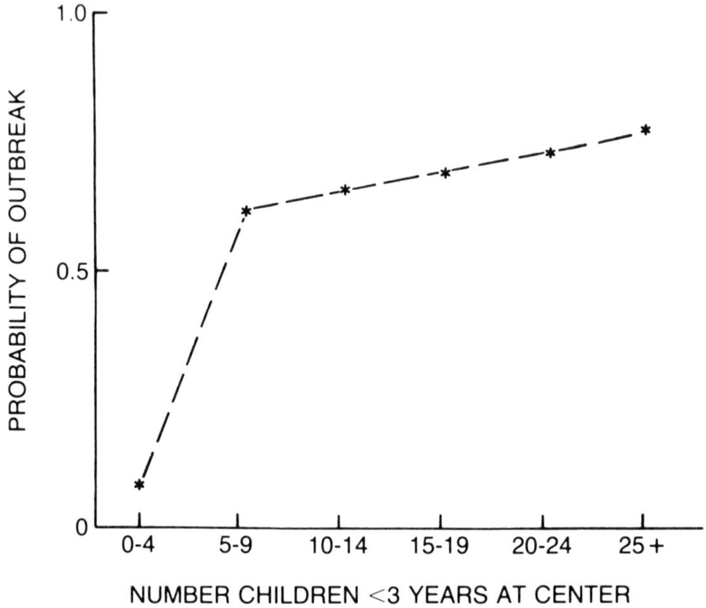

FIGURE 13. Effect of the number of children younger than 3 years of age on the probability of spread of hepatitis A in daycare centers. (From Hadler, S. C., Erben, J. J., Francis, D. P., Webster, H. M., and Maynard, J. G., *J. Infect. Dis.*, 145, 255, 1982. With permission.)

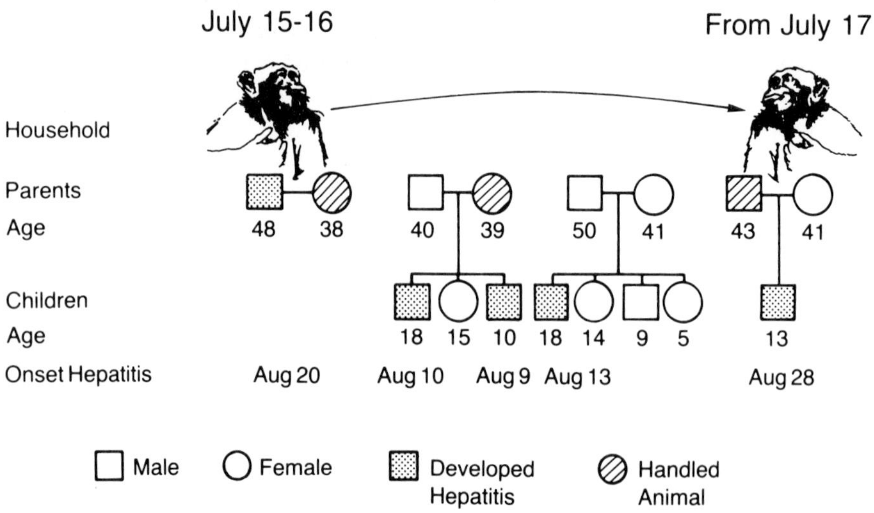

FIGURE 14. Outbreak of chimpanzee-associated hepatitis in Wayne County, Michigan, 1964. (From Davenport, F. M., Hennessy, A. V., Christopher, N., and Smith, C. K., *Am. J. Epidemiol.*, 83, 146, 1966. With permission.)

animal dealer from Liberia. The animal was held in New York and was forwarded to Michigan on July 15th. On arrival in household A, the animal's fur was soiled with excreta and it was immediately bathed and placed in diapers. At night it was caged without diapers and each morning had its haunches sponged to clean them of excreta. From July 15 to 16, both adults in household A cared for the animal; they were assisted by two nephews who lived in household B and a boy from household C. This person only had contact with the chimpanzee on a single occasion, July 16. On the following day the animal was sold to a family that lived 40 mi. away. On arrival in household D, the chimpanzee was noted to be unwell and for several days had rhinorrhea and passed 8 to 10 loose stools per day. It was cared for by a young boy (in whose room it slept) and the boy's father.

Between August 9 and 28, five of the eight people who had direct contact with the chimpanzee developed hepatitis. The order in which hepatitis developed in the four households is shown in Figure 14. Four of the five infections must have occurred between July 15 and July 17, just before the animal became unwell. Infection was presumably acquired by contact with the animal's feces; as none of the handlers wore gloves while handling the animal there were ample opportunities for infected material to be transferred to the handlers' mouths. The chimpanzee presumably acquired his infection while being held in New York.

Another interesting outbreak involving man and two different species of nonhuman primates occurred at the Beardsly Park Zoo in Bridgeport, Connecticut in 1970.[98] In November of that year a young chimpanzee was obtained from Sierra Leone and was caged with a second, adult animal. Upon admission to the zoo, the young chimpanzee was listless and suffering from rhinorrhea and diarrhea. Thirty days later his keeper developed hepatitis and this was followed over the next 2 to 3 weeks by seven additional cases in animal handlers. Serological testing of two wooley monkeys held in the same room as the chimpanzees and cared for by the same attendants revealed evidence of hepatitis in one. Retrospective testing of sera saved from animals and humans involved in both these outbreaks has confirmed both were due to infection with HAV.

While most of the outbreaks which have been documented have been associated with recently captured chimpanzees, other primates (including a gorilla, celebes apes, gibbons, and wooley monkeys) have occasionally been involved. What is not certain is whether the animals were already infected at the time of capture or became infected during the holding period; most of the epidemiological data suggest the latter.

2. *Food Borne*

a. Shellfish

Common vehicle epidemics of hepatitis A due to the consumption of contaminated food have been reported on numerous occasions. In many countries the most important mechanism of transmission appears to be consumption of raw or partially cooked shellfish.

Shellfish-associated hepatitis was first reported from Sweden in 1956 when more than 600 cases occurred among people who had eaten oysters which had been harvested from waters known to have been contaminated with human sewage.[99] In 1961, two further outbreaks involving approximately 400 people were reported from the U.S.; one outbreak was associated with consumption of raw oysters,[100] the other with clams.[101] Since that time many other outbreaks have been documented in the U.S. and elsewhere.[102-104] It is probable that numerous outbreaks and sporadic cases occur but go unrecognized because of difficulties in tracking the source of infection.

An example of a large epidemic is the one which occurred in the southern U.S. in 1973[105,106] involving 278 people in Houston, Texas and Calhoun, Georgia, who had eaten oysters at 11 different restaurants over a 17-day period (see Figure 15). The oysters were supplied by a single distributor who had obtained them from two bays approved for oyster growing. The bays had been closed for oyster harvesting 6 weeks earlier because of pollution following torrential rain and flooding in the Mississippi Valley, but at the time the oysters were collected they were classified as safe. Either the shellfish were able to retain virus for 6 weeks or the standards for detecting fecal contamination were inadequate.

Shellfish are particularly likely to transmit hepatitis A because they must filter large quantities of water in order to obtain adequate supplies of food and oxygen. During this

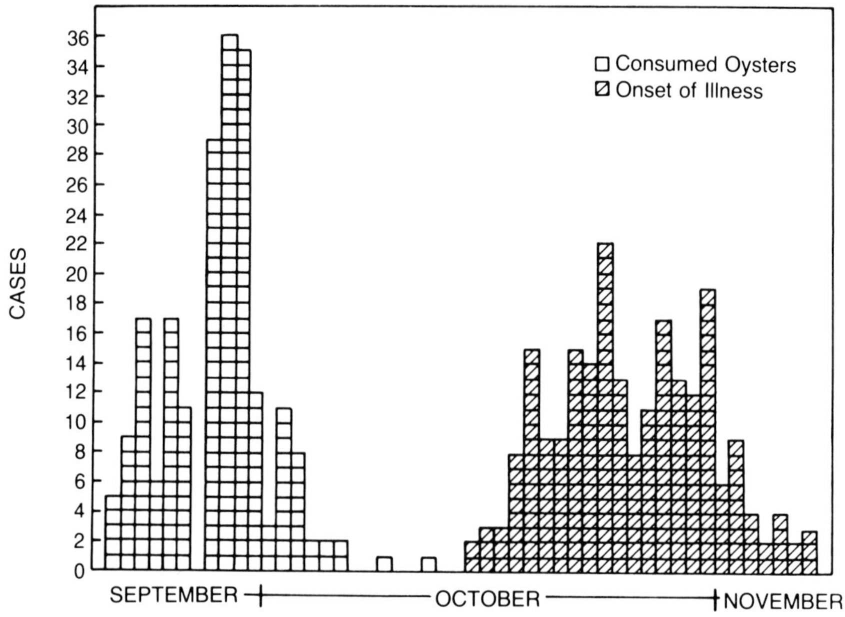

FIGURE 15. Cases of hepatitis A related to oyster consumption at a restaurant in Houston in 1973. (From MacKowiak, P. A., Caraway, C. T., and Portnoy, B. L., *Am. J. Epidemiol.*, 103, 181, 1976. With permission.)

process they also concentrate bacteria, viruses, and certain chemicals which may be present in the water. In many countries shellfish are grown commercially, frequently in shallow waters adjacent to populated areas. If these waters become contaminated with sewage, the potential for transmission of hepatitis A and other human diseases exists.[107,108]

In most countries the cultivation and distribution of shellfish is carefully regulated; however, these regulations are difficult and sometimes impossible to police. If waters become contaminated, the shellfish may be given a fictitious origin or simply transported to adjacent clean waters from which they are sold. The harvest from an individual bed may be sent to a variety of suppliers and then distributed in small quantities to a large number of outlets. These factors coupled with the long incubation period of the disease make shellfish-associated outbreaks very difficult to document.

Hepatitis A is more heat stable than other enteroviruses (see Chapter 3), being stable at 60°C for 60 min but destroyed by boiling for 1 min or more. Shellfish are often eaten raw or after gentle steaming which is sufficient to cause the shell to open but inadequate to inactivate the virus.[109]

The best data on the ability of shellfish to concentrate enteroviruses have been obtained from studies with large volumes of water seeded with vaccine strains of PV. Oysters have been demonstrated to be able to concentrate PV up to 60-fold. Using this system, not only are oysters able to concentrate enteroviruses, but recent studies indicate that they may retain the virus even after 100 hr of depuration.[110] Clams on the other hand, usually eliminate the virus within 24 to 48 hr.[111]

The acceptability of water supplies is usually determined by measuring the quantity of fecal *E. coli* present. As viral contamination can exist in the absence of detectable enteric bacteria, elimination of shellfish-associated hepatitis will require either a sensitive test for the presence of the agent in large volumes of water or development of a means of treating shellfish which retains their flavor but destroys the virus.

b. Other Food

Hepatitis A is an important cause of food-borne viral disease and has the potential to infect large numbers of people. Despite the explosive nature of these outbreaks and the frequency with which food handlers become infected, they are relatively uncommon. For example, in the U.S. approximately 1000 food handlers are thought to develop hepatitis A each year, yet the number of food handler-associated outbreaks reported to the CDC is no more than four or five.[112]

As hepatitis A is inactivated by heating, most outbreaks are associated with uncooked food or food which has been cooked and then is contaminated prior to serving. A wide variety of vehicles have been involved in outbreaks of hepatitis A; these include beverages such as raw milk and orange juice, salads, hamburgers, spaghetti, cold meat, cream, and pastries, and on one occasion a punch made from maotai.[113-119] In many of the outbreaks that have been documented, the majority of those infected have been young adults which probably reflects the type of food being consumed and the low rate of immunity in this group in most developed countries.

The major features of 20 food-borne epidemics of hepatitis A reported to CDC between 1968 and 1980 have recently been reviewed.[112] Incubation periods ranged from 10 to 50 days (means of 26 to 32 days). Attack rates for other restaurant employees ranged from 2.9 to 66.7 per 100 and consumers of the food from 4.8 to 56.0 per 100. The interval between the onset of jaundice in the food handler and the first patron involved in the outbreak ranged from 3 to 24 days (mean 11 days). In none of the outbreaks was there evidence of a secondary wave of infection.

An infected food handler was identified in 18 of the 20 outbreaks and on each occasion uncooked food or food handled after cooking was the presumed vehicle of infection.

The investigation of outbreaks of food-borne hepatitis has been simplified by the development of specific tests for HAV infection. As a result it is now possible to rapidly confirm the diagnosis and to screen suspect food handlers for evidence of infection irrespective of whether they are symptomatic. The availability of assays which determine past exposure to the virus also allow for the selective use of gamma globulin to control an outbreak.

One of the first epidemics in which these techniques were shown to be used was an outbreak among recruits at the Naval Training Center in San Diego which occurred in October 1974.[116] The outbreak was traced to a food handler responsible for the preparation of salads and desserts who contaminated batches of food during the prodromal period of his illness. Subsequently, within 3 to 6 weeks of eating food prepared by the person, 136 recruits developed biochemical evidence of hepatitis. HAV particles were detected in stool specimens from a number of infected people and rising titers of anti-HAV were in 23 of 25 serum pairs examined.[39]

In most outbreaks, transmission of infection can be traced to a food handler with poor standards of hygiene. The most commonly described cause is the failure of the person to wash their hands after defecation. Infection is usually transmitted by a food handler in the late incubation period of the disease, when virus shedding is at its peak and the person may be completely asymptomatic. As public health authorities are not usually notified until a cluster of cases has occurred, it is usually too late for gamma globulin to have a major impact except in reducing the risk of secondary cases.[116]

Investigations of outbreaks of food-borne hepatitis are exercises in classical field epidemiology. Public health officers should interview patients to try and determine a common source of infection through the use of food questionnaires. Publicity is often useful for detecting other exposed but well people. Liver function tests and tests for hepatitis A-specific IgG and IgM are usually performed to confirm evidence of current infection or past immunity, and can be used as a source of reassurance and a means of selecting people who should be offered gamma globulin. In determining the source of infection it is important to question the food handlers individually and also obtain information about the practices of their colleagues. Information on the adequacy and regularity of hand washing procedures and the use of gloves when handling food is particularly important.

Colleagues working with an infected food handler may also be at risk as such people frequently eat together at work. Gamma globulin should be considered for this group. Food handlers should not be allowed to prepare food until they have completely recovered from their illness.

3. Water Borne

The first water-borne epidemic of hepatitis was described in 1920[120] and since then many other episodes have been documented.[121] However, caution needs to be exercised in evaluating some of the early data as it is now recognized that water-borne outbreaks of hepatitis (especially in developing countries) may be caused by other agents[122] (see Chapter 12). Transmission has followed fecal contamination of traditional water supplies and also, ingestion of contaminated water not intended for human consumption. In the U.S. most outbreaks have followed ingestion of private supplies, but municipal supplies have also been implicated.[68]

Water-borne outbreak of hepatitis A are surprisingly uncommon in developed countries in view of the frequency of other forms of water-borne enteric disease. Extensive surveillance in the U.S. revealed six outbreaks of water-borne hepatitis A in 1971, five in 1972, two in 1973, and one each in 1975 and 1977, with none in 1974, 1976, or 1978, comprising less than 6% of all water-borne outbreaks reported during the period.[123] It has been estimated that water-borne transmission accounts for less than 1% of all cases of hepatitis seen in the U.S.[121]

Although it is theoretically possible to acquire hepatitis A by swimming in contaminated pools or sea water, we are unaware of any data which show that this is a real risk.

Control measures for prevention of water-borne hepatitis are largely the responsibility of sanitary engineers. In most countries the safety of public water supplies is regularly monitored and evidence of fecal contamination detected by the presence of enteric bacteria. As these methods do not guarantee the absence of enteric viruses, separate standards may be required.[124] Experience during World War II indicates that even heavily contaminated water can be rendered safe for drinking by the addition of high levels of chlorine (see Chapter 1), a total chlorine residual of 1 mg/ℓ, and a free chlorine residual of 0.4 mℓ/ℓ over 30 min, inactivated virus in the same water only after the water had been treated by coagulation, settling, and filtration. If there is real doubt about the purity of drinking water it should be boiled before use.

In recent years improved methods have been developed for concentrating enteric viruses from drinking water, mainly by utilizing techniques such as absorption and elution from micoporous filters, followed by organic flocculation. Sobsey and his colleagues in the School of Public Health at the University of North Carolina have demonstrated the HAV can be recovered and concentrated more than 100-fold from samples of seeded tap water, with recovery of about 50% of the input virus.[125]

They have used this technique to demonstrate the presence of HAV in samples of fecally contaminated well water, collected during an outbreak of hepatitis A in a rural community.[126] The outbreak took place in a small dairy farming community in Washington County, Maryland from April to June 1981, and involved 14 people in a susceptible population of 120 (an attack rate of 11.7%). Most of the population of this area consumed untreated ground water from their own wells. Epidemiological investigation suggested that the common factor was drinking water obtained from a shallow well or contact with a person who had drunk water from such a well.

In the middle of the outbreak, samples of water (500 to 950 ℓ) were obtained from the wells of five households in which patients with hepatitis lived and a nearby spring. Enteric viruses were concentrated by absorption to electropositive pleeted-sheet cartridge filters and the absorbed viruses eluted with a beef-extracted, glycine buffer at pH 9.5. After adjusting the pH, the viruses were further concentrated by organic flocculation adjusted to pH 3.5 and the floc deposited by centrifugation. Aliquots of the suspended concentrate, plus antibiotics were inoculated into BGMK and MA-104 cells to detect common enteric viruses and into AGMK cells to detect the presence of HAV. Replication of HAV was detected using an RIFA and the specificity of the results confirmed by neutralization tests with appropriate antisera and animal inoculation studies. HAV was isolated from four of the six samples and successfully transmitted to chimpanzees using pools of this material.

Subsequently Sobsey and his colleagues have used these techniques to study the absorption and survival of HAV in ground water, waste water, and soil, and shown that in the cold the virus can remain viable in these settings for up to 12 weeks.

4. Other

There is no evidence that percutaneous transmission of hepatitis A is an important means of transmission. The increased prevalence of the disease among i.v. drug users in some countries is usually ascribed to their poor standard of living and lack of routine hygiene. Similarly, there is no data to support a role for blood-sucking arthropods.

Spread by droplet infection has been postulated by some workers to account for explosive outbreaks in which no common vehicle has been identified,[127] and for the higher prevalence of infection in Israeli communes which used crude oxidant-pond effluents to spray-irrigate their crops.[128]

Although it is theoretically possible that infection could occur in this way the case is still unproven. Certainly the pattern of transmission of HAV in the community is markedly different from the pattern produced by respiratory viruses.

III. ECOLOGY OF HAV

A. Hosts

Man is considered to be the only important reservoir of HAV, however the possibility of extra human reservoirs of infection has not been extensively investigated.

A wide range of nonhuman primates have been shown to be susceptible to infection (see Chapter 4) and some of these animals have been demonstrated to have detectable levels of anti-HAV shortly after capture. Whether these antibodies represent infections acquired at the time of capture or during the holding period, the occurrence of HAV in the wild or the existence of cross-reacting simian enteroviruses has not been resolved. Subclinical infections of nonhuman primates can serve as a link in the chain of infection in man, particularly among primate handlers, experimenters, and their families.

Few attempts have been made to search for evidence of natural infection with HAV in animals other than primates. Several years ago we examined sera from several pigs and dogs which lived in a crowded village on an isolated Pacific island. The animals were tested because they played an important role in the removal of human feces from the village. Apparently, specific antibodies were detected in two of the three dogs but none of the pigs examined. Antibodies in the dog sera were reactive in a standard RIA and caused aggregation of HAV particles by IEM[129] (see Figure 16). Whether dogs can be infected (or immunized by ingestion of infected human feces) or these findings indicate the existence of cross-reacting canine enteroviruses has not been determined.

It has been suggested that under some circumstances shellfish may serve as reservoirs of infection by concentrating virus from regions which are not subject to purging with clean water and depuration.[130] This mechanism could explain the apparent persistence of infection in some isolated Pacific islands where human feces is disposed of into a central lagoon.

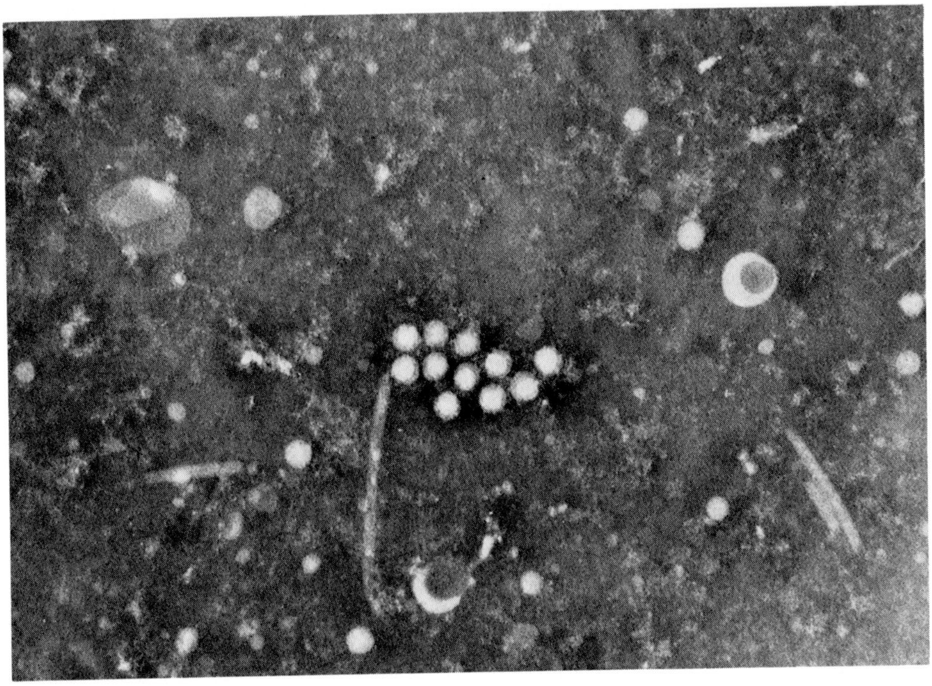

FIGURE 16. An immune aggregate obtained by reacting HAV purified from the feces of a patient with hepatitis A with a serum specimen collected from a dog in a village in Vanuatu. This specimen also appeared to contain anti-HAV in a standard RIA. EM kindly provided by Dr. John Marshall.

Little data exist on the survival of HAV in the environment, however, the relative stability of the agent to heat suggests that it may be capable of retaining infectivity in the environment for several days or weeks, especially if it is contained in feces.[131]

B. Strain Differences

Although individual strains of HAV show differences at the molecular level, there is no convincing evidence for the existence of epidemiologically important differences. Polyclonal antibodies and monoclonals directed against the major antigenic determinant appear to be capable of detecting strains of HAV isolated in different parts of the world. Normal immunoglobulin prepared in different countries appears to be effective wherever it is used and second infections with HAV are extremely rare.

REFERENCES

1. **Raska, K.,** Epidemiology of infectious hepatitis, *J. Ky. Epidemiol. (Prague),* 1, 121, 1957.
2. **Reisler, D. J., Brachott, D., and Mosely, J.,** Viral hepatitis in Israel: morbidity and mortality data, unpublished observations.
3. **Koff, R. S., Chalmers, T. C., Culhane, P. L., and Iber, F. L.,** Under reporting of viral hepatitis, *Gastroenterology,* 64, 1194, 1973.
4. **Mosely, J. W. and Kendrick, M. A.,** Hepatitis as a world problem, *Bull. N.Y. Acad. Med.,* 45, 143, 1969.
5. **Mosely, J. W.,** Hepatitis epidemiology: an overview, *Am. J. Med. Sci.,* 270, 253, 1975.
6. **Bachmann, L.,** Hepatitis epidemica in Europe, in *World Atlas of Epidemic Diseases,* Part 1, Rodenwalt, E., Ed., Falk Verlag, Hamburg, 1952, 67.

7. **Mosley, J. W.,** Viral hepatitis: a group of epidemiologic entities, *Can. Med. Assoc. J.*, 106, 427, 1972.
8. **Gust, I. D., Lehmann, N. I., Lucas, C. R., Ferris, A. A., and Locarnini, S. A.,** Studies on the epidemiology of Hepatitis A in Melbourne, in *Viral Hepatitis,* Vyas, G. N., Cohen, S. N., and Schmid, R., Eds., Franklin Institute Press, Philadelphia, 1978, 105.
9. **Skinhoj, P., Hillinger, F. G., Hoving-Hougen, K., and Lous, P.,** Infectious liver diseases in three groups of Copenhagen workers: correlation of hepatitis A infection to sewage exposure, *Arch. Environ. Health,* 36, 139, 1980.
10. **Corey, L. and Holmes, K.,** Sexual transmission of hepatitis A in homosexual men: incidence and mechanism, *N. Engl. J. Med.*, 302, 435, 1980.
11. **Malkani, P. K. and Grewal, A. K.,** Observations on infectious hepatitis in pregnancy, *Indian J. Med. Res.*, 45 (Suppl.), 77, 1957.
12. **Bornhanmanesh, F., Haghighi, P., Hekmat, K., Rezaizadeh, K., and Ghavami, A. G.,** Viral hepatitis during pregnancy: severity and effect on gestation, *Gastroenterology,* 64, 304, 1973.
13. **Khuroo, M. S., Teli, M. R., Skidmore, S., Sofi, M. A., and Khuroo, M. I.,** Incidence and severity of viral hepatitis in pregnancy, *Am. J. Med.*, 70, 252, 1981.
14. **Krugman, S.,** Infectious hepatitis: detection of virus during the incubation period and in clinically inapparent infection, *N. Engl. J. Med.*, 261, 729, 1959.
15. **Havens, W. P., Jr.,** Elimination in human feces of infectious hepatitis virus parentially introduced, *Proc. Soc. Exp. Biol. Med.*, 61, 210, 1946.
16. **Havens, W. P., Jr.,** Period of infectivity of patients with experimentally induced infectious hepatitis, *J. Exp. Med.*, 83, 251, 1946.
17. **Neefe, J. R.,** Homologous serum hepatitis and infectious (epidemic) hepatitis: studies in volunteers bearing on immunological and other characteristics of the etiological agents, *Am. J. Med.*, 1, 3, 1946.
18. **Havens, W. P., Jr., Ward, R., Drill, V. A., and Paul, J. R.,** Experimental production of hepatitis by feeding icterogenic materials, *Proc. Soc. Exp. Biol. Med.*, 57, 206, 1944.
19. **Neefe, J. R., Stokes, J., Jr., Garber, R. S., Gellis, S. S.,** Studies on the relationship of the hepatitis virus to persistent symptoms, disability, and hepatic disturbance (''chronic hepatitis symdrome'') following acute infectious hepatitis, *J. Clin. Invest.*, 26, 329, 1947.
20. **Stokes, J., Jr., Berk, J. E., Malamut, L. L., Drake, M. E., Barondess, I. A., Baske, W. J., Wolman, I. J., Farquhar, J. D., Bevan, B., Drummond, R. J., Maycock, W.'d'A., Capps, R. B., and Bennett, M. A.,** The carrier state in viral hepatitis, *JAMA,* 54, 1059, 1954.
21. **Bennett, A. M., Capps, R. B., Drake, M. E., Ettinger, R. H., Mills, E. H., and Stokes, J., Jr.,** Endemic infectious hepatitis in an infants orphanage. II. Epidemiologic studies in infants and young children, *Arch. Intern. Med.*, 90, 37, 1952.
22. **Capps, R. B., Bennett, A. M., and Stokes, J., Jr.,** Endemic infectious hepatitis in an infants orphanage. I. Epidemiologic studies in student nurses, *Arch. Intern. Med.*, 89, 6, 1952.
23. **Maynard, J. E.,** Infectious hepatitis at Fort Yukon, Alaska: report of an outbreak 1960—1961, *Am. J. Public Health,* 53, 31, 1963.
24. **Skinhoj, P. and McNair, A.,** An outbreak of infectious hepatitis in South West Greenland, *Am. J. Epidemiol.*, 99, 50, 1974.
25. **Skinhoj, P., Mikkelson, F., and Hillinger, B.,** Hepatitis in Greenland: importance of specific antibody testing in epidemiologic surveillance, *Am. J. Epidemiol.*, 105, 140, 1977.
26. **McCollcom, R. W.,** Hepatitis as it relates to the Arctic, *Arch. Environ. Health,* 17, 529, 1968.
27. **Gust, I. D.,** The Epidemiology of viral hepatitis, in *Viral Hepatitis and Liver Disease,* Vyas, G. N., Dienstag, J. L., and Hoofnagle, J. N., Eds., Grune & Stratton, Orlando, 1984, 417.
28. **Mosley, J. W.,** Epidemiology of HAV infection, in *Viral Hepatitis,* Vyas, G. N.,Cohen, S. N., and Schmid, R., Eds., Franklin Institute Press, Philadelphia, 1978, 85.
29. **Gust, I. D., Lehmann, N. I., Dimitrikakis, M.,** A seroepidemiologic study of infections with HAV and HBV in five Pacific islands, *Am. J. Epidemiol.*, 110, 237, 1979.
30. **Wong, D. C., Purcell, R. H., and Rosen, L.,** Prevalence of antibody to hepatitis A and hepatitis B viruses in selected populations of the South Pacific, *Am. J. Epidemiol.*, 110, 227, 1979.
31. **Ashley, A.,** Use of gamma globulin for control of infectious hepatitis in an institution for the mentally retarded, *N. Engl. J. Med.*, 252, 88, 1955.
32. **Drake, M. E. and Ming, C.,** Gamma globulin in epidemic hepatitis: comparative value of two dosage levels, apparently near the minimal effective level, *JAMA,* 155, 1302, 1954.
33. **Matthew, E. B., Dietzman, D. E., Madden, D. L., Newman, S. J., Sever, J. L., Nagler, B., Bouton, S. M., and Rostafinski, M.,** A major epidemic of infectious hepatitis in an institution for the mentally retarded, *Am. J. Epidemiol.*, 98, 199, 1973.
34. **Lehmann, N. J., Sharma, D. L. B., and Gust, I. D.,** Prevalence on antibody to the hepatitis A virus in a large institution for the mentally retarded, *J. Med. Virol.*, 2, 335, 1978.
35. **Helmsing, P. J., Duermeyer, W., Van Hattem, G. C. A. M., and Wielaard, F.,** An outbreak of hepatitis A in an institution for the mentally retarded, *J. Med. Virol.*, 5, 143, 1980.

36. **Ward, R., Krugman, S., Giles, J. P., Jacobs, A. M., and Bodansky, O.,** Infectious hepatitis: studies of this natural history and presentation, *N. Engl. J. Med.*, 258, 407, 1958.
37. **Szmuness, W., Purcell, R. H., Dienstag, J. L., and Stevens, C. E.,** Antibody to hepatitis A in institutionalized mentally retarded patients, *JAMA*, 237, 1702, 1977.
38. **Dienstag, J. L., Kapikian, A. Z., Feinstone, S. M., Purcell, R. H., Boggs, J. D., and Conrad, M. E.,** Fecal shedding of hepatitis A antigen, *Lancet*, 1, 765, 1975.
39. **Dienstag, J. L., Routenberg, J. A., Purcell, R. H., Hooper, R. R., and Harrison, W. D.,** Food handler-associated outbreak of hepatitis type A, *Ann. Intern. Med.*, 83, 647, 1975.
40. **Rakela, J. and Mosley, J. W.,** Fecal excretion of hepatitis A virus in human, *J. Infect. Dis.*, 135, 933, 1977.
41. **Locarnini, S. A., Gust, I. D., Ferris, A. A., Stott, A. C., and Wong, M. L.,** A prospective study of acute viral hepatitis with particular reference to hepatitis type A, *Bull. WHO*, 54, 199, 1976.
42. **Hall, W. T.,** U.S. Environmental Protection Agency Report, EPA-600/1-77-149, U.S. Government Printing Office, Washington, D.C., 1977.
43. **Coulepis, A. G., Locarnini, S. A., Lehmann, N. I., and Gust, I. D.,** Detection of hepatitis A virus in the feces of patients with naturally acquired hepatitis, *J. Infect. Dis.*, 141, 151, 1980.
44. **Favero, M. S., Maynard, J. E., Leger, R. T., Graham, D. R., and Dixon, R. E.,** Guidelines for the care of patients hospitalized with viral hepatitis, *Ann. Intern. Med.*, 91, 872, 1976.
45. **Mayhall, G. G.,** Isolation techniques for hospital patients with viral hepatitis: new guidelines premature, *Infect. Control*, 1, 71, 1980.
46. **Papavangelou, G., Roumeliotou-Karayannis, A., and Contoyannis, P.,** The risk of nosocomial hepatitis A and B virus infections from patients under care without isolation procedures, *J. Med. Virol.*, 7, 143, 1981.
47. **Favero, M. S.,** Enteric precautions for hepatitis patients, *Am. J. Infect. Control*, 9, 58, 1981.
48. Outbreak of viral hepatitis in staff of a pediatric ward — California, *Morbid. Mort. Wkly. Rep.*, 26, 77, 1977.
49. **Orenstein, W. A., Wu, E., Wilkins, J., Robinson, K., Francis, D. P., Timko, N., and Wayne, R.,** Hospital acquired hepatitis A: report of an outbreak, *Pediatrics*, 67, 494, 1981.
50. **Giles, J. P., Liebhaber, H., Krugman, S., and Lattimer, C.,** Eary viremia and viruria in infectious hepatitis, *Virology*, 24, 107, 1964.
51. **Krugman, S., Ward, R., and Giles, J. P.,** The natural history of infectious hepatitis, *Am. J. Med.*, 32, 717, 1962.
52. **Boggs, J. D., Melnick, J. L., Conrad, M. E., and Felsher, B. K.,** Viral hepatitis: clinical and tissue culture studies, *JAMA*, 214, 1041, 1970.
53. **Francis, T., Jr., Frisch, A. W., and Quilligan, J. J., Jr.,** Demonstration of infectious hepatitis virus in presymptomatic period after transfer by transfusion, *Proc. Soc. Exp. Biol. Med.*, 61, 276, 1946.
54. **Harden, A. G., Barondess, J. A., and Parker, B.,** Transmission of infectious hepatitis by transfusion of whole blood, *N. Engl. J. Med.*, 253, 923, 1955.
55. **Meyers, J. D., Huff, J. C., Holmes, K. K., Thomas, E. D., and Bryan, J. A.,** Parenterally transmitted hepatitis A associated with platelet transfusions: epidemiologic study of an outbreak in a marrow transplantation center, *Ann. Intern. Med.*, 81, 145, 1974.
56. **Meyers, J. D., Dienstag, J. L., Purcell, R. H., Thomas, E. D., and Holmes, K. K.,** Parenterally transmitted non-A, non-B hepatitis: an epidemic reassessed, *Ann. Intern. Med.*, 87, 57, 1977.
57. **Seeberg, S., Brandberg, A., Hermodsson, S., Larsson, P., and Lundgren, S.,** Hospital outbreak of hepatitis A secondary to blood exchange in a baby, *Lancet*, 1, 1155, 1981.
58. **Barbara, J., Howell, D., Briggs, M., and Parry, J. V.,** Post-transfusion hepatitis A, *Lancet*, 1, 738, 1982.
59. **Skidmore, S., Boxhall, E., and Ala, F.,** A case report of post-transfusion hepatitis A, *J. Med. Virol.*, 10, 223, 1982.
60. **Hollinger, F. B., Khan, N. C., Oefinger, P. E., Yawn, D. H., Schmulen, A. C., Dreesman, G. R., and Melnick, J. L.,** Post-transfusion hepatitis type A, *JAMA*, 250, 2313, 1983.
61. **Sheretz, R. J., Russell, B. A., and Reuman, P. D.,** Transmission of hepatitis A by transfusion of blood products, *Arch. Intern. Med.*, 144, 1579, 1984.
62. **Stevens, C. E., Silbert, J. A., Miller, D. R., Dienstag, J. L., Purcell, R. H., and Szmuness, W.,** Serologic evidence of hepatitis A and B virus infection in thalassemia patients: a retrospective study, *Transfusion*, 18, 94, 1978.
63. **Papavangelou, G., Frosner, G. G., Economidou, J., Parcha, S., Roumelitou, A.,** Prevalence of hepatitis A and B infections in multiply transfused thalasemic patients, *Br. Med. J.*, 1, 689, 1978.
64. **Szmuness, W., Dienstag, J. L., Purcell, R. H., Prince, A. M., Stevens, C. E., and Levine, R. W.,** Hepatitis type A and hemodialysis: a seroepidemiologic study in 15 U.S. centers, *Ann. Intern. Med.*, 87, 8, 1977.

65. **Batten, P. J., Runte, V. E., and Skinner, H. G.,** Infectious hepatitis: infectiousness during the presymptomatic phase of the disease, *Am. J. Hyg.*, 77, 129, 1963.
66. **Knight, V., Drake, M. E., Belden, E. A., Franklin, B. J., Romer, M., and Copple, L. O.,** Characteristics of spread of infections hepatitis in schools and households in an epidemic in a rural area, *Am. J. Hyg.*, 59, 1, 1954.
67. **Strom, J.,** A comparison between family infections during epidemics of poliomyelitis and hepatitis in Stockholm, *Acta Med. Scand.*, 165, 49, 1949.
68. **Mosley, J. W. and Galambos, J.,** Viral Hepatitis, in *Diseases of the Liver*, Schiff, L., Ed., Lippincott, Philadelphia, 1975, 525.
69. **Cline, A., Mosley, J., and Scovel, F. G.** Viral hepatitis among American missionaries abroad: a preliminary study, *JAMA*, 199, 119, 1967.
70. **Frame, J. E.,** Hepatitis among missionaries in Ethiopia and Sudan: susceptibles at high risk, *JAMA*, 203, 99, 1968.
71. **Gust, I. D., Lehmann, N. I., and Lucas, C. R.,** Relationship between prevalence of antibody to hepatitis A antigen and age: a cohort effect? *J. Infect. Dis.*, 138, 425, 1978.
72. **Frosner, G. G., Overby, L. R., Fleming, B., Gerth, H. J., Haas, H., Decker, R. H., Ling, C. M., and Frosner, H. R.,** Seroepidemiological investigation of patients and family contacts in an epidemic of hepatitis A, *J. Med. Virol.*, 1, 1963, 1977.
73. **Gust, I. D., Lehmann, N. I., Lucas, C. R., Ferris, A. A., and Locarnini, S. L.,** Studies on the epidemiology of hepatitis A in Melbourne, in *Viral Hepatitis*, Vyas, G. N., Cohen, S. N., and Schmid, R., Eds., Franklin Institute Press, Philadelphia, 1978, 108.
74. **Chaudauri, A. K. R.,** Tourist hepatitis in the west of Scotland, *Scot. Med. J.*, 23, 213, 1978.
75. **Iwarson, S. and Stenquist, K.,** Tourist hepatitis and gamma globulin prophylaxis, *Scand. J. Infect. Dis.*, 8, 143, 1976.
76. **Steffen, R.,** Tourist hepatitis, *Br. Med. J.*, 4, 1543, 1977.
77. **Skinhoj, P., Gluud, C., and Ramsoe, K.,** Travellers hepatitis: origin and characteristics of cases in Copenhagen 1976—1978, *Scand. J. Infect. Dis.*, 13, 1, 1981.
78. **Frosner, G. G., Roggendorf, M., Frosner, H., Gerth, H. J., Borst, V., Blochinger, G., and Schmid, W.,** *Viral Hepatitis*, Szmuness, W., Alter, H. J., and Maynard, J. E., Eds., Franklin Institute Press, Philadelphia, 1982, 157.
79. **Maynard, J. E., Bradley, D. W., Hornbeck, C. L., Fields, R. M., Doto, I. L., and Hollinger, F. B.,** Preliminary studies of antibody to hepatitis A virus in populations in the United States, *J. Infect. Dis.*, 134, 728, 1976.
80. **Dienstag, J. L., Szmuness, W., and Stevens, C. E.,** Hepatitis A virus infection: new insights from seroepidemiologic studies, *J. Infect. Dis.*, 137, 328, 1978.
81. **Papavangelou, G. J., Gourgouli, K. P., Vissoulis, H. G.,** Characteristics of hepatitis A virus infections in Greece, *Am. J. Epidemiol.*, 112, 482, 1980.
82. **Corey, L. and Holmes, K. K.,** Sexual transmission of hepatitis A in homosexual men, *N. Engl. J. Med.*, 302, 435, 1980.
83. **Kryger, P., Pederson, N. S., Mathieson, L., and Neilson, J. O.,** Increased risk of infection with hepatitis A and B viruses in men with a history of syphilis: relation to sexual contacts, *J. Infect. Dis.*, 145, 23, 1982.
84. **Mindell, A. and Tedder, R.,** Hepatitis A in homosexuals, *Br. Med. J.*, 282, 1666, 1982.
85. **Szmuness, W., Dienstag, J. L., Purcell, R. H., Harley, E. J., Stevens, C. E., and Wong, D. C.,** Distribution of antibody to hepatitis A antigen in urban adult populations, *N. Engl. J. Med.*, 295, 755, 1976.
86. **McFarlane, E. S., Embil, J. A., Manuel, F. R., and Thiebaux, E. J.,** Antibodies to hepatitis A antigen in relation to the number of lifetime sexual partners in patients attending an STD clinic, *Br. J. Vener. Dis.*, 57, 58, 1981.
87. **Christenson, B., Brotsom, C., Bottinger, M., Hermanson, J., Weiland, O., Ryd, G., Berg, J. V. R., and Sjoblom, R.,** An epidemic outbreak of hepatitis A among homosexual men in Stockholm. Hepatitis A: a special hazard for the male homosexual subpopulation in Sweden, *Am. J. Epidemiol.*, 116, 599, 1982.
88. **Hoybye, G., Skinhoj, P., Hentzer, B., Faber, V., and Mathieson, L.,** An epidemic of acute viral hepatitis in male homosexuals: etiology and clinical characteristics, *Scand. J. Infect. Dis.*, 12, 241, 1980.
89. **Szmuness, W., Purcell, R. H., Dienstag, J. L., and Stevens, C. E.,** Antibody to hepatitis A in institutionalized mentally retarded patients, *JAMA*, 237, 1702, 1977.
90. **McFarlane, E. S. and Embil, J. A.,** Antibody to hepatitis A virus in young inmates, *J. Infect. Dis.*, 146, 442, 1982.
91. **Siebke, J. C., Degre, M., Ritland, S., and Enger, S. C.,** Prevalence of hepatitis A antibodies in a normal population and some selected groups of patients in Norway, *Am. J. Epidemiol.*, 115, 185, 1982.
92. **Widdell, A., Hansson, B. G., Moestrop, T., and Nordenfelt, E.,** Increased occurrence of hepatitis A with cyclic outbreaks among drug addicts in a Swedish community, *Infection*, 11, 198, 1983.

93. **Storch, G., McFarland, L. M., Kelso, K., Heilman, C. J., and Caraway, C. T.,** Viral hepatitis associated with day-care centers, *JAMA,* 243, 1514, 1979.
94. **Hadler, S. C., Erben, J. J., Francis, D. P., Webster, H. M., and Maynard, J. E.,** Risk factors for hepatitis A in day-care centers, *J. Infect. Dis.,* 145, 255, 1982.
95. **Hillis, W. D.,** An outbreak of infectious hepatitis among chimpanzee handlers at a United States Air Force Base, *Am. J. Hyg.,* 73, 316, 1961.
96. **Deinhardt, F.,** Hepatitis in primates, in *Advances in Virus Research,* Academic Press, New York, 1976, 20, 113.
97. **Davenport, F. M., Hennessy, A. V., Christopher, N., and Smith, C. K.,** A common source multi-household outbreak of chimpanzee-associated hepatitis in humans, *Am. J. Epidemiol.,* 83, 146, 1966.
98. **Dienstag, J. L., Davenport, F. M., McCollum, R. W., Hennessy, A. V., Klatskin, G., and Purcell, R. H.,** Nonhuman primate-associated viral hepatitis type A: serologic evidence of hepatitis A virus infection, *JAMA,* 236, 462, 1976.
99. **Roos, B.,** Hepatitis epidemic conveyed by oysters, *Sven. Lakar,* 53, 989, 1956.
100. **Mason, J. O. and McLean, W. R.,** Infectious hepatitis traced to the consumption of raw oysters, *Am. J. Hyg.,* 75, 90, 1962.
101. **Dougherty, W. J. and Altman, R.,** Viral hepatitis in New Jersey, *Am. J. Med.,* 32, 704, 1962.
102. **Ruddy, S. J., Johnson, R. F., Mosley, J. W., Atwater, J. B., Rossetti, M. A., and Hart, J. C.,** An epidemic of clam-associated hepatitis, *JAMA,* 208, 649, 1969.
103. **Dienstag, J. L., Gust, I. D., Lucas, C. R., Wong, D. C., and Purcell, R. H.,** Mussel-associated viral hepatitis A: a serological confirmation, *Lancet,* 1, 561, 1976.
104. **Goh, K. T., Chan, L., Ding, J. L., and Oon, C. J.,** An epidemic of cockles-associated hepatitis A in Singapore, *Bull. W.H.O.,* 62, 893, 1984.
105. **Portnoy, B. L., Mackowiak, P. A., Caraway, C. T., Walker, J. A., McKinley, T. W., and Klein, C. A.,** Oyster-associated hepatitis: failure of shellfish certification programs to prevent outbreaks, *JAMA,* 23, 1065, 1975.
106. **Mackowiak, P. A., Caraway, C. T., and Portnoy, B. L.,** Oyster-associated hepatitis: lessons from the Louisiana experience, *Am. J. Epidemiol.,* 103, 181, 1976.
107. **Earampamoorthy, S. and Koff, R. S.,** Health hazards of bivalve-mollusc ingestion, *Ann. Intern. Med.,* 83, 107, 1975.
108. **Nickerson, J. T. and Sinskey, A. J.,** in *Microbiology of Foods and Food Processing,* Elsevier, New York, 1972, 197.
109. **Koff, R. S. and Sear, H. S.,** Internal temperature of steamed clams, *N. Engl. J. Med.,* 276, 737, 1967.
110. **Mitchell, J. R., Presnall, M. W., Atkin, E. W., Cummins, J. M., and Liu, O. C.,** Accumulation and elimination of poliovirus by the Eastern oyster, *Am. J. Epidemiol.,* 84, 40, 1966.
111. **Metcalf, T. G., Mullin, B., Erckerson, D., Moulton, E., and Larkin, E. P.,** The accumulation of enteric viruses by the oyster, *Crassostrea virginica, Appl. Environ. Microsc. Biol.,* 38, 275, 1979.
112. **Carl, M., Francis, D. R., Maynard, J. E.,** Food borne hepatitis: recommendations for control, *J. Infect. Dis.,* 148, 1133, 1983.
113. **Cliver, D. O.,** Implications of food borne infectious hepatitis, *Public Health Rep.,* 81, 159, 1966.
114. **Philip, J. R., Hamilton, T. P., Albert, T. J., Stone, R. S., and Pait, C. F.,** Infectious hepatitis outbreak with maotai as the vehicle of transmission, *Am. J. Epidemiol.,* 97, 50, 1973.
115. **Schoenbaum, S. C., Baker, O., and Jezek, Z.,** Common-source epidemic of hepatitis due to glazed and iced pastries, *Am. J. Epidemiol.,* 104, 74, 1976.
116. **Hooper, R. R., Jeuls, C. W., Routenberg, J. A., Harrison, W. O., Kilpatrick, M. E., and Kendra, S. J.,** An outbreak of type A hepatitis in a Naval training center, San Diego: epidemiologic evaluation, *Am. J. Epidemiol.,* 105, 148, 1977.
117. **Zachoval, R., Frosner, G. G., Deinhardt, F., and John, I.,** Hepatitis A transmission in cold meats, *Lancet,* 2, 260, 1981.
118. **Leger, R. T., Boyer, K. M., Pattison, C. P., and Maynard, J. E.,** Hepatitis A: report of a common source outbreak with recovery of a possible etiologic agent. I. Epidemiologic studies, *J. Infect. Dis.,* 131, 163, 1975.
119. **Chaudhuri, A. K. R., Cassie, G., and Silver, M.,** Outbreak of food borne type-A hepatitis in Greater Glasgow, *Lancet,* 2, 223, 1975.
120. **Andersson, O.,** Epidemic of jaundice, *Nord. Hyg. Tidskr.,* 2, 251, 1921.
121. **Mosely, J. W.,** Transmission of viral hepatitis by drinking water, in *Transmission of Viruses by the Water Route,* Berg, G., Ed., Interscience, New York, 1967, 5.
122. **Khuroo, M. S.,** Study of an epidemic of non-A, non-B hepatitis: possibility of another human hepatitis virus distinct from post-transfusion non-A, non-B type, *Am. J. Med.,* 68, 818, 1980.
123. Centers for Disease Control, Water related disease outbreaks surveillance, annual summary, CDC, Atlanta, Ga., 1978, 1.

124. **Melnick, J. L., Gerba, C. P., and Wallis, C.,** Viruses in water, *Bull W.H.O.*, 56, 499, 1978.
125. **Sobsey, M. D., Oglesbee, S. E., and Wait, D. A.,** Evaluation of methods of concentrating hepatitis A virus from drinking water, *Appl. Environ. Microsc.* 50, 1457, 1985.
126. **Sobsey, M. D., Oglesbee, S. E., Wait, D. A., and Cuenca, A. I.,** Detection of hepatitis A virus in drinking water, *Water Sci. Technol.*, 17, 23, 1985.
127. **Aach, R. D., Evans, J., and Losee, J.,** An epidemic of infectious hepatitis possibly due to airborne transmission, *Am. J. Epidemiol.*, 87, 99, 1968.
128. **Katzenelson, E., Bunim, I., and Shuval, H. I.,** Risk of communicable disease infection associated with waste water irrigation in agricultural settlements, *Science*, 194, 944, 1976.
129. **Gust, I. D.,** Comparison of the epidemiology of hepatitis A and B, in Viral Hepatitis, Szmuness, W., Maynard, J. E., and Hoofnagle, J., Eds., Franklin Institute Press, Philadelphia, 1978, 129.
130. **Gust, I. D., Lehmann, N. I., Dimitrikakis, M., and Zimmet, P.,** Seroepidemiology of hepatitis A and B viruses in an isolated Pacific population, *J. Infect. Dis.*, 139, 559, 1979.
131. **McCaustland, K. A., Bond, W. W., Bradley, D. W., Ebert, J. W., and Maynard, J. E.,** Survival of hepatitis A virus in feces after drying and storage for 1 month, *J. Clin. Microsc.*, 16, 957, 1982.
132. **Weisfuse, A.,** personal communication.

Chapter 10

GENERAL PREVENTATIVE MEASURES

I. INTRODUCTION

The tremendous success which has been achieved in the control of virus diseases such as smallpox, poliomyelitis, measles, and rubella by the use of vaccines has blinded many people to the importance of traditional public health approaches. This is particularly relevant for enterically transmitted infections such as hepatitis A. Experience in many industrialized countries has shown that this disease can be controlled and virtually eliminated when economic conditions and living conditions are favorable. A variety of factors are involved, of which the most important appear to be provisions for adequate quantities of clean water for drinking and washing, proper disposal of sewage, adequate housing, and good standards of personal hygiene.

Experience in Scandinavia, Japan, Australia, and several other industrialized countries since World War II indicates that if these conditions are attained hepatitis A ceases to be a major public health problem although cases may continue to occur among individuals or groups at special risk of infection.

II. GENERAL PREVENTIVE MEASURES

A. Personal and Environmental Hygiene

1. In the Home

Good personal hygiene has been shown to reduce the risk of transmission of enteric infections and is particularly important in attempts to reduce transmission of hepatitis A. When infection is spread to family members or to other close contacts, the virus is usually transferred on the hands of the patient or by a person who has been in direct contact with the patient or has touched items which have been contaminated. The risk of transmission is greater when the person who is infected is not obviously ill, because under these circumstances standards of personal hygiene tend to be more relaxed. For this reason most infections appear to be transmitted either by adults who are in the incubation period of their illness (at which time virus shedding is also at its peak), or by children in whom the infection is usually subclinical.

Evidence supporting the importance of poor personal hygiene in transmission of infection has been obtained from a number of studies in which a higher incidence of infection has been demonstrated among families from lower socioeconomic groups, in crowded households, and in institutions for the intellectually impaired (see Chapter 9). Nevertheless, even in developed countries with high standards of hygiene, excellent living conditions and modern plumbing, spread of infection within families is common. In most circumstances the hygienic habits of the infected person are more important in determining the risk of spread than the habits of his or her contacts. Within the home strict observance of basic hygiene, in particular regular washing of hands with soap and water after going to the toilet, careful preparation of food, and adequate disposal of feces and soiled objects, seem to be particularly important in limiting spread of the virus.

Patients with hepatitis A can usually be nursed at home and do not require strict isolation unless they are not toilet trained or are incontinent; their dishes and linen do not require separate handling provided that they can be thoroughly washed with soap and water, and separate toilet facilities are not required. Admission to hospital may be required to establish the diagnosis, because of the severity of the illness, or because the patient is unable to receive adequate care at home. In the case of family outbreaks it may sometimes be necessary to admit the entire family to ensure that the parents receive adequate rest.

Supervised hand washing is sometimes introduced into schools in the face of an epidemic, and while this practice is worthwhile from an educational point of view, there is little evidence that it will slow the course of the outbreak. Similarly, disinfection of the classroom or objects that children with the disease have been in contact with (e.g., desks, door knobs, etc.) is of no proven value.

2. In Hospital

Regular hand washing appears to be highly effective in protecting hospital staff against the risk of infection. Patients with hepatitis A do not need to be isolated unless they are unable to control their bowels or are unreliable. They can usually share communal toilet facilities. Crockery and cutlery should be thoroughly washed before being reused and visitors should wash their hands after contact with the patient. To ensure that people in direct contact with the patient appreciate the risk of infection and do not inadvertently contaminate their clothing, some hospitals also insist that all visitors and staff wear protective gowns which are removed when they leave the room.

Laboratory staff handling specimens from patients with hepatitis A (and patients with other infectious diseases) should be gowned and gloved, the work areas should be covered with impervious disposable paper, and spills should be rapidly disinfected. Care should be taken to limit the production of aerosols by undertaking such procedures as homogenization of fecal specimens in biological containment hoods and using centrifuges with sealed buckets. Special care should also be taken to avoid contamination of the skin or the production of aerosols when HAV is purified from large volumes of feces, liver, or cell culture lysate. To prevent inadvertent ingestion of virus from infected surfaces or gloves, staff should remove their gowns and wash their hands before leaving the laboratory; smoking, eating, and mouth pipetting in the work area should be absolutely forbidden. If specially hazardous procedures are being undertaken, consideration should be given to selecting only immune staff and/or protecting susceptible staff members with regular injections of human normal immunoglobulin (HNIG).

B. Water and Food-Borne Infection

Prevention of water-borne infections is largely an exercise in engineering; the same practices which have virtually eliminated water-borne bacterial infections have proved effective in limiting the spread of enterically transmitted viral infections. Water supplies should be regularly monitored for evidence of contamination with human or animal feces by testing for the presence of fecal microorganisms. Should the coliform count rise to dangerous levels, the water should be treated with chlorine as an emergency measure and steps taken to identify the source of contamination and eliminate it.

In industrialized countries, water-borne epidemics are relatively uncommon and usually result from a once-only contamination of the water supply. Nevertheless it is important that they be rapidly recognized so that sources of persistent or continuing contamination can be identified and remedied. If the epidemic is recognized early, people who have been exposed can be offered gamma globulin. Occasionally water-borne epidemics of hepatitis A are preceded by an outbreak of bacterial or viral gastroenteritis. Although not all epidemics of water-borne gastroenteritis are followed by outbreaks of hepatitis A, they provide evidence that the water supply has been contaminated with human feces. In populations in which the majority of people who drank the contaminated water are likely to be susceptible to infection with HAV, the widespread use of gamma globulin should be considered.

While food-borne epidemics of hepatitis A are well documented (see Chapter 9) they are relatively rare in developed countries because of the high standards of hygiene imposed on food handlers by health authorities and the increasing use of gloves when handling food. In most outbreaks in which an infected handler has been implicated a serious breakdown in hygiene can be documented, usually the result of poor education or carelessness on the part of the food handler or occasionally, deliberate contamination. Food handlers who develop hepatitis A should be prohibited from preparing food until they have fully recovered.

It is not possible to ascertain the importance of contamination of food in the spread of infection in developing countries, however the frequency with which travelers to these areas become infected suggests that contamination is relatively common. The risk of acquiring infection can be reduced by eating only in restaurants with high standards of cleanliness, avoiding uncooked foods, especially fresh washed salads, raw milk, and drinking bottled fluids. People who are known to be susceptible to infection with HAV and plan to spend periods in areas in which hepatitis A is endemic and are forced to eat the local food should consider the prophylactic use of gamma globulin (see below).

Shellfish-associated hepatitis is more difficult to prevent because of the nature of this industry and its complicated channels of supply. Shellfish eaten in a large city may come from a variety of suppliers, some of whom may be located thousands of miles away or in other countries in which the cultivation and harvesting of shellfish is largely unregulated. For example, major outbreaks of hepatitis A have occurred in Singapore when cockles and other shellfish have been imported from neighboring countries and mixed with the local catch.[29] In general, the application of proper environmental standards to the waters in which shellfish are grown, regular sampling of the water, together with licensing of suppliers, and a system of inspection and control should reduce the risk of transmission of disease. In practice the problems of policing this system and preventing illegal harvesting and distribution proved almost insurmountable; under these circumstances the best guarantee of the safety of the product is the reputation of the supplier.

C. Primate-Associated Infection

Because jungle-caught animals occasionally become infected with HAV prior to shipping (see Chapter 4), primate-associated infections and spread of infection within primate colonies can be reduced by quarantining recently imported animals for about 8 weeks. During this period the animals should be separately housed and handled by the minimum number of staff, each of whom is clad in protective clothing. Attendants should be taught to wash their hands after feeding or cleaning the animals and to avoid close contact and play with young animals. It is imperative that there be no smoking, eating, or drinking in areas in which the animals are housed and that these restrictions are rigidly enforced.[1]

Care of newly imported animals should be restricted to the most highly trained and responsible attendants. If these people are seronegative, regular administration of gamma globulin may be useful.

III. PASSIVE IMMUNIZATION

A. General

At present the single most effective measure of protecting an individual against infection with HAV is the administration of HNIG. In the 1940s the introduction of techniques for the large-scale separation of proteins made it possible to prepare concentrated antibody solutions from pools of human plasma. Initially there was some concern that this material might contain infectious agents because of the occasional transmission of hepatitis B by injection of pooled plasma obtained from patients convalescing from measles or mumps (see Chapter 1). These doubts were soon dispelled and the gamma globulin fractions prepared from pools of "normal" human serum, HNIG or immune serum globulin (ISG) as it is known in the U.S., by cold-ethanol fractionation or modifications of this technique,[2] have proven to be among the safest of all biological products.

Most preparations of HNIG which are available are obtained from large pools of plasma collected from paid or volunteer donors, or from outdated units of blood. The immunoglobulins obtained by precipitation in cold ethanol are usually diluted to a final concentration of at least 15 g/100mℓ and after adjustment of the pH and addition of a preservative, ampouled and stored in the cold. The final product is largely made up of IgG antibody with IgM and IgA usually contributing less than 2% of total protein.[3] In most countries HNIG is produced from pools of plasma obtained from more than 1000 donors and hence contains a representative cross-section of the antibodies found in that population. Although most regulatory authorities require batches of HNIG to have minimum titers of diphtheria antitoxin and certain levels of antibody to measles and poliovirus, there is no agreed minimum level of anti-HAV.

B. Protective Efficacy

The first attempt to prevent hepatitis A by the use of HNIG occurred in 1944 when Neefe and Stokes administered it during a common-source epidemic in a summer camp in Pennsylvania.[4] Only 3 of 53 people (5.7%) who received 0.15 cc/lb of the preparation i.m. developed jaundice, compared with 125 of 278 (45%) who did not. These results were rapidly confirmed and after Havens and Paul[5] demonstrated that high levels of protection could be induced with smaller doses (0.08 cc/lb), HNIG was approved for use in American troops fighting in the Mediterranean theater, an area in which hepatitis A was occurring in epidemic form. Some 2700 soldiers received injections of gamma globulin and the protective efficacy over the next 7 weeks, calculated by the reduction in the incidence of jaundice, was estimated to be 89%.[6] The degree of immunity appeared to decline with time and the protective efficacy at 12 weeks had fallen to 73%. Thus, the value of HNIG in prevention of hepatitis A was rapidly established and it has been widely used ever since.

It is assumed that HNIG acts by neutralizing circulating virus and preventing it from reaching susceptible liver cells. If this is so, hepatitis A infection must be or have been, relatively common in those communities in which HNIG is produced and antibodies which are produced as a result of natural infections must be long lived. The efficacy of HNIG produced in any part of the world against infections acquired in other parts of the world implies that there is only one serotype of HAV, or that all strains share a common specificity.[7,8]

Passive immunization if undertaken prior to exposure (pre-exposure prophylaxis), will prevent infection with HAV and if given during the incubation period (postexposure prophylaxis) may reduce the severity of infection, converting potential clinical infections into subclinical infections, which can only be detected by monitoring the patient' liver functions tests or demonstrating the development of long-lasting antibodies. The first evidence that HNIG could suppress clinical expression of infection was obtained in 1954 by Drake and Ming during an institutional outbreak.[9] Two different dose schedules (0.01 mℓ/kg and 0.02 mℓ/kg), were administered. The incidence of clinical hepatitis was reduced in both of the groups receiving HNIG, however the incidence of inapparent infection (as demonstrated by transient elevations in ALT levels) was virtually identical to the incidence of clinical hepatitis among subjects who received no treatment. The duration of protection appeared to be dose dependent, being only 8 weeks in those who received the lower dose, compared with 13 weeks in those receiving the higher dose.

These findings have now been confirmed in both naturally acquired and experimentally induced infections in a variety of settings.[10,11] For example, Schneider and Mosley demonstrated that although the incidence of elevated transaminase levels in family members who received HNIG during a community-wide outbreak was no different from those who did not receive the preparation, although the incidence of icteric disease was significantly reduced.[10] Most of the patients who developed elevated transaminase levels did so within 5 weeks of the onset of illness in the index case in their family, suggesting that HNIG had been given after exposure.

Because some patients who receive HNIG may develop inapparent infections and may excrete HAV while continuing to work or attend school, it has been suggested that they may actually promote the spread of the disease.[12] In practice this does not appear to be the case; administration of HNIG to half a school population has been shown to reduce the incidence of infection among both the inoculated and uninoculated children, suggesting that infections occurring in the presence of circulating antibody reduce the shedding of virus.[13]

When infections are acquired in the presence of circulating antibody, passive/active immunity may result. This phenomenon was first observed during an epidemic of hepatitis in an institution: the children who received HNIG remained free of hepatitis for the next 8 months, whereas numerous cases of the disease were observed among children who had not received HNIG.[14] Passive/active immunity will only occur under circumstances in which the recipient of HNIG is at a high risk of infection — under normal circumstances the protection provided by HNIG lasts for only a few months.[15]

C. Pre-Exposure Prophylaxis

There is clear evidence that HNIG will prevent infection with HAV when administered prior to exposure. For this reason it is commonly administered to troops and groups such as diplomats, Peace Corps workers, and others who plan to spend extended periods in areas in which the disease in endemic. Administration of HNIG to Swedish troops serving in the Gaza Strip in 1957 reduced the incidence among those who received it to 1 per 1000 compared with 90 per 1000 in those who were unprotected.[16]

The duration of immunity has not been adequately defined, partly because of methodological problems and partly because of difficulties in comparing the potency of different batches of HNIG. The half-life of passively administered anti-HAV has not been determined, although studies with antibodies to other enteric viruses have given values of 3 to 6 weeks.[17] While these data are of interest, it is of little help in predicting the duration of immunity to HAV since the minimum protective dose of antibody is not known. Studies in institutions in which hepatitis A was occurring in epidemic form have shown a reduction in the incidence of cases of 4 to 12 months when doses varying from 0.01 to 0.02 mℓ/kg were used.[9,11,14] However these figures probably overestimate the efficacy of HNIG because of the occurrence of passive/active immunization. In adult Peace Corps workers receiving 0.01 mℓ/kg, protection appeared to be significantly decreased after 16 weeks,[18] whereas injections of 2, 5, or 10 mℓ of HNIG in U.S. soldiers in Korea appeared to provide protection for up to 6 months.[8]

D. Post-Exposure Prophylaxis

Administration of HNIG has been found to significantly reduce the incidence of jaundice among household contacts of patients with hepatitis. When administered shortly after disease is recognized in the index case, efficacies of 80 to 90% have been obtained.[9,19] Doses of 0.01 to 0.02 mℓ/kg appear to be adequate provided they are given in the week or two after exposure; the earlier the material is administered, the higher the protection obtained.[20] Administration of HNIG in the second half of the incubation period (i.e., in the last 2 weeks before the onset of jaundice) appears to have little or no effect.[21]

E. Indications for Use

Guidelines for the use of HNIG vary from country to country, with most physicians being guided by the recommendations of the World Health Organization or of the Immunization Practices Advisory Committee of the U.S. Centers for Disease Control.

For pre-exposure prophylaxis of travelers to high-risk areas, a single i.m. dose of 0.02 mℓ/kg is generally recommended. For prolonged travel or residence in such areas, a larger dose is recommended (0.06 mℓ/kg) with the advice that it be repeated every 5 months.

There appears to be no problem in repeated administration of HNIG even over many years.

In all situations where postexposure prophylaxis is indicated (e.g., household contacts of an index case, children and adults associated with outbreaks in day care and other centers, or in common-source outbreaks) a single i.m. dose of 0.02 mℓ/kg is suggested.

In developed countries the rates of secondary infection among household contacts may be as high as 45% in children and 5 to 20% in adults. For this reason it is usual to administer gamma globulin to all household contacts as soon as an index case is diagnosed. While it is possible to bleed each family member and test them for anti-HAV before deciding who to immunize, in practice this is rarely done.

Epidemiological evidence suggests that children are more likely to contract infection from their siblings or playmates than in the classroom. For this reason, when a case occurs in a child of school age, it is not generally considered necessary to administer HNIG to the remainder of the class. Exceptions to this rule may occur, with classroom contacts who are also playmates, and if the source of infection appears to be in the school.

Similarly there is generally no justification for administering HNIG to adults who have casual contact with a workmate with hepatitis, even though they share the same eating and toilet facilities.

Widespread immunization is often recommended for children in nursery schools, child care centers, and institutions in which hepatitis A is occurring.

Passive immunization is usually recommended in the face of common-source, water- or food-borne epidemics; however, by the time the source of infection has been identified, it is usually too late for HNIG to be effective.

Administration of HNIG is not required for hospital or laboratory personnel caring for patients with hepatitis unless accidents occur. Regular administration should be considered for seronegative animal handlers responsible for the care of newly imported nonhuman primates.

F. Potency of Preparations of HNIG

While there is good evidence that preparations of HNIG prepared in different parts of the world using plasma from a variety of different sources are effective in protecting against hepatitis A, there are also data to indicate that different batches may vary in the level of protection afforded.

Mosley and his colleagues have demonstrated that two batches of HNIG produced in Israel by different techniques had markedly different potencies.[7] In a placebo-controlled study in which the two preparations were directly compared, one batch reduced the incidence of icteric hepatitis from 6.4 to 0.8 cases per 100 children, while the other batch only reduced it to 3.4 cases per 100 children. While these differences appear to have been due to the manner in which the preparations were made, the incident highlighted the need for standardization of the antibody titer of each batch.

The development of sensitive assays for the detection of anti-HAV and the demonstration that antibodies detected by these techniques correlate well with the presence of neutralizing activity has enabled various groups to study the titers of anti-HAV in batches of HNIG produced by different manufacturers at different times.

The declining prevalence of anti-HAV in blood donors in many developed countries has led to fears that preparations of HNIG may become ineffective and that it might soon be necessary to produce the material from a selected population of antibody-positive donors. While considerable variation in titers of anti-HAV have been demonstrated in batches of HNIG obtained from different manufacturers,[22] no significant difference was observed in the titer of Danish batches of HNIG manufactured between 1969 and 1977.[23] When U.S. batches of HNIG manufactured between 1967 and 1977 were studied the mean titer of antibody detected was found to have increased rather than decreased[24] (see Table 1). The

Table 1
DETECTION OF ANTI-HAV TITERS OF ≧ 1:100 AND GEOMETRIC MEAN TITERS (GMT) OF BATCHES OF HNIG PREPARED IN THE U.S. BY VARIOUS MANUFACTURERS

	Batches containing anti-HAV/batches tested (GMT)		
Manufacturer	**1967**	**1972**	**1977**
A	NT[a]	8[b]/10 (237)[c]	21/21 (424)
B	5/9 (100)	9/10 (120)	21/21 (254)
C	5/7 (100)	NT	22/22 (329)
D	4/10 (178)	7/10 (158)	21/21 (183)
E	5/10 (158)	3/10 (100)	21/21 (157)
F	4/10 (178)	7/9 (199)	NT

[a] NT = Not tested.
[b] Anti-HAV titer of ≧ 1:100 by competition RIA.
[c] GMT of batches in which antibody titer was ≧ 1:100.

From Smallwood, L. A., Tabor, E., Finlayson, J. S., and Gerety, R. J., *J. Med. Virol.*, 7, 21, 1981.

reason for this appears not to lie in the stability of the antibodies, but in a trend to the preparation of HNIG from plasma collected from paid donors in which the prevalence of anti-HAV is higher than among the general community.[25,26]

In an attempt to assist manufacturers to standardize the antibody content of batches of HNIG, the World Health Organization has developed an International Reference Preparation of anti-HAV which has been arbitrarily assigned a unitage of 100 international units (IU) per ampoule. This preparation was originally obtained from a commercial manufacturer as bulk immunoglobulin in powder form. After preliminary study by the U.S. Office of Biologics, the powder was made into a sterile bulk solution and aliquoted into 8,000 1-mℓ ampoules which were then lyophilized and sealed. Some 17 laboratories with special expertise in hepatitis A serology participated in an international collaborative study to determine the titer of the preparation and to determine the reproducibility of the results from country to country and center to center. The preparation was found to have a titer of 1:500 to 1:1000 by RIA and 1:400 to 1:8000 by IAHA with very similar results being obtained from all laboratories. As accelerated stability tests showed no appreciable decline in antibody titer with time, this preparation was acceptable as an International Reference preparation.

While the minimum titer of anti-HAV which is required to ensure that a preparation of HNIG is effective remains unknown, the titer of anti-HAV in the reference preparation is similar to that found in many batches which have been shown to prevent disease in man. As a consequence, manufacturers are able to define the titer of individual batches of HNIG and by ensuring that it is at least 100 IU/mℓ, be confident that it contains protective levels of antibody.

G. Safety

With the discovery of the human immunodeficiency virus (HIV), concern has been expressed about the safety of preparations of HNIG. This concern arose for three main reasons: first some preparations produced between 1980 and 1985 are known to have contained plasma from patients infected with HIV; second, some viruses (e.g., HNANB) are known to survive the fractionation process; and third, while 25% ethanol has been demonstrated to inactivate HIV at room temperature it appears to be less effective at −5°C, the temperature usually used in Cohn fractionation.

To resolve this issue, the World Health Organization recently convened a meeting of manufacturers of HNIG and other experts in the field to review data on the safety of existing manufacturing techniques.[27] It was found that the fractionation procedures which are employed by manufacturers vary widely with a minimum of three and a maximum of six precipitation steps being undertaken. In addition, some manufacturers introduced additional steps involving filtration, column chromatography, or storage of the final product at 27°C for 24 days before use.

The results of a number of studies showed that following each step in the fractionation HIV is found almost exclusively in the precipitates, most of which are discarded. As each separate step was found to result in a loss of 1 to 4 $\log_{10}$ ID/mℓ of virus, final estimates of the cumulative virus loss through the fractionation procedure ranged from 7 to 15 $\log_{10}$ ID/mℓ. A further 4 $\log_{10}$ reduction could be obtained if the final preparation was stored at 27°C for 24 days prior to use.

The most detailed analysis of the safety of current preparations of immunoglobulin is that of Wells and her colleagues at the U.S. Food and Drug Administration.[28] Using a Cohn-Oncley fractionation procedure which has six partitioning or inactivating steps, they found each step removed from 1 to 4 $\log_{10}$ in vitro infectious units of HIV/mℓ. Assuming that the efficiency of each step is independent of the starting dose, they calculated an overall reduction of 15 $\log_{10}$/mℓ. Given that the mean infectivity of plasma from infected subjects is 0.02 ID/mℓ, that plasma pools are now prepared from donors that are screened for anti-HIV, and that the sensitivity of the existing assays is greater than 98%, they have calculated that a 1000-ℓ plasma pool prepared from screened donors in the U.S. would contain less than 0.13 ID/mℓ of HIV.

While this calculation contains many assumptions, the data suggest that the capacity of the Cohn-Oncley fractionation process to remove and inactivate HIV is vastly greater than any likely virus load — a finding supported by clinical observations and laboratory studies. While many thousands of people have now received batches of HNIG containing antibodies to HIV, there is no evidence that these preparations have ever transmitted infection. Furthermore, many lots of immunoglobulin which contain antibody to HIV have been cultured, without virus being recovered on any occasion.

Thus, a considerable body of data obtained from laboratory studies together with observations in the field support the historical impression that HNIG produced by the Cohn-Oncley process is one of the safest of all biological preparations.

REFERENCES

1. **Krushak, D. H.,** Application of preventive health measures to curtail chimpanzee-associated infectious hepatitis in handlers, *Lab. Anim. Care,* 20, 52, 1970.
2. **Cohn, E. J., Oncley, J. L., Strong, L. E., Hughes, W. L., Jr., and Armstrong, S. H., Jr.,** Chemical, clinical and immunological studies on the products of human plasma fractionation. I. The characterization of the protein fractions of human plasma, *J. Clin. Invest.,* 23, 417, 1944.
3. **Heiner, D. C. and Evans, L.,** Immunoglobulin and other proteins in commercial preparations of gamma globulin, *J. Pediatr.,* 70, 820, 1967.
4. **Neefe, J. R. and Stokes, J., Jr.,** An epidemic of infectious hepatitis apparently due to a water borne agent, *JAMA,* 128, 1063, 1945.
5. **Havens, W. D. and Paul, J. R.,** Prevention of infectious hepatitis with gamma globulin, *JAMA,* 129, 270, 1945.
6. **Gellis, S. S., Stokes, J. Jr., Brother, G. M., Hall, W. M., Gilmore, H. R., Beyer, E., and Morrissey, R. A.,** The use of human immune serum globulin (gamma globulin) in infections (epidemic) hepatitis in the Mediterranean theatre of operations, *JAMA,* 128, 1062, 1945.

7. **Mosley, J. W., Reisler, D. M., Brachott, D., Roth, D., and Weiser, J.,** Comparison of two lots of immune serum globulin for prophylaxis of infectious hepatitis, *Am. J. Epidemiol.*, 87, 539, 1968.
8. Prophylactic gamma globulin for prevention of epidemic hepatitis: effects of U.S. gamma globulin upon the incidence of viral hepatitis and other infections diseases in U.S. soldiers abroad, *Arch. Intern. Med.*, 128, 723, 1971.
9. **Drake, M. E. and Ming, C.,** Gamma globulin in epidemic hepatitis comparative value of two dosage levels, apparently near the minimal effective level, *JAMA*, 155, 1302, 1954.
10. **Schneider, A. J. and Mosley, J. W.,** Studies of variations of glutamic-oxaloacetic transaminase in the serum in infectious hepatitis, *Pediatrics*, 24, 367, 1959.
11. **Krugman, S., Ward, R., Giles, J. P., and Jacobs, A. M.,** Infectious hepatitis: studies on the effect of gamma globulin and on the incidence of inapparent infection, *JAMA*, 174, 823, 1960.
12. **Noble, H. B. and Peterson, D. R.,** Evaluation of immune serum globulin for control of infectious hepatitis, *Public Health Rep.*, 80, 173, 1965.
13. **Paktoris, E. A.,** Control problems in epidemiology and prevention of Botkin's disease, *Vestn. Akad. Med. Nauk.*, 18, 55, 1963.
14. **Stokes, J., Jr., Farquhar, J. A., Drake, M. E., Capps, R. B., Ward, C. S., Jr., Mill, S., and Kitts, A. W.,** Infectious hepatitis: length of protection by immune serum globulin (gamma globulin) during epidemics, *JAMA*, 147, 714, 1951.
15. **Werlhe, P. F. and Hammon, W. M.,** Absence of active immunization against infectious hepatitis, *JAMA*, 167, 2602, 1958.
16. **Kluge, R.,** Gamma globulin in the prevention of viral hepatitis: a study of the effect of medium sized doses, *Acta Med. Scand.*, 174, 469, 1963.
17. **Martin, C. M., Gordon, R. S., Felts, W. R., McCullough, N. B., Kolb, R. W., Kasel, J. A., and Szwed, C. F.,** Studies in gamma globulins. I. Distribution and metabolism of antibodies and gamma globulin in hypogamma globulinemic patients, *J. Lab. Clin. Med.*, 49, 607, 1957.
18. **Woodson, R. D. and Clinton, J. J.,** Hepatitis prophylaxis abroad: effectiveness of immune serum globulin in protecting Peace Corp volunteers, *JAMA*, 209, 1053, 1969.
19. **Fowinkle, E. W. and Guthrie, N.,** Comparison of two doses of gamma globulin in prevention of infectious hepatitis, *Public Health Rep.*, 79, 634, 1964.
20. **Krasna, V. and Radkovsky, J.,** Evaluation of the effectiveness of gamma globulin in the prevention of infectious hepatitis in Prague in 1953—1956, *J. Epidemiol. Microbiol. Immunol.*, 6, 295, 1957.
21. **Pollock, J. M. and Reid, D.,** Assessment of British gamma globulin in preventing infectious hepatitis: a report to the Director of the Public Health Laboratory Service, *Br. Med. J.*, 3, 451, 1968.
22. **Frosner, G. G., Haas, H., and Hotz, G.,** Hepatitis A antibody in commercial lots of immune serum globulin, *Lancet*, 1, 432, 1977.
23. **Mathieson, L. R., Skinhoj, P., and Balstrup, F.,** Antibody to hepatitis A virus in plasma of Danish blood donors and in immune serum globulin produced in Denmark, *Scand. J. Infect. Dis.*, 12, 81, 1980.
24. **Smallwood, L. A., Tabor, E., Finlayson, J. S., and Gerety, R. J.,** Antibodies to hepatitis A virus in immune serum globulin, *J. Med. Virol.*, 7, 21, 1981.
25. **Hall, W. T., Mundon, F. K., Madden, D. L., De Souze, D., and Kane, J. L.,** Distribution of antibody to hepatitis A antigen in a population of commercial plasma donors, *J. Clin. Microbiol.*, 6, 132, 1977.
26. **Hall, W. T., Madden, D. L., Mundon, F. K., Brandt, D. E. L., and Clark, N. A.,** Protective effect of immune serum globulin (ISG) against hepatitis A infection in a natural epidemic, *Am. J. Epidemiol.*, 106, 72, 1977.
27. **Gust, I. D., Deinhardt, F., Hollan, S. R., and Rey, F.,** The Safety of Blood Products, in *Aids: the safety of blood and blood products,* Petricianni, J. C., Gust, I. D., Hoppe, P. A., and Krijnen, H. W., Eds., John Wiley & Sons, Chichester, 1987, in press.
28. **Wells, M., Witteck, A., Epstein, J., et al.,** Inactivation and partition of human T-cell lymphotropic virus, type III, during ethanol fractionation of plasma, *Transfusion*, 26, 210, 1986.
29. **Goh, K. T.,** personal communication.

Chapter 11

ACTIVE IMMUNIZATION

I. INTRODUCTION

The declining incidence of hepatitis A which has occurred in many highly industrialized nations as a consequence of improved living standards has led some people to suggest that hepatitis A vaccines are either unnecessary or likely to be of limited usefulness. A closer examination of the data suggests that the reverse is the case.

While it is true that in areas such as the U.S., Canada, Australia, Japan, Scandinavia, and some parts of Western Europe, hepatitis is a declining public health problem, it is by no means an insignificant problem. In the U.S., where the incidence of the disease had declined steadily since 1970, more than 20,000 cases are still reported to CDC annually, an incidence of 9.2 per 100,000/year. Given the vagaries of the reporting system the true incidence of the disease is likely to be five or ten times higher. No accurate figures on the cost of the disease to the community exist. In 1979 Storch et al.[2] studied 26 patients with hepatitis A in 17 different households, one third of whom required hospitalization, and estimated the total cost per patient to be $1,952 which in 1987 terms would probably be in excess of $3,000. If these figures are accurate the cost of caring for the 20,000 reported cases which occur in the U.S. each year is probably between $60,000,000 and $300,000,000, figures which make control of the disease relevant on economic grounds alone. In highly industrialized countries certain groups of the population are at increased risk of infection and might be considered for targeted use of hepatitis A vaccines. These include children and staff of daycare centers, homosexual men, i.v. drug users, travelers to countries in which the disease is endemic, and people such as armed forces personnel, business men, diplomats, missionaries, and Peace Corps workers who are posted overseas for prolonged periods. Since in the U.S. this would represent a target population of several million people each year and would probably fail to prevent a large number of sporadic cases of the disease, a good argument could be made for inclusion of hepatitis A vaccine in the general program of childhood immunizations.

From a numerical point of view, the major users of hepatitis A vaccine are likely to be health authorities in rapidly developing countries in Asia, Southern and Central America, and the Middle East. In many countries rapid improvements in living conditions have resulted in a decline in infection with HAV and other enterically transmitted organisms. Paradoxically, hepatitis A has emerged as a major public health problem, because although fewer infections are occurring, a higher proportion take place in the late childhood or early adult life when they usually cause clinically apparent disease. In recent years this pattern has been recognized in many parts of Asia (e.g., China, Korea, Singapore), South and Central America (e.g., Chile and Cuba), and Europe (Greece, Italy, Portugal). The incidence of notified cases of hepatitis A in some of these countries is extremely high. For example, in Chile it has increased from 46 per 100,000 in 1975 to 91 per 100,000 in 1985 with even higher rates predicted for 1986[3] (see Figure 1). Similarly, the incidence of hepatitis A notified to public health authorities in Greece, Italy, and Portugal is currently more than 45 per 100,000 per year.[4] Experience in other countries suggests that once hepatitis A emerges on such a scale, it is likely to remain a serious problem for at least two decades. Under these circumstances a safe, effective vaccine which can be administered early in life is urgently needed.

At present the only areas of the world which would have no immediate need for a vaccine are those in which low living standards plus poor hygiene and sanitation lead to almost universal exposure to infection early in life. Under these circumstances clinical disease is

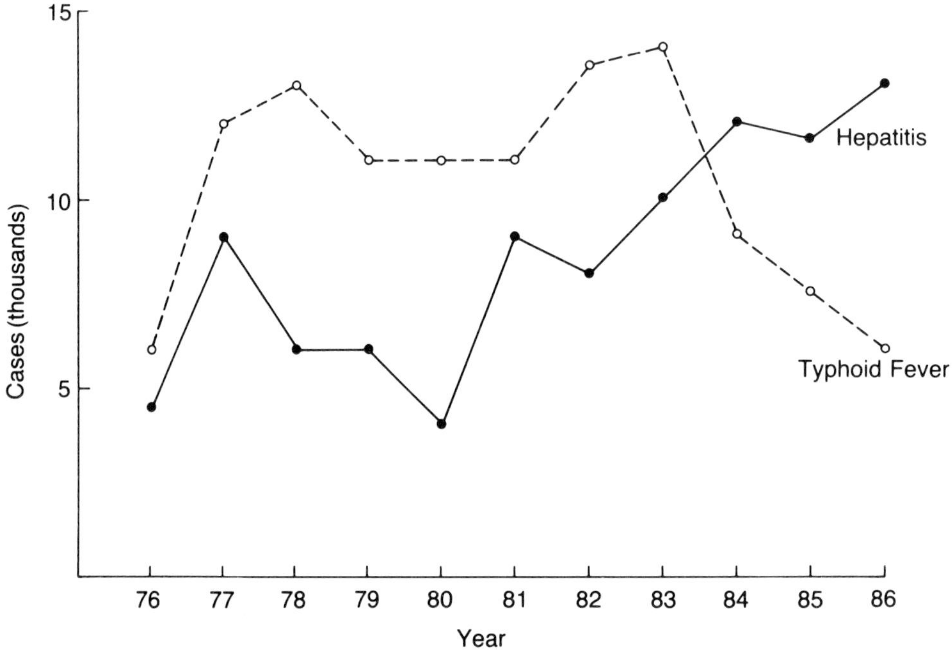

FIGURE 1. The annual incidence of cases of viral hepatitis notified to public health authorities in Chile over the period 1976 to 1986. Rising standards of hygiene and improved sanitation have led to a major reduction in enterically transmitted bacterial infections and a concomitant increase in the incidence of hepatitis A. (From data kindly provided by Professor Marta Velasco, University of Chile.)

virtually nonexistent and since no long-term sequelae of HAV infection are recognized, immunization is probably unnecessary except as part of a coordinated program to control common childhood infections.

II. APPROACHES TO IMMUNIZATION

Traditionally the requirements for developing a human viral vaccine have been an in vitro propagation system, sensitive assays for detecting the presence of the virus and neutralizing antibodies directed against it, and susceptible animals in which candidate vaccines can be evaluated prior to their use in man.

All of these requirements can now be met for hepatitis A. As outlined in previous chapters, HAV has been adapted to growth in a variety of cell cultures some of which are suitable for vaccine production (see Chapter 6). A variety of assays have been developed for detecting infectious virus, viral antigens, and viral nucleic acid (see Chapters 2, 6, and 8) and infections have been transmitted to several species of nonhuman primates (see Chapters 4 and 5). In addition, the relatively new techniques of molecular biology and recombinant DNA technology have made it possible to develop a range of alternative approaches to vaccine development such as the production of synthetic peptide, chimeric, and idiotype/anti-idiotype vaccines.

There has long been argument as to whether killed or live attenuated vaccines provide better protection against infection with enteric viruses. Most of the debate has hinged on the relative merits of the Salk and Sabin vaccines and has not always been confined to scientific issues.

Live attenuated vaccines are usually produced by passaging a virulent virus through an unnatural host and selecting progeny in which successive mutations have resulted in a loss of virulence for man. The genetic basis for attenuation is generally not known and may be multifactorial.

The major advantages of live enterovirus vaccines are ease of administration, relatively low cost (because the antigenic mass of the virus is increased by replication in the gut), and the fact that such vaccines induce a balanced immune response which mimics the natural infection. Local and systemic antibodies stimulated by live vaccines are generally long lasting and directed against all the major epitopes of the virus.

The disadvantages of live vaccines are that they may be contaminated with other viruses which are present in the cell cultures in which the vaccine is produced, they may retain a low level of virulence (which can be a problem if they are administered to immunocompromised patients), they may occasionally revert to virulence either in the immunized subject or in a close contact to whom the virus has spread, and they may lose their activity under the storage conditions available in the field.

These disadvantages are relatively minor and many of them are readily overcome. For example in the case of the live PV vaccine, sensitive assays now exist for detecting evidence of viral contamination, reversion to virulence is estimated to occur with a frequency of less than one in several million, and stabilizers have been developed which greatly improve survival of vaccines in the field. For these reasons oral polio vaccines have become widely used throughout the world and have proved to be extremely effective.

Inactivated vaccines offer a potential advantage in safety in that, provided the inactivation procedures are properly carried out, the final product can be guaranteed to be free of live viruses. Killed vaccines tend to be more stable than live vaccines and offer greater safety when administered to the immunocompromised (a feature which may become of increasing importance as infection with the HIV spreads). They are usually excellent at stimulating circulating antibodies but poor at stimulating secretory antibody which is responsible for protection against reinfection. As a result, wild-type virus can continue to circulate in a community which has received killed vaccine. However, killed PV vaccines have been extensively used in three countries in Scandinavia with outstanding results.

It is of interest that while most workers feel that the ideal hepatitis A vaccine would be a live attenuated preparation that could be administered orally, a considerable amount of effort has also been expended in attempts to develop killed vaccines.

A. Killed Vaccines

1. Marmoset Liver

In 1973 Mascoli, Provost, and their colleagues at the Merck Institute for Therapeutic Research, Pennsylvania, successfully infected marmosets with HAV.[5] Subsequently they demonstrated that considerable quantities of infectious virus or viral antigen could be extracted from the livers of infected animals and that the infectivity could be destroyed by treatment with formaldehyde.[6] In a further series of studies they were able to show that injections of formalin-inactivated virus obtained from infected livers would stimulate antibodies in susceptible marmosets and protect them against challenge with live virus.[7]

To prepare this vaccine, a rufiventer *(Saguinus labiatus)* marmoset was infected i.v. with the CR326 strain of HAV which had previously been passed 25 times in marmosets. The animal was sacrificed 14 days after inoculation at the first sign of biochemical hepatitis, the liver perfused *in situ,* exised, homogenized in PBS, and clarified by low-speed centrifugation. After heat treatment (60°C for 30 min) and further clarification, the final 50% extract was treated with formalin (final concentration 1:4000) for 4 days at 35.5°C. The residual formalin was neutralized with sodium bisulfite and the preparation stored at -40°C. The final concentration of viral particles was estimated by EM to be of the order of $1.4 \times 10^{10}/m\ell$. Control antigens were prepared by processing the liver of an uninfected, susceptible marmoset in the identical fashion.

Table 1
RESULTS OF CONTROLLED STUDY OF HEPATITIS A VACCINE PRODUCED IN MARMOSET LIVER IN PROTECTING MARMOSETS AGAINST HEPATITIS A

	After inoculation of 1 cc s.c. at 14-day intervals (× 8)		After i.v. challenge with 10^3 MID_{50} of HAV	
Preparation	Antibody response	Enzyme elevation	Antibody response	Enzyme elevation
Normal liver vaccine	0/8	0/8	8/8	8/8
Hepatitis A vaccine	8/8	0/8	0/8	0/8

The two preparations were inoculated s.c. into susceptible rufiventer marmosets in 1-mℓ volumes, every 14 days for 14 weeks (a total of eight injections). The presence of residual live virus was sought by injecting six susceptible animals with a single 1-mℓ dose of vaccine i.v. All animals were challenged 17 weeks after the first injection with 1 mℓ of a 10^{-6} dilution of CR326 virus estimated to contain about 10^3 marmoset ID_{50} of virus (see Table 1).

All eight marmosets which received the vaccine developed anti-HAV within 12 weeks of the first dose compared with none of eight that received the control preparation.

Following i.v. inoculation of live virus, all the marmosets which received the control preparation developed hepatitis within 4 to 6 weeks. By contrast, none of the vaccinated animals developed hepatitis.

None of six marmosets that received a single dose of HAV vaccine i.v. as a safety test developed hepatitis, indicating that the virus had been completely inactivated. Interestingly, only one of these animals developed detectable levels of antibody (By IAHA). Following challenge, two of the seronegative animals developed hepatitis while the other four remained well. These data suggest that protection may exist even in the absence of detectable antibodies. On the other hand, it is possible that low levels of IgM antibody were present but not detected by IAHA which detects mainly IgG.

While clearly demonstrating the feasibility of developing a conventional killed hepatitis A vaccine, this approach is unsuitable for production of large quantities of vaccine as marmosets are in short supply and frequently infected with other agents. As a result attention turned to the production of cell-culture derived vaccines and the use of adjuvants to increase their antigenicity.

2. *Cell Culture*

In 1986 two groups, Provost and his colleagues[8] at Merck, Sharpe and Dohme, and Binn and his colleagues[9] at the Walter Reed Army Institute for Research, reported the production of prototype hepatitis A vaccines from cell-culture derived virus.

a. Merck, Sharpe and Dohme

The Merck group used the CR326 strain of HAV which had originally been isolated and passaged in FRhK6 cells and then adapted to LLC-MK2 cells (a transformed line of African green monkey kidney [AGMK] cells) and passaged 15 times. The virus for the experimental batch of vaccine was produced by inoculating $10^{7.7}$ $TCID_{50}$ of passage 15 virus into confluent LLC-MK2 cells and harvesting the infected cells 21 days later. The cell pellets from 24,850 cm^2 roller bottles were lysed and sonicated in hypotonic buffer containing detergent, clarified by low-speed centrifugation, and pelleted through a 20% sucrose cushion. The resuspended

Table 2
ANTIGENICITY OF KILLED HEPATITIS A VACCINE IN MICE AND MARMOSETS

	Detection of anti-HAV			
	Mice		Marmosets	
Dose (μg)	Standard[a]	Modified[b]	Standard	Modified
1.0	ND[c]	ND	4/4	4/4
0.25	10/10	10/10	ND	ND
0.1	10/10	10/10	4/4	4/4
0.01	1/10	5/10	3/4	4/4
0.001	0/10	0/10	0/4	2/4
None	0/10	0/10	0/4	0/4

[a] Standard RIA for anti-HAV (Ausab).
[b] Modified RIA estimated to be 10 times more sensitive than the standard assay.
[c] ND = not done.

pellets were further purified by centrifugation on a 25 to 50% CsCl gradient after which the virus containing fractions were pooled, pelleted, resuspended in PBS, sonicated, and filtered. Analysis of the final material by SDS-PAGE showed a high degree of purity and a final yield of about 100 μg of antigen. Purified antigen diluted to contain 10 μg/mℓ was treated with 1:4000 formalin at 37°C for 72 hr after which thiomerosal was added as a preservative. This material, containing 5 μg/mℓ of HAAg was absorbed to aluminum hydroxide and diluted to give a final product containing 1 μg/mℓ of HAAg.

The efficacy of the inactivation process was demonstrated by the failure of the vaccine virus to propagate in susceptible LLC-MK2 cells or to induce disease when inoculated i.v. into seronegative marmosets.

The antigenicity of the vaccine was studied by injecting graded doses into groups of 5-week-old mice and young *S. labiatus* marmosets. The mice received a single dose of vaccine and were bled 4 weeks later. The marmosets received three doses at 4-week intervals and were bled weekly for 12 weeks and again at 14 weeks. In mice the lowest dose of vaccine which induced an antibody response after a single i.p. dose was 0.01 μg. (This dose induced antibody in half the animals studied.) Whereas in marmosets, which received three injections of vaccine, the 50% effective dose was 0.001 g. It is noteworthy that with lower doses of vaccine, anti-HAV could not be detected in some animals with a standard RIA (Abbott). Antibodies could only be detected in these animals when a modified (ten times more sensitive) assay was employed (see Table 2).

Fifteen weeks after receiving their first dose of vaccine, representative marmosets were inoculated i.v. with 1 cc of virulent HAV strain CR326 (containing about 10^5 marmoset infectious doses) and observed closely for 12 weeks. Serum samples collected at weekly intervals were assayed for total and class-specific anti-HAV and for elevated ICD levels.

Three animals which had received alum diluent alone and two animals which had received 0.001 μg of vaccine and failed to seroconvert, developed typical acute hepatitis A on challenge. In these animals ICD elevations were noted usually within 3 weeks of inoculation, and these were followed by rising levels of anti-HAV IgM and later by anti-HAV IgG. Two other animals which had received 0.001 μg of vaccine and had developed low levels of anti-HAV, detectable only by the modified RIA, were protected against disease. Although no elevations in enzyme levels were observed in these animals, transient rises in anti-HAV IgM and total antibody suggest that limited replication of virus may have occurred. These

Table 3
RESULTS OF CHALLENGE OF IMMUNIZED MARMOSETS WITH $10^5 MID_{50}$ OF CR 326

Dose	Animal	Post vaccination anti-HAV		Post challenge		
		Standard[a]	Modified[a]	ICD elevated (week)	Anti-HAV IgM peak (week)[b]	Anti-HAV standard[a]
1.0	1	96	99	—	—	97
	2	96	99	—	—	96
0.1	3	92	98	—	—	96
	4	93	98	+(5—9)[c]	—	96
0.01	5	91	98	—	3(6)	87
	6	91	99	—	—	88
	7	94	99	—	—	95
	8	29	81	—	8(4)	82
0.001	9	21	69	—	3(4)	80
	10	−2	23	+(3,4)	34(4)	93
	11	27	73	—	6(6)	79
	12	10	20	+(2,3)	38(4)	97
None (alum alone)	13	1	26	+(2—4)	29(4)	96
	14	−13	5	+(4,5)	17(4)	95
	15	9	5	+(2—4)	32(4)	97

[a] % Inhibition.
[b] Expressed as P/N ratio (cpm test sample per cpm negative control).
[c] This animal died at week 9 of an illness unrelated to hepatitis A.

findings have important implications for future vaccination programs, since they imply that subjects who have encountered HAV and mounted an immune response are likely to be immune even if their circulating antibody levels have become undetectable by conventional assays.

Of the three animals which received three doses of 0.01 μg of vaccine and were challenged, one that had a poor antibody response developed low levels of anti-HAV IgM and a boost in total anti-HAV, but no biochemical evidence of hepatitis. The other two animals which had developed high levels of anti-HAV after vaccination were also completely protected against disease and showed no increase in anti-HAV titers. Identical findings were obtained in marmosets injected with higher doses (0.1 and 1.0 μg) of vaccine (see Table 3).

These findings indicate that not only is it possible to produce a killed hepatitis A vaccine from cell-culture derived virus, but that provided the vaccine is adjuvanted, protective levels of antibody can be induced in labortory animals with relatively low doses. Coupled with improved methods of growing the virus and obtaining higher yields in supernatant fluid, commercial production of killed vaccines appears feasible. The substrate used to produce this experimental lot of vaccine (LLC-MK2) is a transformed cell line and thus unlikely to be acceptable to licensing authorities. However both CR326 and other strains of HAV have been adapted to growth in human diploid fibroblasts[10,11] with little loss in yield. Preliminary indications are that hepatitis A vaccines produced in these cells have similar properties to those produced in LLC-MK2 and should be acceptable for use in man.[8]

b. Walter Reed Army Institute for Research

Similar data on the safety, antigenicity, and efficacy of a candidate-killed hepatitis A vaccine, has recently been provided by Binn and his colleagues[9] at the Walter Reed Army

Institute for Research. Three lots of vaccine were produced in B-SC-1 cells maintained in serum-free medium and infected with the HM175 strain of HAV. Roller bottle cultures were incubated for 3 to 4 weeks and the infected cells collected and lysed by sonication. HAV was purified from the clarified cell lysates and supernatant fluid, inactivated with 1:4000 formalin at 35°C for 96 hr (lots 1 and 2) or 10 days (lot 3) after which residual formalin was inactivated with sodium bisulfite and the product stored at −60°C. By contrast with the Merck studies no adjuvant was used. Lots differed from the other two not only in the duration of formalin treatment, but in being filtered prior to inactivation.

The harvested pools of supernatant fluid and cell lysate used to prepare the vaccine contain $10^{6.8}$ to $10^{7.4}$ RIFU of HAV/mℓ. The immunogenicity of the vaccine was evaluated in young seronegative guinea pigs (Hartley strain) each of which received three 1.0-mℓ doses i.m. at 3-week intervals (days 0, 21, and 42). Antibody was detected in all animals within 1 week of receiving the third dose of vaccine, irrespective of which lot was used. Injection of lots 1 and 2 which had been stored for 9 to 12 months in the cold gave similar results.

A total of 16 seronegative, colony-bred owl monkeys *(Aotus trivirgatus)* were used for the vaccine studies and to provide controls. Baseline blood specimens and liver biopsies were obtained over a 3-month period prior to inoculation, after which six animals were given 1.0 mℓ of vaccine (lot 3) 1/*M* on days 0, 28, and 58, the remaining animals receiving a saline control. Weekly blood tests were performed on all animals and fecal samples collected four times a week from the vaccinated group. On day 136, 78 days after receiving the third dose of vaccine, the monkeys were challenged with a virulent strain of HAV (PA33), three by nasogastric tube and three i.v. Two control animals were given the challenge virus by each route. The titer of the challenge pool was estimated by oral administration of 10^{-2}, 10^{-4}, and 10^{-6} dilutions to additional pairs of animals.

The six monkeys which received three doses of vaccine remained well during the following 18 weeks. None of the animals developed biochemical or biopsy evidence of hepatitis and none shed HAAg in the feces. All six animals developed anti-HAV (detectable by RIA and neutralization assays) after two doses of vaccine with only slight increases in titer being produced by the third dose (see Figure 2).

Following challenge with PA33 by either the oral or i.v. route, none of the vaccinated animals showed evidence of disease and only one shed HAAg in its feces. Although no significant increases in antibody titer were detected by RIA, anti-HAV levels detected by neutralization increased fourfold in monkeys challenged orally and up to 64-fold in animals challenged i.v. As expected, control animals which had received saline instead of the vaccine developed detectable levels of circulating anti-HAV. All of the monkeys receiving dilutions of PA33 developed hepatitis, indicating that the challenge pool contained at least 10^6 ID/mℓ.

These studies support the conclusion that it is now feasible to produce killed hepatitis A vaccine for use in man. The substrate used, B-SC-1, like LLC-MK2 is a heteroploid line which has not previously been licensed for use in man. Recently, however, heteroploid lines have been approved for production of inactivated polio vaccine in some countries,[12] so that the substrate in which the vaccine is produced may not be an insurmountable barrier to having the product licensed.

While the results of the Merck, Sharpe and Dohme and Walter Reed Army Institute for Research studies are encouraging, the yields of virus obtained in cell culture remain relatively low, suggesting that until this problem is overcome, the vaccine is likely to be modertely expensive and to require several doses, perhaps with an adjuvant. For these reasons, increasing emphasis has been placed on the development of live attenuated vaccines.

B. Live Attenuated Vaccines

In comparison with killed vaccines, live attenuated vaccines have a number of attractive

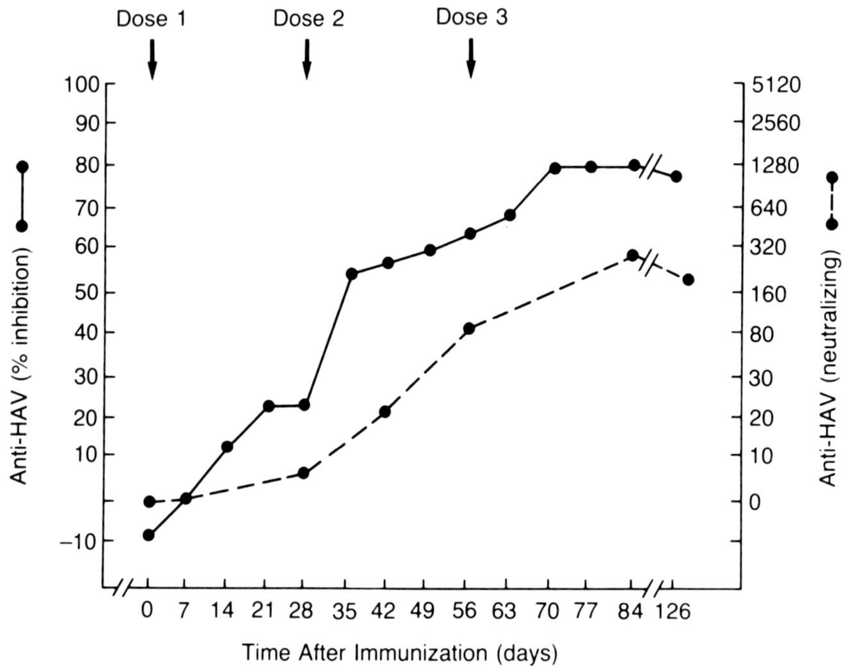

FIGURE 2. Development of anti-HAV in owl monkeys who had received 1, 2, or 3 doses of the W.R.A.I.R. inactivated hepatitis A vaccine. (From Binn, L. N., Bancroft, W. H., Lemon, S. M., Marchwicki, R. H., LeDuc, J. W., Trahan, C.J., Staley, E. C., and Keenan, C. M., *J. Infect. Dis.*, 153, 749, 1986. With permission.)

properties such as the ability to administer them orally, thus avoiding the use of needles and syringes. In addition, live vaccines are often cheaper to produce, and may result in longer lasting immunity as virus replication in the gut multiplies the amount of antigen present. Induction of gut immunity can also be expected to reduce the circulation of wild-type virus in the community and thereby further reduce the risk of disease.

In 1982 two groups of workers, Provost and his colleagues at the Merck Institute[11] and Feinstone and his colleagues at the National Institutes of Health (NIH),[13] produced evidence that strains of HAV which had been passaged many times in cell culture, retained their ability to infect marmosets and chimpanzees but appeared to have lost their ability to produce disease.

In the first study, the CR326 strain of HAV which had been adapted to growth in FRhK6 cells (after 31 passages in the marmoset) and to human diploid lung fibroblasts (WI-38 and MRC5) was used.

Two particular variants were prepared and evaluated for their degree of attenuation, antigenicity, and protective efficacy while several more variants were evaluated simply for their degree of attenuation. The first variant (no. 1) was produced by 31 passages in marmosets, 10 passages in FRhK6 cells followed by a further nine passages in MRC5 cells at 32°C. To ensure that the final pool was homogenous and contined only HAV particles, the final three passages in the series were performed with terminal dilutions of the inoculum.

The second variant (No. 2) was produced by 31 passages in marmosets followed by 16 passages in FRhK6 and eight passages in human diploid cells (two in WI-38, six in MRC5) at 35°C. As with the previous strain, three of the later passages (five, six, and seven) were carried out by limiting dilutions.

HAV was more readily adapted to growth in human diploid fibroblasts at 32°C than 35°C; however, once adaptation had been achieved the two variants had similar growth characteristics and produced similar yields of infectious virus ($10^{7.3}$ and $10^{6.6}$ $TCID_{50}$) and viral antigen (P/N 7.1 and 5.3, respectively).

Four *S. labiatus* marmosets were inoculated i.v. with 1 mℓ of 1:10 dilutions of both variants (in PBS) and observed for 8 months. During this period the animals were bled weekly and their plasma tested for ICD and anti-HAV. ICD levels of more than 1500 sigma units, on more than one occasion, were considered to be elevated. Using this definition, only one animal (no. 4) developed a significant rise in ALT — which was seen on a single occasion only, 35 days after inoculation just prior to the development of anti-HAV. Another animal developed an elevated ICD (1040 units) on a single occasion, 46 days after inoculation, also just preceding the development of antibody. It seems reasonable to attribute both of these rises to the vaccine. All four animals developed circulating antibody to HAV within 46 to 90 days of immunization.

None of the four animals which received variant 2 developed significant elevations in ICD, however all four showed minor elevations (550 to 710 units) on two consecutive samples collected between 25 and 68 days of inoculation, usually coinciding with the development of antibody.

At 32 weeks after inoculation, animals that had been immunized with variant 1 were challenged with a single i.v. dose of a 10^{-5} dilution of a pool of CR326 extracted from infected marmoset liver and known to have a titer of about 10^9 MID_{50}. The animals appeared to be completely protected, with none developing hepatitis or a boost in antibody titer. Similarly, all four animals immunized with the second variant and challenged 20 weeks later were fully protected. By contrast, all the control (unimmunized) animals which received live virus developed acute hepatitis with typical rises in ICD and anti-HAV.

These data indicate that is is possible to produce strains of HAV which are attenuated for marmosets, but suggest that even with these strains, virus replication is occurring in the liver. Histological study of biopsy tissue obtained from animals receiving both variants were reported as either normal or showing minor periportal imflammation, mononuclear infiltration, and occasional focal necrosis of hepatocytes. Direct evidence of virus replication in the liver (by passage in animals, inoculation of cell culture, or IF) was not reported.

In addition to the two major variants (1 and 2), a number of other variants were produced and inoculated into marmosets to evaluate the degree of attenuation. Strains which had been passaged in FRhK6 for 25 passages or more appeared to have lost their virulence but retained their capacity to infect the animals. Variants produced by further passage in diploid fibroblasts were either fully attenuated or over attenuated (i.e., had lost their ability to infect marmosets).

Similar data were reported by Feinstone and his colleagues[13] with the HM175 strain of HAV which had been passaged more than 20 times in AGMK and evaluated in chimpanzees.

The HM175 strain of HAV was originally obtained from the feces of a patient with acute hepatitis A in Melbourne,[14] Australia, and was directly isolated in AGMK without prior passage through marmosets.[15] After a further 25 passages, the parent wild-type virus and cell culture passage 10 and 20 material were evaluated for their infectivity, virulence, and immunogenicity in susceptible chimpanzees.[16] One milliliter of wild-type virus induced hepatitis in chimpanzees at a dilution of 10^{-6} but not at 10^{-7}. By contrast, virus that had been passaged 10 or 20 times in AGMK remained fully infectious for chimpanzees at dilutions as high as 10^{-5} and 10^{-6}, respectively, but produced no evidence of liver disease. Viral replication and virus shedding in the stools were markedly reduced in animals receiving either the P10 or P20 material.

While these findings are encouraging it should be noted that the HM175 strain, although highly infectious when administered parenterally, is of relatively low infectivity when given orally. In addition, P10 and P20 HAV was not attenuated when evaluated in marmosets (see Table 4).

These early findings indicated that a live attenuated vaccine against hepatitis A was feasible

Table 4
ATTENUATION OF THE HM175 STRAIN OF HAV IN CHIMPANZEES AND MARMOSETS

	Chimpanzees		Marmosets	
Strain of virus	**No.**	**ALT (mean maximum)**	**No.**	**ALT (mean maximum)**
Wild type	7	314	10	4777
P10	7	62	NT[a]	NT
P20	5	48	7	2936

[a] NT = not tested.

but left a number of unresolved questions and problems requiring further study. It is still not clear how to determine the proper level of attenuation for a strain before evaluating it in man. Some workers have argued that experimental batches of hepatitis A vaccine should not induce even minor abnormalities in liver function in laboratory animals, whereas others regard such changes as inevitable (because the virus which replicates in the liver) and evidence that the vaccine has taken. In this regard, while an ideal live hepatitis A vaccine would not produce any alterations in liver function and no cases of jaundice, it should be remembered that most attenuated viral vaccines (such as measles, rubella, mumps, and polio) have a slight risk of producing symptoms and this risk needs to be balanced against the costs of failing to control the disease.

While animal studies are useful guides to the behavior of candidate vaccine in man, neither marmosets nor chimpanzees can be expected to exactly mimic man. In addition, the results obtained in these species may differ. There is already evidence from several laboratories that prolonged passage of HAV in cell culture may attenuate it for chimpanzees but have little effect on its pathogenicity for marmosets.[17] In these situations the ultimate test of the value of each candidate live hepatitis A vaccine will be its performance in man.

Additional areas which require further study are the development of easily detected markers of attenuation which will enable laboratory workers to rapidly identify vaccine strains from wild-type virus, and the development of strains which are well adapted to growth in the alimentary tract and can be given orally rather than by injection.

In the last 4 years, groups at Merck and NIH have undertaken a number of further studies aimed at development of a live attenuated vaccine available for use in man. The NIH has carried out extensive studies with two high-passage variants of HM175 produced after 21 and 32 passages in AGMK, each of which had been triply cloned by limiting dilution[18] and passed standard tests for safety, sterility, and potency. The titer of virus in the vaccine lots, determined by inoculating serial dilutions in AGMK cells, was $10^{6.5}$ (P21) and $>10^{5.5}$ (P32), respectively.

Wild-type P21 and P32 virus were administered to chimpanzees and marmosets i.v. in doses of 10^{-3} CID_{50} and 10^{-3} or 10^{-5} $TCID_{50}$, respectively. Animals were observed daily and had serum samples and liver biopsies collected weekly. Sera were tested for ICD, ALT, and anti-HAV. In addition, weekly fecal samples were assayed for HAAg by RIA and HAV RNA by cDNA hybridization (see Table 5).

In chimpanzees, the virus was clearly attenuated on passage, with decreasing numbers of animals showing biochemical or histological evidence of hepatitis with increasing passage of the virus. The geometric mean of the peak ATL levels detected in chimpanzees receiving P21 and P32 virus were only 66 and 50 MU/mℓ. In keeping with these findings, none of the animals receiving P21 or P32 virus showed histological signs of disease, compared with

Table 5
RESULTS OF INOCULATION OF SUSCEPTIBLE ANIMALS WITH WILD TYPE AND P21 AND P32 STRAINS OF HAV HM175

		Evidence of Hepatitis		Detection of HAAg in liver [duration in weeks (mean)]	Detection of HAV in feces [duration in weeks (mean)]	
	Inoculum	Biochemical[a]	Histological[b] (severity)	IF	RIA	cDNA hybridization
Chimpanzees	NT	3/4	2/4 (0-3+)[c]	1—3 (1.7)	0—2 (0.5)[d]	0—3 (1.7)
	P21	2/6	0/6 (0)	0—2 (0.5)	0 (0.0)	0—2 (0.3)
	P32	1/6	0/6 (0)	0.3 (0.6)	0 (0.0)	
Marmosets	WT	4/4	4/4 (1—2+)	10—14 (12.2)	1—2 (2)	2 (2)
	P21	8/8	8/8 (1—4+)	0—10 (3.5)	0—3 (1.5)	1—2 (1.7)
	P32	1/5	2/5 (0—1+)	0—7 (2.4)	0 (0.0)	0—3 (1)

[a] ALT and/or ICD levels 2 times mean preinoculation levels within 4 weeks of seroconversion.
[b] Typical histological changed developing within 6 weeks of seroconversion.
[c] Scoring used: 1+ mild hepatitis; 2+ moderate hepatitis; 3+ bridging necrosis; 4+ massive necrotic hepatitis.
[d] P/N.

50% of chimpanzees receiving wild-type virus. Evidence of virus replication in the liver was obtained with the vaccine strains but this was markedly reduced in comparison with the pattern seen with wild-type virus. In addition, shedding of virus into the feces was markedly reduced, being undetectable with both strains by RIA and only occasionally detected by cDNA hybridization in animals infected with the P21 virus. The mean duration of excretion (detectable by this technique) fell from 2 weeks with wild-type virus, to 2 days with P21 virus, to zero with P31 virus.

The situation in marmosets was less dramatic. Animals which received P21 virus all developed biochemical and histological hepatitis and although there was some reduction in the extent and duration of replication in the liver, the duration and mean-maximum shedding of virus in the feces were largely unchanged. By contrast P32 virus was markedly, but not completely attenuated for marmosets — while four or five animals inoculated remained well, one developed a marked peak of ICD (12 times mean-preinoculation value). Marmosets receiving the P32 virus not only showed diminished replication of the virus in the liver but shed very little virus into the feces (mean maximal titers being at least 100-fold lower than with the wild-type virus).

All animals developed anti-HAV detectable by a standard RIA and by neutralization assays. In chimpanzees the anti-HAV titers detected by RIA in animals receiving P21 or P32 virus were vitrually identical to those detected in animals receiving the wild type. By contrast the titers of neutralizing antibodies in animals receiving vaccine virus were about two logs less than in animals receiving the wild type.

By contrast, marmosets generally developed higher levels of antibody than chimpanzees and only minor differences in titer were observed between animals receiving wild-type or vaccine virus. Provided that protection can be demonstrated, these strains, especially P31, have the potential to be used in man. The P31 strain is well attenuated in chimpanzees, produces minimal liver damage, and is poorly excreted in the feces. Preliminary data suggest that virus shed into the feces retains its attenuated characteristics.

Recently Provost and his colleagues reported additional data on the candidate live-hepatitis A vaccines being developed at Merck, Sharp and Dohme.[19] Two variants known as F and F^1 have been studied extensively. The CR326F strain of HAV, derived from the feces of the child from whose serum CR326 was originally isolated, was used. The variant F was obtained after passaging CR326F, 15 times in FRhK6 cells and 8 times in MRC5 cells, while F^1 was obtained by passaging the virus 15 times in FRhK6 and 16 times in MRC5, a process which took more than 2 years. Prior to administration to animals and man, each lot of virus was purified by three limit-dilution steps and tested for safety, sterility, and potency by standard techniques. Representative variants of HAV with different passage histories were inoculated into seronegative marmosets to determine whether attenuation had occurred (see Table 6).

The virus remained virulent for marmosets after 10 passages in FRhK6 cells but by the 15th passage was beginning to show evidence of attenuation. Variant F master seed and the vaccine lot stimulated high titers of antibody in all animals, but produced mild and short-lived enzyme elevations in 3 of 12 marmosets. After further passages in MRC5 cells the virus become completely attenuated for marmosets. The F^1 master seed and vaccine lot produced no evidence of biochemical hepatitis in 12 of 12 animals and stimulated anti-HAV in 10. All of the vaccinated animals were challenged with a single i.v. dose of 10^5 ID_{50} HAV (CR326F) and none developed disease (see Table7).

Both variants stimulated antibodies without biochemical evidence of hepatitis when inoculated into seronegative chimpanzees. Some animals which received the F variant, however, showed biopsy evidence of mild hepatits. Low levels of HAV (10^3 to 10^4 $TCID_{50}/g$) were detected by cell culture in the feces of both animals that received variant F for 2 weeks prior to development of anti-HAV. Further passage of this virus to seronegative chimpanzees also produced silent infections with no evidence of reversion to virulence.

Table 6
EVALUATION OF DEGREE OF ATTENUATION OF HAV STRAIN CR326F AT DIFFERENT PASSAGE LEVELS

	Passage	Evidence of infection: Elevated ICD	Evidence of infection: Seroconversion	Anti-HAV titer (IAHA)
	FRhK6 × 5	4/4	4/4	≧ 2560
	FRhK6 × 15	2/4	4/4	≧ 2560
(F)	FRhK6 × 15, MRC5 × 8	1/6	6/6	293
	Vaccine lot	2/6	6/6	293
	FRhK6 × 15, MRC5 × 12	0/6	6/6	94
(F^1)	FRhK6 × 15, MRC5 × 16	0/6	4/6	120
	Vaccine lot	0/6	6/6	160

Table 7
PROTECTIVE EFFICACY OF HAV VARIANTS F AND F^1 IN MARMOSETS

	Pre-challenge Anti-HAV		Post-challenge Anti-HAV	
Inoculum	Detected	GMT (IAHA)	ICD elevation	GMT (IAHA)
F	4/4	80	0/4	240
F^1	6/6	27	0/6	67
Controls	0/7	<5	7/7	≧1280

Table 8
RESULTS OF TRIAL WITH F AND F^1 VACCINES IN MAN

Vaccine	Dose	ALT elevation	Seroconversion
F	$10^{5.6}$	3/9	9/9
	$10^{5.6}$	2/11	6/11
	$10^{4.6}$	3/9	6/9
	$10^{3.6}$	1/8	2/8
F^1	$10^{6.3}$	0/11	6/11[a]

[a] 10/11 positive when more sensitive "modified" RIA used.

Because of these encouraging findings, both the F and F^1 variants were administered to human volunteers in 1-mℓ doses by the subcutaneous route. The doses used were $10^{5.6}$ $TCID_{50}$/mℓ, $10^{4.6}$ or $10^{3.6}$ $TCID_{50}$/mℓ (F) and $10^{6.3}$ $TCID_{50}$/mℓ (F^1). The major findings are summarized in Table 8.

The F variant was quite well attenuated for man, producing single abnormal ICD levels only in 9 of the 25 recipients who seroconverted, with the frequency diminishing with lower doses. The most encouraging results were obtained with the F^1 variant. Of 11 subjects who received this vaccine, none developed elevated ALTs and all but one developed anti-HAV, although in four the levels were so low that they could only be detected by a modified RIA.

It appears that doses of ≧ 10^5 $TCID_{50}$ of virus will be needed per vaccine dose in man and since titers of 10^7 or 10^8 $TCID_{50}$ are readily obtained in cell culture, large-scale production of live attenuated hepatitis vaccines is now feasible.

C. Other Approaches

A detailed account of alternative approaches to development of hepatitis A vaccines has recently been provided by Anderson and his colleagues[20] from which some of the following information is drawn. Additional information can be found in Chapter 7.

1. Recombinant DNA

Alternative viral vaccines can now be produced by the application of the relatively new disciplines of molecular biology and recombinant DNA technology. In the simplest case, a segment of viral DNA, or a complementary DNA (cDNA) copy of viral RNA, is ligated into a vector to form a molecular hybrid and introduced into a prokaryotic or eukaryotic cell. All progeny of the original bacterial cell will carry identical (cloned) copies of the viral DNA or cDNA, which can be characterized further by restriction endonuclease-mapping and nucleotide-sequencing analyses. A variety of computer programs to facilitate these analyses are available, and most of these allow for the determination of the amino-acid sequences which would be expected to be encoded by either strand of the cloned DNA. Again, in the simplest case, the cloned DNA can be genetically engineered to a position downstream from, and in appropriate reading frame to, a promoter such that the peptide specified by the cloned DNA is produced within the cell. After lysis of the cell the resulting peptide can then be purified and used for vaccine purposes. Of the many examples of the successful use of this approach, and the one the most relevant to the production of a hepatitis A vaccine, is the production of virion protein 1 (VP1) of FMDV in *Escherichia coli,*[21] and its use to immunize animals and protect them against challenge by FMDV. To date this approach has been relatively unrewarding with other picornaviruses such as PV[22] because they express their proteins as a single large polyprotein which is later cleaved into several small functional peptides. In addition epitopes present on at least three of the four major structural proteins appeared to be involved in stimulating neutralizing antibodies.[23]

More commonly, stable expression of cloned genes cannot be obtained or maintained in bacteria, necessitating the recloning of the DNA into other host/vector systems. This approach is often necessary if the final antigenic products embody posttranslational modifications such as specific intrachain polypeptide processing.

Racaniello and Baltimore[24] found that a complete copy of the PV genome cloned as cDNA in the plasmid vector pBR322 was infectious when transfected into eukaryotic cells in culture, yielding intact, infectious PV virions. This result has important implications for the development of vaccines against picornaviruses such as HAV as it renders feasible the generation of attenuated viral strains by genetic engineering of cDNA in vitro.

2. Chimeras

An alternative approach to vaccine development is to splice DNA coding for the antigenic determinants of HAV into cDNA copies of existing attenuated PV variants to generate polyvalent chimeric strains. Extreme care would be necessary with such studies in view of the possibility of generating hepatotropic strains of PV.

Examples of the use of a viral genome to carry a DNA-segment encoding an antigen of a different virus have been reported by Smith and his colleagues,[25] who inserted DNA encoding the hepatitis B virus surface antigen (HBsAg) into the DNA genome of vaccinia virus. The HBsAg was subsequently expressed to a level of 2.6 μg per 5 million cells infected with the variant virus, and rabbits inoculated with the live variants developed antibodies to HBsAg within 9 days. Smith, Mackett, and Moss[26] together with Panicali et al.[27] have extended this work by demonstrating, in animals, that vaccinia virus variants carrying the influenza virus hemagglutinin gene are as effective as live attenuated vaccines in prevention of infection. Smith and Moss[28] have estimated that the vaccinia virus genome can accommodate up to 25,000 base pairs of foreign DNA (almost 20-fold more than the length of HBsAg DNA), raising the possiblity of constructing infectious polyvalent vaccines against a number of different pathogenic agents.

Ticehurst et al.,[29] Von der Helm and Deinhardt,[30] and Najarian et al.[31] have published accounts of cloning HAV cDNA into plasmid vectors in *E. coli,* the first step in the development of recombinant DNA-derived HAV vaccines. In the latter case more than 99% of the HAV genome was represented among a series of overlapping cloned DNA fragments, and some preliminary nucleotide sequence information was reported.[32] The sequence information has been extended to all 7478 bases.[31] To date, there have been no confirmed reports of stable expression of antigens from cloned HAV fragments, although this is just a matter of time.

3. Synthetic Polypeptides

A relatively new approach to the development of viral vaccines is to make use of the peptide sequences predicted from the nucleotide sequences determined from the cloned viral DNA or cDNA. Oligopeptides likely to be antigenic can be chemically synthesized, coupled to carriers mixed with adjuvants, and tested in animals to determine whether they are immunogenic, and if the immune response is virus-neutralizing, or protective. This approach was employed successfully by Bittle et al.,[33] who synthesized peptides of the VP1 protein of FMDV and used this material to stimulate a neutralizing antibody response in animals. Linemeyer et al.[34] have used DNA sequence data from HAV cDNA clones to compare predicted amino-acid sequences with known sequences of cyanogen bromide-cleaved fragments of VP1 and VP3 of HAV to locate the cDNA sequences coding for these structural proteins. In the near future, it may be possible to use these data to synthesize a series of oligopeptides which stimulate antibody to major antigenic site(s) of HAV.

There are several potential advantages in the use of polypeptide vaccines, including the low cost of syntheses, the absence of adventitious agents, and the relative ease of varying the antigenic content. Before this approach is practical for hepatitis A, more information is needed on the epitopes which stimulate neutralizing antibodies, in particular their three-dimensional structure.

III. DISCUSSION

A number of approaches are now available for the production of hepatitis A vaccines and several candidate vaccines are undergoing preliminary evaluation in man. It seems likely that safe, effective vaccines will be developed and licensed by the early 1990s, with perhaps the first being a conventional killed vaccine which can be used in combination with other early childhood vaccines (tetanus, diphtheria, pertussis) which are given by injection. If live attenuated vaccines are developed it would seem logical to incorporate them with the three oral polio strains and produce a quadrivalent vaccine which could be administered to babies early in life.

The strategy of administration is likely to vary from country to country depending on the types of vaccine available, their cost, and the type of delivery system available. In developed countries with high standards of living and a good health infrastructure, the vaccine is likely to be restricted to groups at increased risk of infection. Where such groups constitute a significant proportion of the population (e.g., 20 to 30%) it will probably be easier to include the vaccine into the regular schedule of childhood immunizations.

For the vaccine to have a major impact in developing countries it will need to be widely administered early in life, either as part of the routine childhood immunization program or as a prerequisite to school entry. It should be stressed that the duration of immunity with these vaccines and the need for booster doses is unknown. Long-term follow up of recipients will be required to determine whether waning antibody levels are associated with an increased risk of infection and disease. This will be particularly important in those countries in which improvements in sanitation and public health are delayed and wild-type viruses continue to circulate (or are regularly introduced) in the community.

REFERENCES

1. *Morbid. Mortal. Wkly. Rep.*, Summary of diseases, 1983.
2. **Storch, G., McFarland, L. M., Kelso, K., Heilman, C. J., and Caraway, C. T.,** Viral hepatitis associated with day-care centers, *JAMA*, 242, 1514, 1979.
3. **Rayo, M. V.,** Epidemiologia de la hepatitis viral en Chile, Second Meeting of the PAHO Ad Hoc Committee on Viral Hepatitis, Caracas, 1986, in press.
4. **Deinhardt, F. and Zuckerman, F.,** Immunization against hepatitis B: report of a WHO meeting on viral hepatitis in Europe, *J. Med. Virol.*, 17, 209, 1985.
5. **Mascoli, C., Ittensohn, O. L., Villarejos, V., Arguedas, G., Provost, P. J., and Hilleman, M. R.,** Recovery of hepatitis agents in the marmoset from human cases occurring in Costa Rica, *Proc. Soc. Exp. Biol. Med.*, 142, 276, 1973.
6. **Provost, P. J., Wolanski, B. S., Miller, W. J., Ittensohn, O. L., McAleer, W. J., and Hilleman, M. R.,** Physical, chemical and morphologic dimensions of human hepatitis A virus strain CR326, *Proc. Soc. Exp. Biol. Med.*, 148, 532, 1975.
7. **Provost, P. J. and Hilleman, M. R.,** An inactivated hepatitis A vaccine prepared from infected marmoset liver, *Proc. Soc. Exp. Biol. Med.*, 159, 201, 1978.
8. **Provost, P. J., Hughes, J. V., Miller, W. J., Giesa, P. A., Banker, F. S., and Emini, E. A.,** An inactivated hepatitis A viral vaccine of cell culture origin, *J. Med. Virol.*, 19, 23, 1986.
9. **Binn, L. N., Bancroft, W. H., Lemon, S. M., Marchwicki, R. H., Le Duc, J. W., Trahan, C. J., Staley, E. C., and Kennan, C. M.,** Preparation of a prototype inactivated hepatitis A virus vaccine from infected cell cultures, *J. Infect. Dis.*, 153, 749, 1986.
10. **Gauss-Muller, V., Frosner, G. G., and Deinhardt, F.,** Propagation of hepatitis A virus in human embryo fibroblasts, *J. Med. Virol.*, 7, 233, 1981.
11. **Provost, P. J., Banker, F. S., Giesa, P. A., McAleer, W. J., Buynak, E. S., and Hilleman, M. R.,** Progress towards a live attenuated human hepatitis vaccine, *Proc. Soc. Exp. Biol. Med.*, 170, 8, 1982.
12. **Montagenon, B. J., Fanget, B., and Vincent-Falquet, J. C.,** Industrial-scale production of inactivated poliovirus vaccine prepared by culture of Vero cells on microcarrier, *Rev. Infect. Dis.*, 6, 341, 1984.
13. **Feinstone, S. M., Daemer, R. J., Gust, I. D., and Purcell, R. H.,** Live attenuated vaccine for hepatitis A, *Dev. Biol. Stand.*, 54, 492, 1983.
14. **Gust, I. D., Lehmann, N. I., Crowe, S., McRorie, M., Locarnini, S. A., and Lucas, C. R.,** The origin of hepatitis A strain HM175, *J. Infect. Dis.*, 151, 365, 1985.
15. **Daemer, R. J., Feinstone, S. M., Gust, I. D., and Purcell, R. H.,** Propagation of human hepatitis A virus in African green monkey kidney cell culture: primary isolation and serial passage, *Infect. Immun.*, 32, 388, 1981.
16. **Daemer, R. J., Feinstone, S. M., Gust, I. D., and Purcell, R. H.,** Titration in cell culture and chimpanzees of hepatits A isolated and serially passaged in primary African green monkey kidney, in *Viral Hepatitis*, Szmuness, W., Alter, H., and Maynard, J., Eds., Franklin Institute Press, Philadelphia, 1981, 755.
17. **Purcell, R. H., Feinstone, S. M., Ticehurst, J., Daemer, R. J., and Baroudy, B. M.,** Hepatitis A virus, in *Viral Hepatitis and Liver Disease*, Vyas, G. M., Dienstag, J. L., and Hoofnagle, J., Eds., Grune & Stratton, Orlando, Fla. 1984, 9.
18. **Karron, R. A., Daemer, R., Ticehurst, J., D'Hondt, E., Popper, H., Mihalik, K., Feinstone, S., and Purcell, R. H.,** Studies of prototype live HAV vaccines in primate models, *J. Infect. Dis.*, in press.
19. **Provost, P. J., Bishop, R. P., Gerety, R. J., Hilleman, M. R., McAleer, W. J., Scolnick, E. M., and Stevens, C. E.,** New findings in live, attenuated hepatitis A vaccine development, *J. Med. Virol.*, 20, 165, 1986.
20. **Anderson, B. N., Coulepis, A. G., and Gust, I. D.,** Towards a hepatitis A vaccine, *J. Hyg.*, 93, 269, 1984.
21. **Kleid, D. G., Yansura, D., Small, B., Dowbenko, D., Moore, D. M., Grubman, M. J., McKercher, P. D., Morgan, D. O., Robertson, B. H., and Bachrach, H. L.,** Cloned viral protein vaccine for foot and mouth disease: responses in cattle and swine, *Science*, 214, 1125, 1981.
22. **Emini, E. A., Jameson, B. A., and Wimmer, E.,** Priming for and induction of anti-poliovirus neutralizing antibodies by synthetic peptides, *Nature (London)*, 304, 699, 1983.

23. **Almond, J. W., Cann, A. J., Minor, P. D., Reeve, P., Schild, G. C., Hauptmann, R., and Stanway, G.,** Nucleotide sequence from neurovirulent and attenuated strains of poliovirus type 3, *Rev. Infect. Dis.*, 6, 5487, 1983.
24. **Racaniello, V. R. and Baltimore, D.,** Cloned poliovirus complementary DNA is infectious in mammalian cells, *Science*, 214, 916, 1981.
25. **Smith, G. L., Mackett, M., and Moss, B.,** Infectious vaccinia virus recombinants that express hepatitis B virus surface antigen, *Nature (London)*, 302, 490, 198.
26. **Smith, G. L., Mackett, M., and Moss, B.,** Construction and characterization of an infectious vaccinia virus recombinant that expresses the influenza virus hemagglutinin gene and induces resistance to influenza virus in hamsters, *Proc. Natl. Acad. Sci. U.S.A.*, 80, 7155, 1983.
27. **Panicali, D., Davis, S. W., Weinburg, R. L., and Paoletti, E.,** Construction of live vaccines by using genetically engineered poxviruses: biological activity of recombinant vaccinia virus expressing influenza virus hemagglutinin, *Proc. Natl. Acad. Sci. U.S.A.*, 80, 5364, 1983.
28. **Smith, G. L. and Moss, B.,** Infectious poxvirus vectors have capacity for at least 25,000 base pairs of foreign DNA, *Gene*, 25, 135, 1983.
29. **Ticehurst, J. R., Racaniello, V. R., Baroudy, B., Baltimore, D., Purcell, R. H., and Feinstone, S. M.,** Molecular cloning and characterization of hepatitis A virus cDNA, *Proc. Natl. Acad. Sci. U.S.A.*, 80, 5885, 1983.
30. **Von der Helm, K. and Deinhardt, F.,** Cloning of hepatitis A virus genome, in *Advances in Hepatitis Research*, Chisari, P., Ed., Masson, New York, 1984, 263.
31. **Najarian, R., Caput, D., Gee, W., and ,** Primary structure and gene organization of human hepatitis A virus, *Proc. Natl. Acad. Sci. U.S.A.*, 82, 2627, 1985.
32. **Baroudy, B. M., Ticehurst, J. R., Miele, T., Maizel, J. V., Purcell, R. H., and Feinstone, S. M.,** Sequence analysis of cloned hepatitis A virus (HAV) cDNA, in, *Viral Hepatitis and Liver Disease*, Vyas, G. N., Dienstag, J. L., and Hoofnagle, J., Eds., Grune & Stratton, New York, 1984, 625.
33. **Bittle, J. L., Houghton, R. A., Alexander, H., Shinnick, T. M., Sutcliffe, J. G., and Lerner, R. A.,** Protection against foot and mouth disease by immunization with a chemically synthesized peptide predicted from the viral nucleotide sequence, *Nature (London)*, 298, 30, 1982.
34. **Linemeyer, D. L., Mltra, S. W., Hughes, J., Bennet, C. D., Menka, J. G., and Young, A.,** Molecular cloning of hepatitis A virus, in, *Viral Hepatitis and Liver Disease*, Vyas, G. N., Dienstag, J. L., and Hoofnagle, J., Eds., Grune & Stratton, New York, 1984, 629.

Chapter 12

ENTERICALLY TRANSMITTED NON-A, NON-B HEPATITIS

I. INTRODUCTION

In recent years it has become apparent that a significant proportion of epidemic and sporadic hepatitis in developing countries is enterically transmited but is caused by an agent other than HAV. It is impossible, therefore, to discuss strategies for the control of enterically transmitted hepatitis without considering this form of the disease.

Following the identification of the hepatitis A and B viruses, it became possible to test large panels of sera which had been collected from patients with acute viral hepatitis in different parts of the world and determine the prevalence of these infections. To most peoples' surprise, in a significant proportion of cases no evidence of infection with either agent was obtained; even more surprising was the observation that several common-source outbreaks of hepatitis, which had been regarded as text book examples of hepatitis A, were not due to infection with HAV! The first suggestion that acute viral hepatitis was caused by more than two agents came from the work of Fred Prince and his colleagues at the New York Blood Center.[1] In 1974, they noted that some recipients of large numbers of units of blood developed a form of post-transfusion hepatitis which was not hepatitis B and had a longer incubation period than hepatitis A. They called this disease hepatitis C, assuming it to be due to a single agent. Shortly afterward Feinstone and his colleagues[2] were able to use the newly developed technique of IEM to demonstrate that this disease was definitely not hepatitis A and gave it the more accurate name of non-A, non-B hepatitis (HNANB), a term which despite its length soon found general acceptance. It soon became apparent that non-A, non-B hepatitis existed in two different forms which had distinctive epidemiological features and were caused by different etiological agents. The first type, which was first recognized as a complication of blood transfusion and is now common in i.v. drug users in many countries, resembles hepatitis B. This disease has a long incubation period, is parenterally transmitted, and appears to be associated with a chronic carrier state: as with hepatitis B, chronic infection this form of HNANB may result in chronic liver disease.[3] The second type of HNANB resembles hepatitis A. This disease has a short incubation period, is spread enterically, occurs in both sporadic and epidemic forms, is not associated with a chronic carrier state, and does not appear to result in long-term sequelae. The main features of these diseases are shown in Table 1.

An account of our evolving understanding of the epidemiology, clinical features, pathology, and etiology of the enterically transmitted form is presented below.

II. EPIDEMIOLOGY

A. The New Delhi Outbreak

In retrospect the first well documented outbreak of HNANB appears to have taken place in New Delhi, India in 1955 to 1956.[4] At the time this was thought to be a major epidemic of water-borne hepatitis A with some unusual features; however, when specific serologic tests become available it was found to have been an outbreak of HNANB.[5,6] Because it illustrates many features of the epidemiology of the disease, this outbreak is worth recapitulating in some detail.

In the first week of December 1955, public health authorities in New Delhi noted an increase in the number of patients with viral hepatitis presenting to local hospitals and clinics. The number of cases rose rapidly, peaking in the last week in December and not declining

Table 1
MAJOR DISTINGUISHING FEATURES OF THE ENTERICALLY AND PARENTERALLY TRANSMITTED FORMS OF HNANB

	Enterically transmitted	Parenterally transmitted
Incubation period	2—6 Weeks	6—8 Weeks
Epidemiology	Causes sporadic disease and occasional outbreaks in many developing countries.	Has been recognized following transfusion of blood products or use of shared syringes in developed countries.
Clinical features	Often produces icteric disease in adults.	Frequently subclinical even in adults.
Case fatality rate	High, especially in pregnant women in second or third trimester (10—20%).	Extremely low.
Outcome	Self-limiting disease — if patient survives acute illness. No long-term sequelae recognized.	Significant risk of developing biochemical and/or histological evidence of chronic liver disease.
Carrier state	No	Yes
Transmitted to	Human volunteers, and several species of nonhuman primates.	Chimpanzees
Etiological agent	Unknown. Best candidate is 22—30 nm particle identified in the feces of some patients and experimentally infected animals.	Unknown. Some evidence to suggest more than one agent exists.

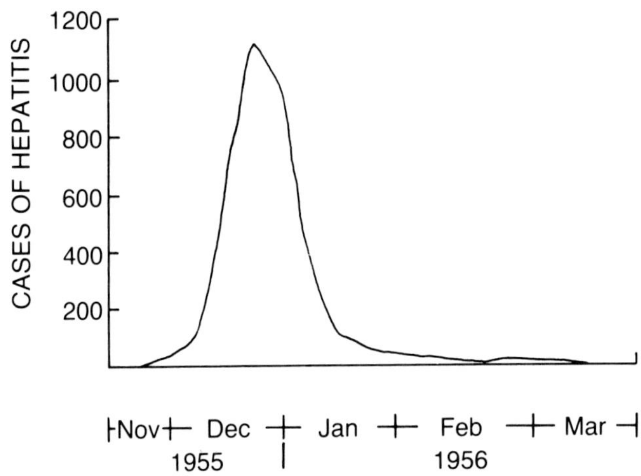

FIGURE 1. The number of cases of viral hepatitis reported to public health authorities during the New Delhi outbreak of 1955 to 1956.

to normal levels until mid or late January, with most cases occurring within a 6-week period (see Figure 1). The peak incidence of the disease was in the 14 to 39-year age group with males being more frequently affected than females. The attack rates in residents less than 14 years of age, from 14 to 39 yr and over 40 years of age were estimated to be 1.3%, 2.9%, and 2.0%, respectively.[32] It is of interest that in the Indian Army, officers were affected four times as commonly as troops, who were in turn affected twice as frequently as sweepers. The precise number of people who were affected is unknown, however, extrapolations from some areas which were carefully studied suggest that more than 29,000 cases of jaundice occurred, with perhaps three times as many people showing milder expressions of disease and up to 60% of the population (1.1×10^6) being infected. Very few secondary cases were recognized.

Table 2
DETECTION OF CHLORIDE IN WATER FROM THE WAZINABAD PUMPING STATION FROM NOVEMBER 10 TO 20, 1955

Date	Chloride level	
November 10	6	
11	13	
12	25	Period of heavy fecal contamination (November 12–16)
13	60	
14	80	
15	90	
16	61	
17	6	
18	6	
19	6	
20	6	

In young adults the clinical, biochemical, and histological picture was typical of acute viral hepatitis although some patients were noted to have evidence of mild cholestasis. While in most people the disease was relatively mild and was followed by a complete recovery, 73 deaths were recorded with a disturbingly high case-fatality rate occurring among pregnant women, especially those who were in the third trimester of pregnancy.

The epidemiology of the outbreak suggested that infection was acquired from a common source. As the highest incidence of the disease was in a section of the city which obtained drinking water from a common source and no cases were seen among embassy staff who used the same supply but boiled their water prior to use, suspicion arose that this supply had become contaminated. Detailed investigation revealed that raw sewage was discharged into the Jumna River just beyond the site of the pumping station which pumped domestic water into the municipal system. Sewage was collected from open drains which ran through the city and were known to contain both human and animal excrement.

A review of weather patterns revealed that unusually heavy rains in October had been followed by severe flooding. In November as the water level fell rapidly, the course of the river changed dramatically with the result that raw sewage was drawn into the water supply for a short period. The existence of gross contamination was rapidly recognized and large doses of chlorine (90 parts per million — superchlorination) were added to the water (see Table 2). These measures appear to have been successful, and were followed by major earth-moving works to prevent future mingling of the sewage outflow and water intake.

Studies of the occurrence of cases among residents of New Delhi and visitors indicate that the incubation period of the disease (from the time the water was noted to be contaminated with sewage to the onset of jaundice) was 4 to 6 weeks. A limited followup of patients involved in the epidemic showed no evidence of prolonged abnormalities of liver function, no excess chronic liver disease, and no excess mortality (except among pregnant women).

The occurrence of a large outbreak of hepatitis A was recognized to be unusual (because it was assumed that most residents of New Delhi would already be immune to the virus) and interpreted as being due either to the very large dose of virus involved or perhaps to a different strain of virus. It was not until many years later, when sera collected from patients were examined for markers of hepatitis A and B infection, that it was realized that this outbreak was due to an enterically transmitted form of HNANB.

B. Further Outbreaks on the Indian Subcontinent

Since 1955 a number of additional outbreaks of hepatitis with similar features have been

Table 3
AGE AND SEX DISTRIBUTION OF PATIENTS WITH HEPATITIS IN THE 1978 TO 1979 EPIDEMIC IN THE KASHMIR VALLEY

	Male		Female		Total	
Age (yr)	No.	Hepatitis	No.	Hepatitis	No.	Hepatitis
<10	2754	17	2615	10	5369	27
11—20	1955	52	1834	36	3789	88
21—30	1270	41	1289	48	2559	89
31—40	1152	34	1049	25	2201	59
41—50	774	7	657	2	1431	9
51—60	502	1	380	0	882	1
>60	233	1	156	1	389	2
	8640	153	7980	122	16620	275

reported from several parts of India including Aurangabad (1961),[7] Ahmedabab (1975 to 1976),[8] Pune (1978 to 1979),[9] the Kashmir Valley (1978 to 1979),[10] Uttar Pradesh (1979 to 1980),[11] and Bihar (1986); not all of these have been described in the scientific literature.

In Aurangabad[7] a sharp outbreak of hepatitis occurred in June 1961 in a small section of a city with a population of about 100,000 people. Detailed investigation revealed that all the families involved were supplied by a tributary line which had numerous breaks and was frequently contaminated with sewage from cess pits and surface drains. As in New Delhi, the outbreak was most severe in young adults and there was a high case-fatality rate among pregnant women. Second cases were seen in about 25% of affected households, but whether these were secondary cases or primary is not clear.

In recent years as knowledge and interest in this disease has increased several outbreaks have been thoroughly investigated. One of the best documented areas is the Kashmir Valley where outbreaks occurred in 1978, 1979, 1980, and 1981.[10] Following a newspaper report of an outbreak of hepatitis in the Kashmir Valley, in early November 1978, Khuroo and his colleagues visited the area and undertook a detailed survey. At the outset every house was visited, questionnaires were completed by the head of the household, and all family members were examined by a doctor or health worker. Each household was revisited three times at 6 to 8 week intervals to determine the condition of patients seen at previous visits and the occurrence of further cases. Blood specimens were collected from patients with hepatitis (or symptoms suspicious of infection) and tested for serum bilirubin, SGOT, and SGPT. Liver biopsies were obtained from selected patients who had provided informed consent.

The population surveyed comprised 16,620 people living in 1435 houses in 15 villages (1603 families). In 11 villages, water was obtained from a local stream; the remaining four had alternative water supplies. Sewage disposal was by pit latrines. All but four cases of hepatitis came from the 11 villages which used stream water for drinking. The stream is fed by melting snow, so that in the summer it flows rapidly and is of good quality; in winter, however, the level of water falls abruptly and the stream is heavily contaminated. Regular testing of the stream during November revealed gross fecal contamination.

Over the period of investigation 275 inhabitants (1.65%), of which 153 (55.6%) were male, developed hepatitis. The highest attack rates were in the 11 to 40-year age group (see Table 3). The majority of cases occurred over a 7-week period (the second week of November to the last week of December, 1978), usually as single cases in a household. In 20 households, two or three cases were detected. Some of these occurred one incubation period after the main epidemic, suggesting that infection was being spread by personal contact. The mean incubation period was estimated to be 15 days with a range of 10 to 50 days.

Table 4
INCIDENCE OF HNANB AND OUTCOME AMONG PREGNANT WOMEN, NONPREGNANT WOMEN, AND MEN INVOLVED IN THE 1978 TO 1979 EPIDEMIC IN THE KASHMIR VALLEY

		Hepatitis		
	Subjects	No.	Fulminant	Deaths
Men (15 — 45 years)	3822	107	3	3
Women (15 — 45 years)	2350	71	0	0
Pregnant women (15 — 45 years)				
First trimester	34	3	0	0
Second trimester	77	15	0	0
Third trimester	97	18	9	6

Twelve patients (four male, eight female) developed fulminant hepatitis and of these, ten (four males, six females) died. All of the women who died were in the third trimester of pregnancy. A total of 128 family contacts of index cases were studied, of these 39 (30%) had abnormal liver function tests suggestive of subclinical infection.

The typical clinical picture was of mild hepatitis which usually resolved within 10 days of the onset of jaundice. In about 20% of cases cholestatic features (clay-colored stools and mild pruritus) were noted. Serological studies revealed no evidence of recent infection with HAV or HBV.

Biopsy findings in patients with uncomplicated hepatitis revealed portal inflammation, ballooning degeneration of hepatocytes, some evidence of biliary stasis, and various levels of hepatocyte necrosis. Follow-up biopsies during the convalescent phase of the illness showed an almost complete recovery of liver architecture. One post-mortem biopsy was obtained from a patient with fatal hepatitis — this showed submassive hepatic necrosis. Prolonged follow up of patients in the Kashmir Valley who developed HNANB in 1978 or 1979 failed to provide evidence of long-term liver disease.[12]

To further evaluate the impact of the disease in women, the data were reanalyzed to determine whether the incidence or outcome was affected by gender, whether or not the patient was pregnant, and stage of pregnancy.[13]

The incidence of hepatitis was much higher in pregnant women (17.3%) than nonpregnant women (2.8%) and men (2.1%) (see Table 4). In addition, a higher proportion of pregnant women with hepatitis developed fulminant disease (22.2%) than nonpregnant women (0.0%) or men (2.8%) — all the fulminant cases of hepatitis in pregnant women occurred in the third trimester. While nonfulminant hepatitis in pregnant women did not have an adverse effect on their pregnancy, fulminant hepatitis resulted in three premature deliveries (with one neonatal death) and five deaths of mother and fetus. These findings confirm and extend observations on the severity of this form of hepatitis in pregnant women that had been documented in isolated case reports over a period of years.[14-19]

To evaluate the role of immune serum globulin (ISG) in prophylaxis of the disease, during one of the outbreaks Khuroo administered locally produced ISG to some 200 pregnant women. Although no cases of hepatitis were observed in these people, the relatively low attack rate at the time the ISG was given and the absence of controls makes these data hard to evaluate.

Khuroo and his colleagues wondered whether the agent responsible for the outbreak was endemic in the Kashmir Valley or had been inadvertently introduced. To study this question they undertook a prospective study of sporadic hepatitis in the area.[20] From April 1979 to

December 1981, 293 additional cases of hepatitis were detected in the population, of which 155 (53%) were demonstrated to be HNANB. As with the epidemic disease, sporadic HNANB was predominately a disease of young adults with males being more frequently affected than females (1.3:1). Interestingly, 19 women with HNANB were pregnant and of these six developed fulminant hepatitis.

No further outbreaks were reported in India until July 1979 when numerous cases were reported in the Azamgarh district of Uttar Pradesh.[11] Because of the relatively high case-fatality rate, the outbreak drew the attention of the All Indian Institute of Medical Sciences (AIIMS) which quickly dispatched a team of investigators to the area. The outbreak appeared to have started in June 1979 and increasing number of cases were seen over the next few months, reaching a peak in May 1980. Overall more than 700 people from a population of about 70,000 were admitted to hospital of which 12% died. The case-fatality rate was highest in pregnant women (39%) followed by nonpregnant women (13.4%) and men (8.4%).

During the 7-week visit of the AIIMS team, 152 patients (105 males, 47 females) were studied. As in other outbreaks, the majority of cases were in the 11 to 49-year age group with men being significantly more frequently involved (2:1). The outbreak was confirmed as HNANB on the basis of typical clinical, biochemical, and histological data supported by negative tests for recent infection with HAV or HBV.

This epidemic has some features which distinguish it from other outbreaks in the Indian subcontinent. First, the prolonged outbreak suggests that a considerable amount of person-to-person spread was occurring. This is borne out by the fact that history of contact with another jaundiced patient was obtained from 70% of cases. Clinically the disease appears to have been similar to other outbreaks with the exception that cholestatic features were observed in less than 10% of patients. Locally produced HNIG was given to 500 individuals including pregnant women and hospital staff. Although the follow up of recipients of gamma globulin was inadequate, none of these people were admitted to hospital with hepatitis over the next 3 months. As the epidemic was declining at that time and only two cases of hepatitis could have been expected in this group by chance, it is not possible to determine whether passive immunization protected against the disease.

In a subsequent study by the same group, the value of locally produced ISG was evaluated during an outbreak of HNANB which took place in troops in Northern India.[21] During this outbreak, which occurred in September 1981, 206 troops in a battalion of 845 men developed acute hepatitis within a 4-week period.

The remaining 639 contacts were examined and 33 with abnormal liver function tests were excluded from study. The 606 men remaining were divided into two groups: group 1 (316 men) received 2 mℓ of a 16.5% solution of commercial ISG (Curewell, India), while group 2 (290) men received no treatment. Both groups were followed for 6 months.

The incidence of biochemical or ecteric hepatitis in the two groups was 0.6% and 2.5%, respectively. This difference is not statistically significant; however as ISG was given late in the epidemic and secondary cases are uncommon, these data do not eliminate the possiblity that ISG may be of value if given prior to exposure. Further studies with a range of ISGs prepared in different locations are required to resolve this issue.

In addition to the outbreaks of HNANB which have been reported from India, epidemics have also been reported in the neighboring countries of Pakistan, Nepal, and Burma. In mid 1982 when it became clear that an epidemic of hepatitis was occurring in the Kathmandu Valley, joint investigation was undertaken by the Nepalese Ministry of Health, the World Health Organization, and the Centers for Disease Control in Atlanta.[22] Inpatient and out-patient data were collected from every hospital in the valley and from local doctors, and was supplemented by information obtained from house-to-house surveys in several affected areas. Analysis of this information revealed that the outbreak had begun in May to June 1981, had a sharp peak in July to August, and then declined to a plateau (about ten times the normal number of cases) until the middle of 1982 when a second large peak (June to August) occurred. Males accounted for 75% of cases with 70% being in the 15 to 34-year age group. Of 119 pregnant women who developed hepatitis during the outbreak 25 (21%) died; the case-fatality rates among nonpregnant women and men were 5.4% and 7.5%, respectively.

A survey of 290 households revealed that 7.6% had at least one case of hepatitis and that the overall attack rate in the population was between 1% and 2%. In contrast with some other outbreaks the secondary attack rate was low (2 to 9%, depending on location).

The piped water provided to the villages flowed for only a few hours per day causing the population to use local rivers and ditches for bathing, drinking, and washing food and clothing. These sources were frequently contaminated with feces, especially during the monsoon rains.

C. Other Outbreaks

In addition to the well-documented outbreaks of HNANB on the Indian subcontinent, epidemics have also been reported in Western China, Mongolia, Burma,[23] the Indochinese peninsula, Afghanistan, the southern U.S.S.R. (1982 to present),[24,25] Algeria,[26] Tunisia, Ethiopia, Sudan, Somalia, the Ivory Coast, and Mexico.[33] In addition there are several reports of outbreaks of hepatitis with a high case-fatality rate in Europe.

The major features of most of these outbreaks are similar to those reported on the Indian subcontinent (see Table 5). For example, Favorov and his colleagues[25] have recently described an epidemic which occurred in the Central Asian region of the U.S.S.R. in 1984.

Of 130 patients with acute viral hepatitis studied during the outbreak, 110 had serological evidence of HNANB infection. The peak incidence of the disease was in young adults between the ages of 15 and 29 years (50.9% of cases). Careful evaluation of the patients revealed a prodromal illness in 83%. This illness which lasted for 3 to 4 days was characterized by weakness, nausea and vomiting, and discomfort or pain in the hypochondrium. In contrast to hepatitis A the onset of jaundice was not accompanied by a rapid improvement in symptoms, most patients continuing to feel weak, to have a poor appetite, or feel nauseated for 5 to 7 days. Pain in the right upper quadrant continued to be a common symptom and was sometimes quite severe. Jaundice was observed for about 2 weeks (14 ± 5 days) and was accompanied by a mild (1 to 5 cm) enlargement of the liver.

As in other outbreaks the disease had a high case-fatality rate among pregnant women. From August to December, 70 pregnant women in the region developed hepatitis, 16 in the first half of their pregnancy, 54 in the second half. No severe illnesses occurred in the former group. By contrast, in 90% of women infected in the second half of pregnancy, the illness was judged to be moderately severe or severe. Eleven of these women died, for a case-fatality rate (in the second half of pregnancy) of 20.4%.

Not all epidemics of HNANB share the same clinical and epidemiological features, suggesting that some may be due to other (perhaps blood-borne) agents. For example, the outbreak which occurred in the Okitsu area of Shimizu City, Japan from April 1980 to November 1981[27] differed from enterically transmitted outbreaks in its duration, the lack of an identifiable source of infection, the long incubation period (2 to 5 months), the absence of excess mortality in women, and the high proportion (58.2%) of patients who had persistently abnormal liver function tests for at least 6 months after the onset of jaundice.

D. Sporadic Disease

Sporadic cases of HNANB in which no evidence of parenteral transmission can be documented were first recognized in Costa Rica in the early 1970s.[28] In their classical studies in Costa Rica, Villarejos and his colleagues obtained serum samples and detailed histories from villagers who developed acute viral hepatitis. Over several years (mid 1960s to early

Table 5
DATA OBTAINED FROM SOME OF THE MAJOR OUTBREAKS OF HNANB

Outbreak	Year	Duration	Population	No. cases (attack rate %)	Secondary attack rate (%)	M/F ratio	Peak age (yr)	Case-fatality rate (%)		
								Preg. female	Nonpreg. female	Male
New Delhi	1955—1956	6 weeks	1,100,000	29,300[a] (2.6)	—	3	15—34			
Aurangabad	1961	7 weeks	100,000	173 (0.2)	19	1.8	21—35	20	0	0
Mandalay, Burma	1976—1977	15 months		20,000			20—29	18		
Kashmir Valley	1978—1979	6 months	16,620	275 (1.6)	27.3	1.2	11—40	16.7[b]	0	2.8
Azamgarh	1979—1980	9 months	70,000	7,700 (71.0)	70	2.2	11—39	39	13.4	8.4
Kathmandu Valley, Nepal	1981—1982	18 months	NR[c]	NR (1.4)	2—9	3:1	15—34	21	5.4	7.8
Rangoon, Burma	1982—1983			399			20—40	12		
Algeria	1980—1981	4 months	68,000	788 (1.2)						
Central Asia	1984	5 months	NR	110			15—29	15.7		

[a] Figures estimated from sample survey. At least three times as many people had mild symptoms consistent with infection.
[b] All in third trimester.
[c] NR = not recorded.

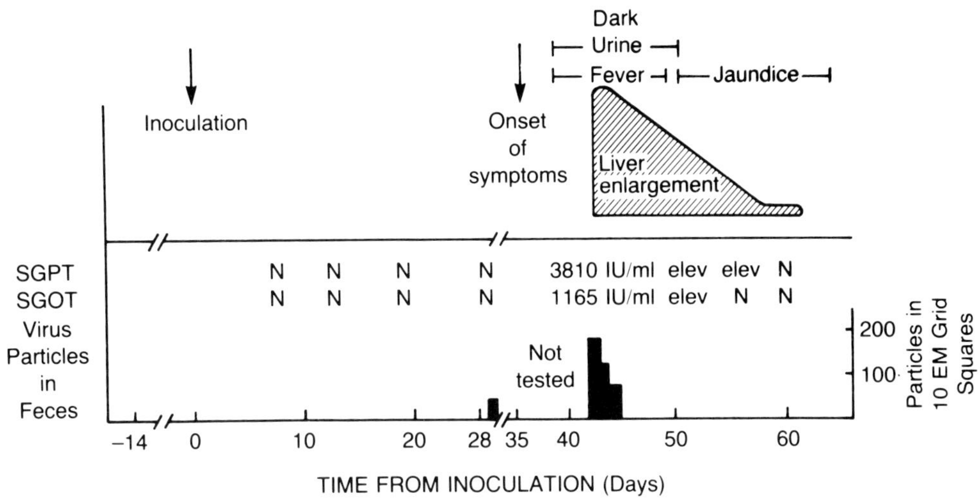

FIGURE 2. Major features of an episode of viral hepatitis which occurred in a human volunteer who was fed an extract of feces from a patient with sporadic HNANB. (From Balayan, M. S., Andzhaparidze, A. G., Swinskaya, S. S., Braginsky, D. M., Savinov, A. P., and Poleschuk, V. F., *Intervirology*, 20, 23, 1981.)

1970s) 103 villagers developed hepatitis of which 12 had no serological evidence of infection with HAV, HBV, CMV, or EBV. In each case the illness was mild, with most patients having fully recovered within 3 weeks. No evidence of the mode of transmission was obtained, although the occurrence of four cases in one family suggested spread from person to person.

Similar data have been reported from other parts of the world indicating that HNANB is widely distributed; indeed, in Southeast Asia it accounts for more than 50% of all hospitalized cases of viral hepatitis. How much of this disease is the enterically transmitted form remains to be demonstrated and will not be known until specific diagnostic tests become available.

III. ATTEMPTS TO IDENTIFY THE ETIOLOGICAL AGENT

Several groups appear to have succeeded in transmitting infection with the enteric form of HNANB to human volunteers or nonhuman primates and reported the detection of virus-like particles or apparently specific viral agents in materials known to be infectious.

None of those particles or antigens has yet satisfied the strict criteria for an infectious agent, i.e., that it is regularly detected in patients (or experimentally infected animals) with the disease, stimulates a specific antibody response, and after isolation in cell culture or purification from infected tissue or other materials, regularly induce disease in laboratory animals or human volunteers.

In 1983, Balayan and his colleagues[24] reported the detection of 27 to 30-nm virus-like particles in fecal samples obtained from a volunteer who ingested a 10% filtered aqueous extract prepared from a pool of 14 fecal samples collected from nine patients with sporadic HNANB. Each of the patients whose specimens were included in the pool had acquired his illness in Tashkent, and all samples were collected within 1 to 4 days of the onset of symptoms. The volunteer, who was known to be immune to hepatitis A, developed symptoms of acute viral hepatitis and elevated transaminase levels 36 days after inoculation. Virus-like particles, measuring 27 to 30 nm in diameter, which were aggregated by his convalescent sera (but not by preinfection sera or animal sera containing anti-HAV) were detected by IEM in fecal samples collected on days 28, 43, 44, and 45 post inoculation, but not in samples collected earlier or later in the illness (see Figure 2). These particles had a buoyant density of 1.35 g/cm^3. Sera collected during the convalescent phase of the patient's illness reacted with these particles by IEM, as did convalescent sera from 23 of 23 patients suspected on epidemiological grounds to have HNANB. Specific antibody was not detected in convalescent sera from 32 patients with acute hepatitis A (who had no history of other attacks of hepatitis).

Two cynomolgus monkeys were inoculated i.v. with 1 cc of a 10% fecal extract which had been collected from the volunteer 42 days after inoculation. Both animals developed mild elevations of SGPT and SGOT 4 to 5 weeks after inoculation and shed virus transiently into their feces. Liver biopsies obtained at the time the animals became ill showed typical acute viral hepatitis.

While these findings were most encouraging it has proved difficult to transmit the disease to other animals or to isolate the agent in cell culture. Because the quantities of virus present in the original samples were small, only limited serological tests have been possible and ELISA assays (for detection of viral antigens and antibodies) have provided equivocal results.

Recently this group has reported the results of further attempts to transmit infection to monkeys.[29] Two species of monkey (Jayan macaques — *Macaca fascicularis* and African green monkeys — *Cercopithecus aethiops)* were studied.

Initially 1 mℓ of a 10% fecal extract obtained from a patient with acute HNANB was inoculated i.v. in two African green monkeys and six macaques. Evidence of infection (shedding of virus-like particles in the feces, elevated AST or ALT levels, typical histological changes, or a rise in antibody titer) was obtained in five of the six macaques.

Feces collected from one of these animals (JM1191) during his illness which was known to contain particles was processed and a filtered extract inoculated i.v. into two additional macaques. While both shed particles into their feces only one (JM1192) developed biochemical and histological evidence of hepatitis. A stool extract from this animal was inoculated i.v. into a further three animals. In this experiment particles were detected in the feces of only one macaque; this animal had mildly elevated liver enzymes and no histological evidence of hepatitis.

Independent confirmation of these observations has recently been reported from a second group in the U.S.S.R.[25,30] Virus-like particles measuring 27 to 30 nm in diameter were detected by IEM in fecal samples obtained from 4 of 40 patients studied during a seasonal outbreak in one of the Central Asian republics. These particles were aggregated by human convalescent sera but not by sera containing anti-HAV alone. Infection was transmitted to five of eight *Macaca fascicularis* monkeys and could apparently be transmitted with infection materials obtained from these animals.[29]

Recently Kane and his colleagues[22] reported the detection of 27-nm virus-like particles in one of eight fecal specimens obtained from patients in the Kathmandu Valley outbreak. These particles were aggregated by convalescent sera collected during the outbreak but not by sera containing anti-HAV alone. Attempts to isolate the agent in cell culture were unsuccessful; however, inoculation of two chimpanzees and four marmosets with the particle containing extract resulted in the induction of mild hepatitis in three of the marmosets. Virus-like particles (27 to 30 nm) were recovered from the feces of these animals 18 to 36 days after inoculation with maximum shedding preceding the onset of biochemical hepatitis.

It has not proved possible to passage this effect in marmosets nor has it proved possible to demonstrate a rising titer of antibodies in patients or experimentally infected animals.

Further studies by the CDC group appeared to have resulted in the transmission of an infectious agent to marmosets and cynomolgus monkeys. One inoculum has now been passaged three times in cynomolgus monkeys.[34] These animals develop abnormal liver function tests from 20 to 47 days after inoculation and appear to shed virus-like particles in their feces during the acute phase of the illness. On passage the severity of the infection appears to decline and the incubation period is shortened. While infected animals and patients convalescing from HNANB using have antibody in their serum which can be detected by IEM, acute phase sera have also been found to contain antibody and seroconversions have rarely been demonstrated. Although the CDC agent is clearly the best candidate to be the etiologic agent, further studies are needed to confirm its role. In particular pre- and postinfection human sera need to be obtained and tested to determine whether seroconversions routinely occur. Such studies are under way.

Particles with a similar morphology have been detected in the feces of three patients in California who developed acute HNANB shortly after returning from visits to Pakistan.[31] The particles varied in size from 17 to 29 nm and had a slightly dimpled appearance. Unusually acute phase sera were found to aggregate the particles more effectively than convalescent sera. Aggregation was also noted when the particles were reacted with convalescent sera obtained from a marmoset infected with virus obtained from the Kathmandu outbreak, suggesting that the two outbreaks were caused by the same or similar agents.

Thus, while several candidate agents have been identified, none of these meets all the criteria required for acceptance as the etiologic agent. Most of the difficulty in confirming the role of these agents lies in the limited quantity of clinical material available for study and, in particular, in the absence of serum samples collected prior to infection. In an attempt to overcome these problems, several prospective studies are being underatken in endemic areas. It is hoped that these studies will generate well-characterized panels of serum and feces which will be as useful for confirming putative etiologic agents as the materials generated at Willowbrook State School were for identifying the causes of hepatitis A and B.

REFERENCES

1. **Prince, A. M., Brotman, B., Grady, G. F., Kuhns, W. J., Hazzi, C., Levine, R. W., and Millian, S. J.,** Long-incubation post-transfusion hepatitis without serologic evidence of exposure to hepatitis B virus, *Lancet,* 2, 241, 1974.
2. **Feinstone, S. M., Kapikian, A. Z., Purcell, R. J., Alter, H. J., and Holland, P. V.,** Transfusion-associated hepatitis not due to viral hepatitis type A or B, *New Engl. J. Med.,* 292, 767, 1975.
3. **Alter, H. J. and Hoofnagle, J. H.,** Non-A, non-B: observations on the first decade, in *Viral Hepatitis and Liver Disease,* Vyas, G. N., Dienstag, J. L., and Hoofnagle, J. H., Eds., Grune & Stratton, Orlando, 345, 1984.
4. **Viswanathan, R.,** Epidemiology, *Indian J. Med. Res.,* Suppl 1, 1, 1957.
5. **Pavri, K.M., Niphadkar, K. B., and Sheikh, B. H.,** Retrospective studies on Australia antigen in sera collected during the epdiemic of viral hepatitis at Delhi in 1956, *Indian J. Med. Res.,* 60, 1575, 1972.
6. **Wong, D. C., Purcell, R. H., Sreenivasan, M. A., Prasad, S. R., and Pavri, K. M.,** Epidemic and endemic viral hepatitis in India: evidence for a non-A, non-B hepatitis virus etiology, *Lancet,* 2, 882, 1980.
7. **Dhamdhere, M. R. and Nadkarni, M. G.,** Infectious hepatitis at Aurangabad: report of an outbreak, *Indian J. Med. Res.,* 1, 1577, 1962.
8. **Sreenivasan, M. A., Banerjee, K., and Pandya, P. G.,** Epidemiological investigations of an outbreak of infectious hepatitis in Ahmedabab City during 1975 — 1976, *Indian J. Med. Res.,* 67, 197, 1978.
9. Annual Report of the National Institute of Virology, Pune, India, 1983.
10. **Khuroo, M. S.,** Study of an epidemic of non-A, non-B hepatitis: possibility of another human hepatitis virus distinct from post-transfusion non-A, non-B, *Am. J. Med.,* 68, 818, 1980.
11. **Tandon, B. N., Joshi, Y. K., Jain, S. K., Gandhi, B. M., Mathieson, L. R., and Tandon, H. D.,** An epidemic of non-A, non-B hepatitis in north India, *Indian J. Med. Res.,* 75, 739, 1982.
12. **Khuroo, M. S., Saleem, M., Teli, M. R., and Sofi, M. A.,** Failure to detect chronic liver disease after epidemic non-A, non-B hepatitis, *Lancet,* 2, 97, 1980.
13. **Khuroo, M. S., Teli, M. R., Skidmore, S., Sofi, M. A., and Khuroo, M. F.,** Incidence and severity of viral hepatitis in pregnancy, *Am. J. Med.,* 70, 252, 1981.
14. **Maklani, P. K. and Grewal, A. K.,** Observations on infectious hepatitis in pregnancy, *Indian J. Med. Res.,* Suppl 45, 71, 1957.

15. **Naidu, S. K. and Viswanathan, R.,** Infectious hepatitis in pregnancy during the Delhi epidemic, *Indian J. Med. Res.,* Suppl 45, 71, 1957.
16. **Borhanmanesh, F., Haghighi, P., and Hekmat, K.,** Viral hepatitis during pregnancy: severity and effect on gestation, *Gastroenterology,* 64, 304, 1973.
17. **D'Cruz, I. A., Balani, S. G., and Iyer, L. S.,** Infectious hepatitis and pregnancy, *Obstet. Gynecol. Col.,* 31, 449, 1968.
18. **Saran, M., Gupta, N. A., and Bhagwat, A. G.,** Hepatitis in pregnancy, in *Hepatitis,* Vakij, B. J. and Shah, S. C., Eds., Adoim Publishers, Bombay, 1975, 41.
19. **Kamat, S. K.,** Prognosis of infective hepatitis in pregnant women, in *Hepatitis,* Vakij, B. J. and Shah, S. C., Eds., Adoim Publishers, Bombay, 1975, 50.
20. **Khuroo, M. S. Duermeyer, W., Ahanger, M. A., and Shah, M. A.,** Acute sporadic non-A, non-B hepatitis in India, *Am. J. Epidemiol.,* 118, 360, 1983.
21. **Joshi, Y. K., Babu, S., Sarin, S., Tandon, B. N., Gandhi, B. M., and Chaturverdi, V. C.,** Immunoprophylaxis of epidemic non-A, non-B hepatitis, *Indian J. Med. Res.,* 21, 18, 1985.
22. **Kane, M., Bradley, D. W., Shrestha, S. M., Maynard, J. E., Cook, E. H., Mishra, R. P., and Joshi, D. D.,** Epidemic non-A, non-B hepatitis in Nepal: recovery of a possible etiologic agent and transmission studies in marmosets, *JAMA,* 252, 3140, 1984.
23. **Myint, H., Soe, M. M., Khin, T., Myint, T. M., and Tin, K. M.,** A clinical and epidemiological study of an epidemic of non-A, non-B hepatitis in Rangoon, *Am. J. Trop. Med. Hyg.,* 34, 1183, 1985.
24. **Balayan, M. S., Andzhaparidze, A. G., Savinskaya, S. S., Braginsky, D. M., Savinov, A. P., and Poleschuk, V. F.,** Evidence for a virus in non-A, non-B hepatitis transmitted via the fecal-oral route, *Intervirology,* 20, 23, 1981.
25. **Favorov, M. O., Khukhlovich, P. A., Zairov, G. K., Listovskaya, E. K., Arakelov, S. A., Vzorov, A. N., Stakhanova, V. M., Ketiladze, E. S., and Zhdanov, V. M.,** Clinical and epidemiological features and diagnosis of viral non-A, non-B hepatitis with fecal-oral transmission mechanism, *Vopr. Virusol.,* 1, 65, 1986.
26. **Belabbes, H., Benatallha, A., and Bourgemouh, A.,** Non-A, non-B epidemic viral hepatitis in Algeria: strong evidence for its water spread, *Viral Hepatitis and Liver Disease,* Vyas, G. N., Dienstag, J. L., and Hoffnagle, J. H., Grune & Stratton, Orlando, Fla., 1984, 637.
27. **Yamauchi, M., Nakajima, H., Kimura, K., Fujisawa, K., Kameda, H., Nakahara, M., Stoh, Y., and Kawamura, H.,** An epidemic of non-A, non-B hepatitis in Japan, *Am.J. Gastroenterol.,* 78, 652, 1983.
28. **Villarejos, V. M., Visona, K. A., and Eduarte, C. A.,** Evidence for viral hepatitis other than type A or type B among persons in Costa Rica, *New Engl. J. Med.,* 173, 1350, 1975.
29. **Andzhaparidze, A. G., Balayan, M. S., Savinov, A. P., Braginsky, D. M., Pokeschuk, V. F., and Zamyatina, N. A.,** Non-A, non-B hepatitis transmitted by the fecal oral mode experimentally produced in monkeys, *Vopr. Virusol.,* 1, 73, 1986.
30. **Zairov, G. K., Stakhanova, V. M., Listovskaya, E. K., Konstantinova, L. A., Favorov, M. O., Ketiladze, E. S., and Zhdanov, V. M.,** Electron microscopic investigations in non-A, non-B viral hepatitis with fecal oral transmission, *Vopr. Virusol.,* 1, 175, 1986.
31. **De Cock, K. M., Bradley, D. W., Sandford, N. L., Maynard, J. E., and Redecker, A. G.,** Non-parenterally transmitted non-A, non-B hepatitis imported from Pakistan, in press.
32. **Melnick, J. L.,** personal communication.
33. **Maynard, J. E.,** personal communication.
34. **Bradley, D. W.,** personal communication.

INDEX

H

I

J

K

L

M

N

O

P

R

S

T

V

W

X

Y